AF522661

Pvt. Ltd.

To

Dr. L.M. Joshi

for his inspiration as a teacher

Cover Photograph

Mosaic of Uredosorus of Wheat Rust by W. Mulder (I.P.O. Building, Institute of Phytopathological Research, Wageningen, Netherlands).

Acknowledgement

Catalyzed and Supported by the Department of Science and Technology, under its Utilization of Scientific Expertise of Retired Scientists Scheme.

Biology and Management of
Wheat Pathogens

K.D. Srivastava
Former Head* and *Professor
Division of Plant Pathology
Indian Agricultural Research Institute
New Delhi - 110012

2010

Studium Press (India) Pvt. Ltd.

Biology and Management of Wheat Pathogens

©2010

This book contains information obtained from authentic and highly regarded sources. Reprinted material is quoted with permission, and sources are indicated. A wide variety of references are listed. Reasonable efforts have been made to publish reliable data and information, but the author and the publisher cannot assume responsibility for the validity of all materials or for the consequences of their use.

All rights are reserved under International and Pan-American Copyright Conventions. Apart from any fair dealing for the purpose of private study, research, criticism or review, as permitted under the Copyright Act, 1956, no part of this publication may be reproduced, stored in a retrieval system or transmitted, in any form or by any means–electronic, electrical, chemical, mechanical, optical, photocopying, recording or otherwise–without the prior permission of the copyright owner.

ISBN: 978-93-80012-22-3

Correct Citation:

Srivastava, K.D. 2010. Biology and Management of Wheat Pathogens. Studium Press (India) Pvt. Ltd., New Delhi, pp. 270.

Published by:

Studium Press (India) Pvt. Ltd.
4735/22, 2nd Floor, Prakash Deep Building
(Near Delhi Medical Association),
Ansari Road, Darya Ganj, New Delhi-110 002
Tel.: 23240257, 65150447; Fax: 91-11-23240273;
jngovil@gmail.com; jngovil@hotmail.com

Printed at:

***India Offset Press,** A-1, Mayaprui Indl. Area, Phase-1*
New Delhi-110064, Tel.: 28116494; Fax: 91-11-28115486

भारत सरकार
पौधा किस्म और कृषक अधिकार संरक्षण प्राधिकरण,
एन.ए.एस.सी. काम्प्लेक्स, (टोड़ापुर के सामने),
डी.पी.एस. मार्ग, नई दिल्ली - 110012 (भारत)

अध्यक्ष
CHAIRPERSON

Government of India
Protection of Plant Varieties and Farmers' Rights Authority,
N.A.S.C. Complex, (Opp. Village Todapur),
DPS Marg, New Delhi-110012 (India)
दूरभाष/Telphone : 25848127
फैक्स/Fax : 011-25840478
Website : www.plantauthority.gov.in
E-mail : chairperson-ppvfra@nic.in

Foreword

Wheat grown during the mild winter months produces good yields thereby sustaining the food security system of the country. The quantum jump achieved by the nation in the 1960's in wheat production through increased productivity and matching agronomic practices is referred to as the "Green Revolution". However, stability in wheat production was achieved by proper understanding of the biology of the wheat pathogens, their epidemiology and management.

Major Sleeman of the East India Company, stationed at Jabalpur, got stem rust samples identified as *Puccinia* in 1828. From then on, a series of investigations were made on the nature and recurrence of cereal rusts. The first scheme to be funded by the ICAR on cereal rusts was awarded to Prof. K.C. Mehta, an eminent pathologist. Based on nearly twenty years of investigations, Prof. Mehta published two volumes on the epidemiology of wheat rusts in 1940 and 1955. He also initiated race analysis work at Flowerdale, Shimla. The cereal rust investigations were subsequently taken up at IARI by Dr. L.M. Joshi, whose contributions to wheat pathology is monumental. The author of this book, Dr. K.D. Srivastava and I had the privilege of working with Dr. Joshi, especially in explaining the "Puccinia Path" and developing a gene deployment strategy over central India to contain stem rust, and along the Gangetic plains to check leaf rust. I am immensely pleased that the author has dedicated this book to Dr. Joshi.

Yellow rust is primarily a closed circuit between North Western Plains Zone (NWPZ) and the adjoining Himalayas. Even though Ug99 concern was expressed by international experts, India did not press the panic button since stem rust is not considered a serious problem in NWPZ.

The contribution of Dr. Manoranjan Mitra, a leading wheat pathologist, towards the understanding of Karnal bunt is worth recollecting. At national level, several stalwarts spent their entire carrier in investigating specific diseases and their management in wheat through breeding disease resistant varieties. The trend set by Dr. B.P. Pal, the renowned plant breeder, and Dr. Mehta in breeding for disease resistance resulted in the development of several disease resistant varieties. This team effort saves each year about six million tonnes of wheat, which otherwise would have been lost to diseases. This book covers an array of plant diseases caused by fungi, viruses, nematodes and other organisms. An exhaustive effort has been made by the author to cover several aspects of wheat pathology. At a time when climate change is posing a threat, and the dynamics of plant diseases are changing, the effort of the author in presenting this comprehensive volume to students, researchers and those involved in wheat improvement is, indeed, commendable.

November 3, 2009

(S. Nagarajan)

About the Author

Dr. Krishna Dutt Srivastava, former Professor and Head of the Division of Plant Pathology, Indian Agricultural Research Institute, New Delhi, obtained Ph.D. degree from Agra University, Agra in 1982. He served the Rockefeller Foundation in Indian Wheat Improvement Programme during 1968–70 and since 1970 Dr. Srivastava devoted his entire scientific carrier of 38 years in investigating the diseases of wheat at IARI. The major area of his research work has been epidemiology of wheat rusts and Karnal bunt, disease resistance, fungicidal and biological control of pathogens and integrated disease management. He has published 130 research papers in reputed national and international journals. He has also contributed several conceptual review articles in edited books and popular articles in reputed magazines. He is co-author of 4 books and a number of manuals, bulletins and technical reports. In the capacity as Head of the Publication Unit of IARI, he coordinated the publication of annual reports, number of success stories, bulletins, folders and newsletters. In addition to research, he was doing post graduate teaching at IARI. Twelve students have obtained M.Sc. and Ph.D. degree under his guidance. P.G. School of IARI has recognized his contributions in teaching and research and conferred Best Teacher Award. Dr. Srivastava is joint recipient of Rafi Ahmad Kidwai Memorial Award instituted by Indian Council of Agricultural Research, New Delhi. He was awarded Sharda Memorial Award by Plant Protection Association, Hyderabad and Prof. M.J. Narasimhan Medal Award by Indian Phytopathological Society, New Delhi. He is the Fellow of the Indian

Phytopathological Society and served as its Joint Secretary and Treasurer. Dr. Srivastava has advanced training on pesticide utilization for plant protection at Japan International Cooperation Agency, Tokyo (Japan) and system analysis and crop growth simulation in agriculture organized by IARI and International Rice Research Institute, Philippines.

Preface

The pests and diseases are serious threat to the successful cultivation of wheat resulting in substantial losses in crop yield. In India, several diseases of fungal, viral, bacterial and nematode origin have been recorded. Physiological disorders due to nutrient deficient soil and few phanerogamic parasites are new emerging problems of wheat. Much research has been done on these topics in the Indian sub-continent but the information so generated is available in various journals, periodicals magazines, research reports etc. and as such it is not easily available in the form of a book. The present work attempts a critical and comprehensive review of literature on wheat diseases.

The book is intended to provide precise information pertaining to diagnosis of wheat diseases, nature and extant of damage due to pathogens, dynamics of disease appearance, ecological factors governing disease development and epidemiology and the variability existing in pathogens. Strategic management planning in tackling the wheat health through integration of host resistance, genetic manipulations using biotechnology, gene deployment, forecasting, chemical, biological and cultural control approaches have been discussed in different chapters of the book.

I hope that the compilation will be of immense use to students offering diseases of field crops at undergraduate and postgraduate levels and teachers of plant pathology, botany and biology. It will also be useful to all those workers involved in crops protection.

I am fully conscious that in a work of this nature, some mistakes might have crept in the text inadvertently and for these I own responsibility and request readers to communicate to me the defects so that I could attempt to rectify them in future edition.

I wish to express my gratefulness to Dr. S. Nagarajan, Chairperson, PPV & FR Authority, New Delhi, for critical appraisal of the manuscript and in giving a foreword to this document. I thankfully acknowledge the encouragement and guidance given to me by friends and colleagues especially Drs. D.V. Singh, Former Head (Plant Pathology), Rashmi Aggarwal, National Fellow, R.K. Jain, Head of the Division of Plant Pathology and D. Prasad, Professor of Nematology, IARI, New Delhi. I am indebted to Drs. A.K. Sharma and Dr. D.P. Singh, Principal Scientists, DWR, Karnal; Indu Sharma and R.C. Sharma, PAU, Ludhiana and Praveen Verma, Staff Scientist, NIPGR, New Delhi who have readily furnished me the information and the literature.

Author is thankful to Drs. S.K. Singh, SKUA & T, Jammu, Ajay Kumar Singh and Amit Dixit, PPV & FR Authority, Robin Gogoi and R.K. Sharma, IARI, who have spared time for providing me the photographs of disease symptoms and making drawings. Mrs. Aditi Verma, Mr. A.K. Rai, Amit Kumar and Ms. Sapna Sharma helped me in more than one way in preparing the typescript and I am greatly indebted to them. Constant inspiration and help provided by my wife Mrs. Usha Rani Srivastava during the preparation of this book is thankfully acknowledged.

I am thankful to the Director, Indian Agricultural Research Institute, New Delhi who permitted me to compile this book. My thanks are also due to Dr. A.K. Ganguly, Principal Scientist (PPI, IARI) who always readily helped me in administrative matters.

The assignment for writing the book on Biology and Management of Wheat Pathogens was catalysed and supported by Department of Science and Technology under its Utilization of Scientific Expertise of Retired Scientists Scheme. I acknowledge unstinting help received from Late Dr. R.C. Srivastava, S.S. Kohli and J.B. Reddy, DST for official support .

I also express gratitude to Mr. Shray Jain and Dr. J.N. Govil of Studium Press LLC who extended valuable help in processing the manuscript and getting the book printed in stipulated time using the latest printing technology.

K.D. SRIVASTAVA

Contents

List of Plates

Chapter

Introduction

Wheat is nature's unique gift to mankind. It is world's most widely cultivated cereal and is consumed in various forms by more than 1000 million people over 60 countries. This cereal is one of the life sustaining food crops which contributes towards food front to the tune of 40% of world's population and provides 20% total global calories to human race (Hanson *et al.*, 1982). The wheat, being high in complex carbohydrate, insoluble fiber, protein, lipids, minerals, vitamins, antioxidants and phytosterols is nutritious, convenient and economic source of food. These components have a significant role in shaping wheat products but it is gluten, the wheat protein that ultimately determines the quality of end product.

The wheat is converted into a number of products, *viz.*, *Atta, Suji, Dalia* and *Maida*, which are either directly used for household consumption or processed to manufacture a number of by products. Bread, sandwiches, cakes, cookies, biscuits, pasta, buns, pizza, noodles, baguettes, spring rolls, dumplings, doughnuts, ice-cream cones, baby food, snack food etc. have found their way in Asia Pacific region. In India, traditionally bread wheat is the staple and it is mainly used for preparation of *Chapatti*, *Tandoori Roti,* and *Nan*. Where as durum wheat is used for preparation of traditional break fast as well as snack foods *viz.*, *Uppuma*, *Kesaribhath*, vermicelli and alike. It is also used in gravies, sauces, soups, candies and beverages.

The bran, husk etc. are very valuable as a feed for livestock. The wheat straw is an important ingredient of dry fodder for cattle in India. It is also utilized in manufacturing mattresses, straw hats, paper and art objects. The wheat straw burried in the soil acts as a source of humus. These are some of the special attributes which signify the great importance of wheat in agriculture, trade and nutrition. In the industry the wheat is used for manufacturing paste, alcohol, oil and gluten.

1.1 WHEAT CLASSIFICATION

Botanically wheat belongs to tribe Triticeae of the grass family Gramineae. The genus *Triticum* is a member of subtribe Triticinae (Fig. 1)

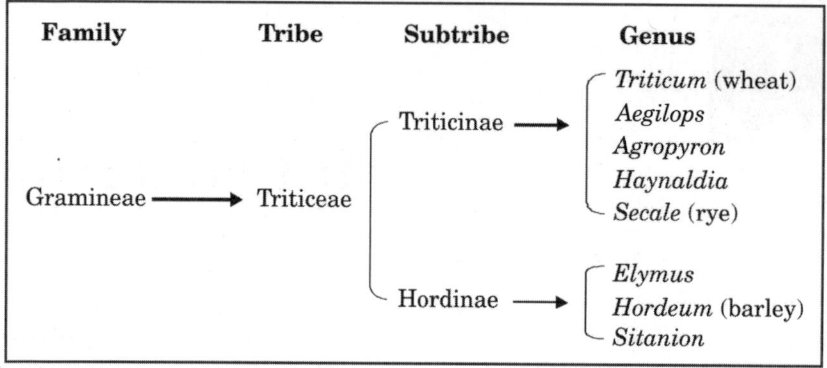

Figure 1: Members of the tribe Triticeae of the family Gramineae

Botanists classify wheat species according to the number of genomes *i.e.* sets of chromosomes. Wild wheats, called einkorn (T*riticum monococcum*) with one genome (7 pairs of chromosomes) are diploid type. Tetraploid wheats (*T. turgidum*) with two genomes (14 pairs of chromosomes) include several cultivated species, where as hexaploid wheats with three genomes having 21 pairs of chromosomes consist of six subspecies of *Triticum aestivum*. Among these, *T. aestivum* is the main subspecies and therefore, it is called common wheat and has economic significance (Martin *et. al.*, 1976). Thus, with in the genus *Triticum* species and a number of subspecies are recognized (Table 1).

1.2 CENTER OF ORIGIN OF WHEAT

It is believed that wheat was derived from wild ancestors through a process of domestication by the hunter gatherers probably in the Neolithic period (the period relating to the period of stone age which began about 10,000 BC *i.e.* 12000 years ago and ended with the rise of agriculture and polished stone tools from about 7500 BC to 1710 BC (*i.e.* about 10,000 to 3700 years ago). A very likely place of origin of wheat is considered to be a hilly region of south-western Asia called the Fertile Crescent (Fig. 2).

Table 1. Species and subspecies of genus *Triticum*

Type	*Triticum* species	Subspecies	Common name	Chromosome number (n)	Genome
Diploids	*T. monococcum*	–	Einkorn	7	AA
Tetra-ploids	*T. turgidum*	spp. *dicoccum*	Emmer	14	AABB
		spp. *durum*	Durum	14	AABB
		spp. *turgidum*	Rivet	14	AABB
		spp. *polonicum*	Polish	14	AABB
		spp. *carthlicum*	Persian	14	AABB
		spp. *turanicum*	Khorasan	14	AABB
	T. timopheevii	–	–	14	AABB
Hexa-ploids	*T. aestivum*	spp. *spelta*	Spelt	21	AABBDD
		spp. *macha*	Macha	21	AABBDD
		spp. *vavilovii*	Vavilov	21	AABBDD
		spp. *aestivum*	Vulgare	21	AABBDD
		spp. *compactum*	Club	21	AABBDD
		spp. *sphaero-coccum*	Shot	21	AABBDD

The area is a rich soil region in the upper reaches of Tigris – Euphrates drainge basin (ancient Mesopotamia) bordered on the other side by the hilly regions embracing Israel, Jordon, Turkey, Iran and Syria. In this area, number of archeological excavations have shown the existence of permanent settlement. This is the same area in which einkorn or emmer, two wild varieties of wheat, are found. Evidence suggests that cereal grains possibly *T. monococcum* or *T. dicoccum* were the principal type of wheats domesticated in this area by the early farmers (Gill, 1979). Meanwhile,

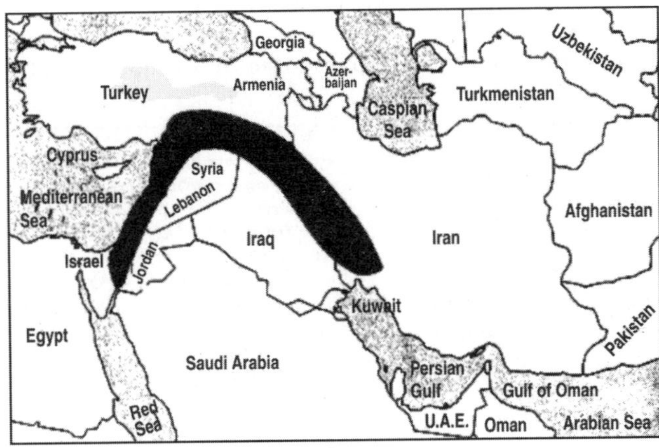

Figure 2: The Fertile Crescent – place of origin of wheat (shaded area)

some of them carried seeds and traveled eastward and introduced wild relatives of wheat in China, India and Japan. It is evident from Chinese classics of Chou that wheat was grown in China about 3000 BC (Pal, 1966). Egyptians believed that Egypt was the early home of wheat and it is a gift form goddess Demeter. The mural paintings in ancient tombs and pyramids (Fig. 3) have established the existence of wheat in Egypt during early period of its origin (Joshi *et al.*, 1986, Gill *et al.*, 1993).

Figure 3: A relief depicting the old age importance of wheat in ancient times showing wheat ears being carried in ceremonial procession to king Araras (800 B.C.)

Vedas are the secred scriptures of Hindus. Max Müller had arbitrarily fixed the age of Vedas as 1500–2000 BC. The Indus civilization was in continuation during Vedic period. One finds a mention of awned and awnless wheats in Atharva Veda showing that this cereal was cultivated in Indus Basin more than 5000 years ago. It has a reference in vedic vulgate "Manasolhas" (Fig. 4). The recovery of carbonized wheat (*Triticum sphaerococcum*) grains from the excavation of Mohenjedaro and Harappa indicated that this species was extensively cultivated during Indus civilization. It is supposed to have originated in the North-western areas of

गोधूमाः क्षलिता शुभ्राः शोषिता रविरश्मिभिः।७५
घरट्टैश्चूर्णिताःश्लक्ष्णाश्चालन्या वितुषीकृताः।
गोधूमचूर्णकं किञ्चित्घृत विमिश्रितम।७६
लवणेन च संयुक्तं क्षीर नीरेणपिण्डितम्।
सुमहत्यां काष्ठपात्र्यां करास्फालैर्विमर्दयेत्।७७
मर्दितं चिक्कणीभूतं गोलकान् परिकल्पयेत्।
स्नेहाभ्यक्तैः करतलैः शालिचूर्णैर्विरूक्षितान्।७८
प्रसारयेत गोलकांस्तान् करसञ्चारयत्नतः।
विस्तृतामण्डकाःश्लक्ष्णाः सितपट्टसमप्रभाः।७९

Wheat was washed, dried in the sun, ground, and cleaned in a sieve. The flour was mixed with clarified butter and salt and made into balls. The balls were turned into cakes with the palms of hands and were cooked in a potsherd. They were baked on live charcoals before eating. Sometimes a wooden roller and a piece of stone were used to change the ball into circular cakes before baking.

Manasolhas III, 1375-79

Figure 4: A vedic description about wheat in Manasolhas

the Indian sub-continent between the Himalayan range and the Hindukush Mountain (Pal, 1966, Wheeler, 1968, Shaktawat and Singhi, 2000). More recent excavations from Chirand in Bihar in Indo Gangetic plains throw light on the cultivation of *T. sphaerococcum* to about 300 BC (Vishnu Mittre, 1971, 1981). Mehra (2000) also made a reference which indicates that *T. dicoccum* was grown is Punjab as early as 2300–1400 BC and *T. aestivum* in Kashmir in 2600–1500 BC., Bihar in 2000–1200 BC., Gangetic plains in 2200–800 BC and Madhya Pradesh and Maharashtra in 2000–800 BC.

1.3 EVOLUTION OF WHEAT

The evolution of modern wheat (*T. aestivum*) is a fascinating genetic origin. It is classic examples of how closely related species can evolve in nature from three genomes A, B and D through several polyploidization. Wild wheat einkorn (*T. monococcum*) is supposed to be the donor of A chromosome where as wild grass *Aegilops speltoides* contributed the B chromosome and another wild grass *Aegilops squarrosa* provided D chromosome to this evolutionary scheme (Jain, 1978). Diploid einkorn wheat (2n = 14) is believed to have hybridized with *Ae. speltoids* to produce tetraploid emmer wheat. This strain now common as durum wheat has AABB genome having chromosome number 2n = 28. The intergeneric cross of tetraploid with *Ae.*

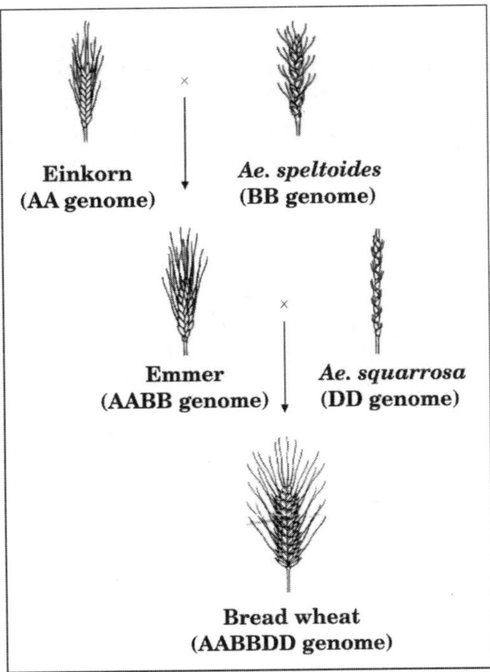

Figure 5: The evolution of wheat from its wild relatives through polyploidization

squarrosa led to the evolution of modern hexaploid (AABBDD) bread wheat (Tomar *et al.*, 2004) which has chromosome number 2n = 42 (Fig. 5).

Two main species, *viz.*, *T. aestivum* (bread wheat) and *T. durum* (macroni wheat) are cultivated on a large scale. Nearly 90% of total wheat area in India is under bread wheat and the macroni or durum wheat is confined to central and southern parts of the country (Hanchinal, 1999).

1.4 ORIGIN OF SEMI DWARF WHEAT

In the early post World War II years, the traditional tall wheat varieties were not favourable to higher doses of fertilizers. Nitrogen applied to tall plants causes them to grow even taller, and they tend to lodge. Wheat breeders had not discovered the genes that would create plants with strong stem to resist lodging and also with the potential to yield benefit from increased nutrients. It was the effort of Japanese scientists who selected semi dwarf wheat, named Daruma which became the ultimate source of dwarfing genes for the new wheats. In 1917, Japanese breeders combined Daruma with Fultz, a high yielding winter wheat imported from United States. Then in 1925, the breeders crossed Fultz-Daruma with Turkey Red, a Russian land race. These combinations gave rise to a new Japanese variety Norin 10, which had a height of only 52–55 cm and was exceptionally high yielding. The seeds of Norin 10 reached United States in 1945 where breeders made crosses of Norin 10 with their locals, in order to confer on its genes that could adapt to local conditions. As a result dwarf, high yielding Gaines wheat, well adapted to the heavy rainfall emerged. However, Gaines seeds were susceptible to rust fungus. In 1953 N.E. Borlaug, utilized Gaines at CIMMYT, Mexico and made crosses with many varieties of wheat collected

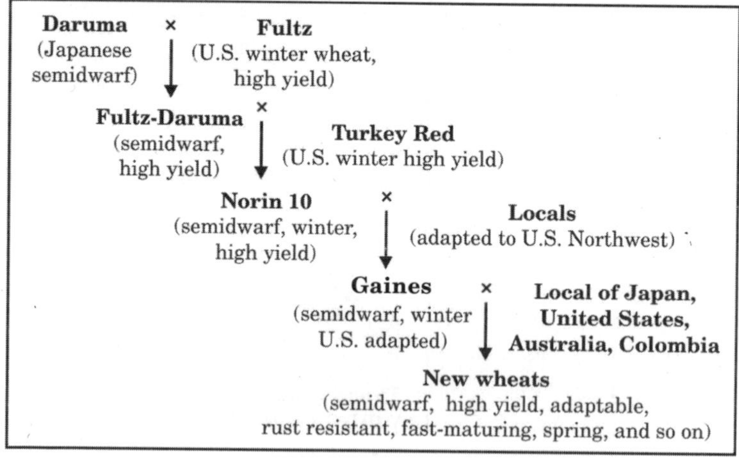

Figure 6: Origin of new wheat

from all around the world. By 1962, the Mexico breeding project had produced new, high yielding, disease resistant semi dwarf varieties with important new characteristics (Fig. 6). These semidwarf wheat varieties have capacity to mature quickly and are insensitive to photo-period. They are also well adapted to a variety of warm climate and are widely cultivated in South and Southeast Asia.

1.5 GROWTH HABITS OF WHEAT

Wheat is annual grass that is grown under a wide range of environmental conditions. It is cultivated between the altitude of 30° and 60° north and between 27° and 40° south. In the northern hemisphere wheat cultivation is practically from Arctic border to almost near equator. In Soviet Union, this cereal is cultivated near archangle at altitude 60° N and even in Tanana Valley, Alaska at 64° N. It is grown in the Tibetan plateau and higher elevation of Himalayas covering India, Nepal and Bhutan (Joshi *et al.*, 1986).

Since the wheat is cultivated under a wide range of environmental conditions, the nature has evolved two growth habits – winter habit wheat and spring habit wheat. The terms winter wheat and spring wheat have a broader meaning than the season in which the crop is grown. Winter-habit wheats require vernalization (exposure to at least several weeks of temperature between 1 and 5°C) to initiate the formation and growth of their reproductive organs. They remain in a vegetative state throughout the winter period and mature in the summer after a total growing period of 9–11 months. They are to a certain extant tolerant to frost and low temperature. In contrast, spring wheat sown in spring matures in the same growing season and do not require vernalization to initiate reproductive growth. They have a continuous growth cycle, generally 3–6 months and no inactive period. They can not survive sustained low temperature.

1.6 MORPHOLOGY OF WHEAT PLANT

The wheat plant emerges from its seed which is technically a caryopsis. The seed consists of an embryo and endosperm enclosed in a mature ovary, that becomes the pericarp. The embryo is a dormant plant comprises of scutellum, coleoptite (plumule), foliage leaf, growing point, first node, radicle and coleorhiza. The seed bears a brush of persistent hair like cell at its terminus.

When planted in well drained clay- loam soil, the young embryo with in dry seed absorbs moisture and swells. As a result the pericarp ruptures to initiate the growth of radicle and plumule. In the process of germination an embryonic seed leaf unfolds breaking through the coleoptile and primary root arises from the radicle. Thereafter, the young seedling makes secondary growth from the crown just beneath the soil surface,. The crown is compacted series of nodes from which clums or tillers and root develop. Clums arise as an elongation from the upper crown internodes. The elongated clum is termed as pseudostem. The leaf sheath encloses the pseudostem and a parallel veined leaf having a ligule and auricle is attached with each node of the clum (Fig. 7).

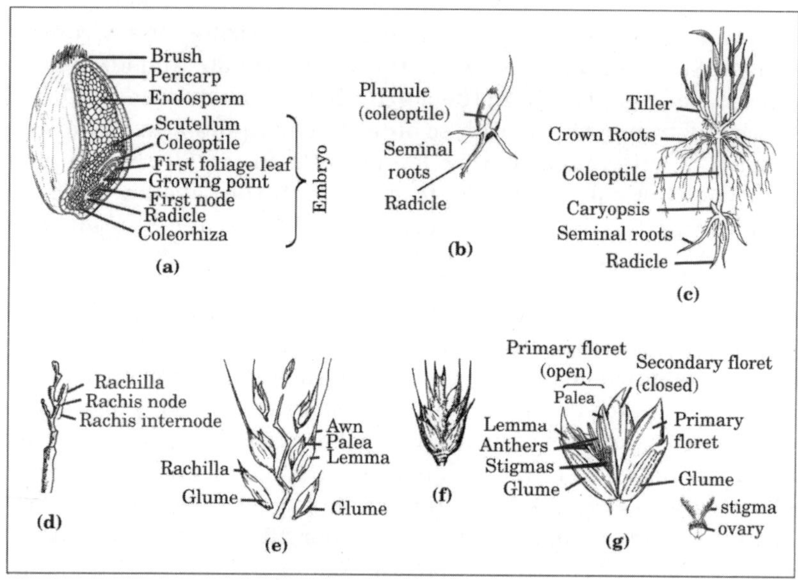

Figure 7: Morphological features of wheat (a) Longitudinal section of seed (b) Seed germination (c) Secondary growth of seedling (d) Rachis (e) Disected spikelet (f) Spikelet and (g) Floret

The ear-head of the wheat emerges during last phase of clum elongation. The upper most nodes are transformed into a twisted axis of the head which is termed as rachis. A number of spikelets are arranged at rachilla along both sides of the rachis. Each spikelet is comprosed of a number of individual florets (flower). The floret is composed of the pistil and the stamens enclosed within glume and protected with outer coverings, the lemma and palea. Initially the developing head is swollen and is sheathed by a leaf, the stage called booting. The self pollination of the flower leads to production and development of the seeds in a spike. The growth of the wheat thus, can be divided into a number of stages. These have been codified by Zadoks *et al.* (1974) into different stages (Figure 8).

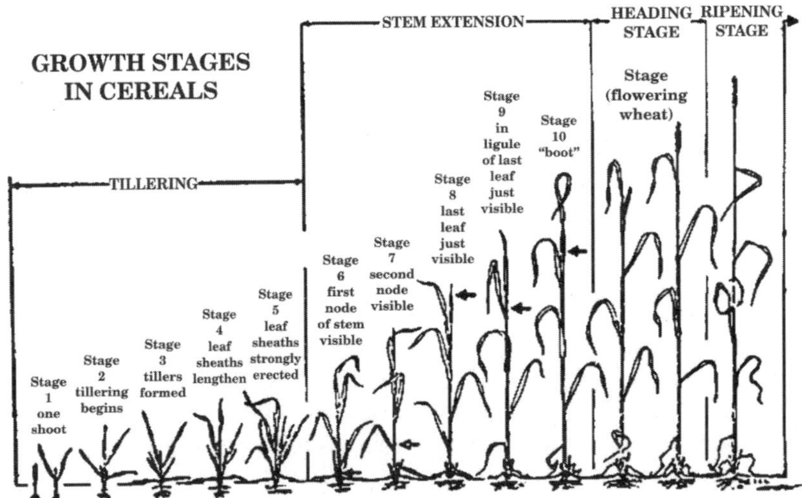

Figure 8: Growth stages of wheat

1.7 WHEAT GROWING ZONES IN INDIA

The wheat growing regions of India differ considerably in soil type, temperature and moisture regime. Hence, wheat is cultivated under diverse environment. Therefore, All India Wheat Improvement Project has identified following six zones under which wheat is grown (Nagarajan, 2001).

Northern Hill Zone (NHZ)

This zone covers the foot hills, mid and higher hills of the Himalayan chain stretching from Kashmir to Arunachal Pradesh in the east. NHZ accounts for 0.7 to 0.8 million ha of wheat.

North Western Plain Zone (NWPZ)

This NWPZ is the fertile, canal and tube well irrigated tract of Gangetic plain with alluvial soil. This zone consists of flat lands of Jammu Kashmir, Punjab, Haryana, western Uttar Pradesh and northern Rajasthan. Here wheat is cultivated over 0.9 million ha. Because of more grains per spike, the yields are very high in this zone.

North Eastern Plain Zone (NEPZ)

The NEPZ covers the eastern part of Uttar Pradesh, Bihar, Jharkhand, West Bengal, Assam, parts of Orissa and small NE Indian states. It has 10 million ha under wheat. The NEPZ has a network of rivers and drain, highly humid, high soil pH and tossing topography with an unreliable irrigation system. Wheat matures in NEPZ in 125 days.

Central Zone (CZ)

The CZ is predominantly high land areas of Madhya Pradesh, Gujarat and parts of Rajasthan and Bundelkhand region of Uttar Pradesh. The lack of assured winter rains discourages wheat cultivation except in deep black soil of high moisture retentivity, where rainfed wheat can be grown. In this tract wheat covers about 5 million ha. Here wheat matures in 100–110 days and is exposed to high temperature.

Peninsular Zone (PZ)

The PZ has at best 1 million ha under wheat in Maharashtra and Karnataka along the elevated areas of western ghats in deep black soil under both conserved moisture and irrigated conditions. The wheat crop matures in 105 days and is subjected to moisture and temperature stress.

South Hill Zone (SHZ)

Wheat in this zone is restricted to high land of Nilgiri and Palni hills of Tamil Nadu. Weather in this zone is alike all the year round. Generally, two crops are taken in SHZ, one between September and April and the other between May and September. Although this zone accounts only for few hundred hectares under wheat but it is important because it serves as the main focus of stem rust inoculum as well as one of the two foci of leaf rust inoculum.

1.8 PRODUCTION AND PRODUCTIVITY OF WHEAT

In India, traditional tall wheat varieties were cultivated prior to 1967. These varieties prone to lodging, also had poor tillering, low sink capacity and longer crop duration exposing plants to the climatic variations, higher sensitivity to thermo and photo variations and vulnerability to diseases. As a result the production and productivity remained low. In 1947, the production of wheat crop was just 6.46 million tonnes and the productivity was merely 663kg/ha. It was not sufficient to feed the Indian population. Therefore, India imported wheat in large quantities from USA under PL-480. However, a major break through in wheat productivity and production started to be visualized in the sixties. Incorporation of 'Norin-10' dwarfing gene from the semi dwarf varieties such as S-227, S-308, S-307 etc. gave rise to dwarf varieties and expanded use of irrigation, fertilizers and pesticides and significantly improved production and productivity of wheat. With the identification and development of dwarf genotypes like Kalyansona, Sonalika, Safed Lerma and Chhoti Lerma in 1967, India got a real momentum and ushured in an era of the Wheat Revolution. A postal stamp was released on July 17, 1968 to commemorate the break through in production. Thereafter, a large number of high yielding, input responsive

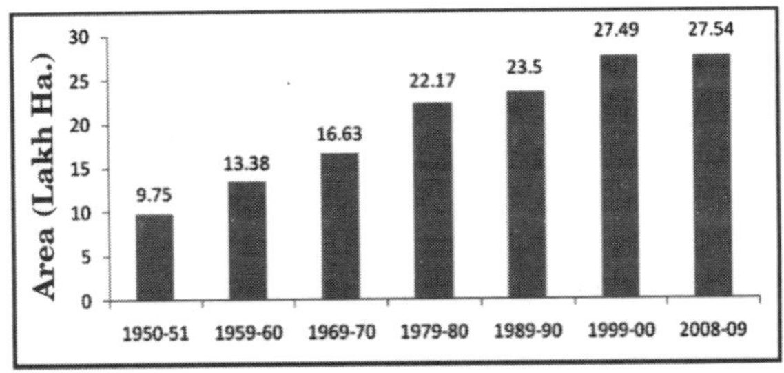

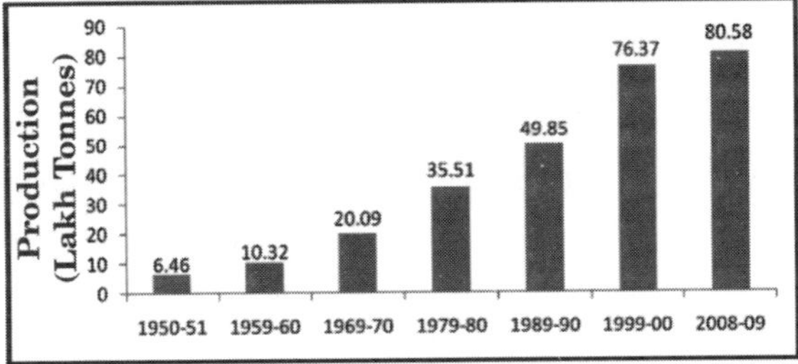

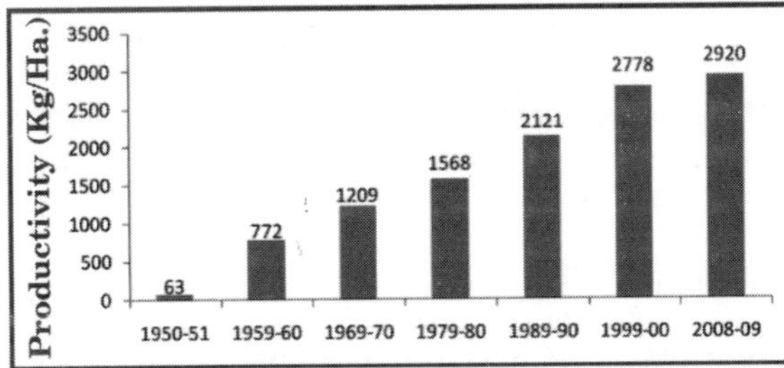

Figure 9: Annual average wheat area, production and productivity in India (10 years period)

and disease resistant wheat varities like UP 262, Arjun, WL 711, HP 1102, HUW 206, HUW 234, HD 2189, HD 2204, HD 2285, VL 616, VL 421, HS 42, WH 147, WH157, WH 542, HD 2329, UP 2003, UP 2338, LOK-1, Raj 1555, Raj 3765, Raj 3077, UP 2425, PPRW 154, PBW 343, PBW 443, PBW 373, HI 8381, HI 8498 HD 2687, KRL 19, HUW 468, GW 273 etc. were developed by agriculture institutes and state agriculture universities. The development of high yielding varieties created an environment for an increase in area. Resultantly an all round increase in production has been recorded. A close analysis shows that the production level and yield of wheat in India have increased progressively. The production of the crop in 1999–2000 remained 76.37 million tones with productively level 27.78 kg/ha. It is worth mentioning that India has registered highest record production of 80.58 million tonnes in 2008–2009. At present, India is second largest producer of wheat in the world after China with about 12.5% world wheat area, whereas it occupies 12% share in the total world wheat production.

In spite of considerable increase in wheat production, India faces different kinds of challenges on several fronts. The anticipated population by the year 2020 AD will be around 1.4 billion with projected demand of about 25m/MT more wheat. The average wheat yield is still much lower than other countries and the decline in productivity growth is a matter of concern. The cultivated area and water availability will further shrink. As these areas are supposed to be the base of the nation's food security, hence, we have to adapt corrective measures. The production can be augmented by enhancing yield potential of wheat varieties through biotechnological interventions. Efforts are required to avoid post harvest losses due to insect pests and diseases. Essentially, the integrated nutrient and water management is needed to meet a certain yield level.

1.9 WHEAT PRODUCTION TECHNOLOGIES

Even with spectacular achievement in increasing the total wheat production, there exists yield gap (1.5 to 2.0 t/ha) between the actual realized yields of the state and the ones achieved in the large number of demonstrations on farmers' fields. It is important that there is little scope for area availability for agriculture owing to other requirements of the economy and crop comptetions within the agriculture. This yield gap can be narrowed by proper resources management and adoption of newly developed cost effective production technologies (Mishra *et al.*, 2005) as described below-

Selection of Variety

An appropriate variety is the first and foremost as part of wheat production technology. At present wide range of high yielding and disease resistant wheat varieties are available for cultivation in different agroclimatic conditions Some of the semi dwarf wheat varieties such as PBW 343, PBW 373, PBW 434, PBW 502, HD 2643, HD 2687, HD 2733, HD 2824, WH 542, WH 869, UP 2338, UP 2425, RAJ 1555, RAJ 3077, RAJ 3765, PDW 233, PDW 291, HUW 468, NW 1012, HP 1731, HP 1761, DL 784-3, K 8804, K 9107, HI 8381, HI 8498, VL 738, VL 804, HS 240 have been recommended for cultivation in the country.

Sowing Period

Wheat crop requires cool and moist climate during its early part of growth. The optimum mean temperature range for germination of wheat seed is 20–25°C. If the temperature goes beyond 25°C at germination phase, the growth of the seedlings is adversely affected, fewer tillers are produced, and heading is accelerated and finally yield is reduced. The desired temperature in northern parts of the country is found in the first fortnight of November. Longer duration dwarf wheats are seeded between November 1–15 and shorter duration dwarf between November 15th and 30th.

Seed rate and Spacing

The requirement of seed rate is based on seed size, germination percentage, time and method of sowing. In general a seed rate of 100 kg/ha at a seed weight of 38 g/100 seed is needed under irrigated and timely sown conditions. Under late sown conditions seed rate should be increased to 125 kg/ha. In case of FIBRS (Furrow Irrigated Raised Bed System) seed rate may be reduced to 75 kg/ha.

The seed is placed about 5–6 cm deep with row spacing of 20–23 cm. In case of late sown wheat 15–18 cm line spacing is adequate.

Fertilization

Wheat is highly responsive to fertilizers under assured irrigated conditions. Nitrogen (N) @ 80–120 kg/ha, phosphate (P) @ 60 Kg of P_2O_2/ha and potash (K) 40 kg/ha should be applied. However, all the fertilizers are applied on basis of soil test and the application depends upon previous crop taken. Wheat following maize, bajra, jowar and rice should be given 120 kg N/ha. Wheat crop following legume may be given only 80 kg N/ ha under irrigated conditions. 1/3rd N with full dose of P and K are to be applied as basal dose and the rest 2/3rd N at first node stage at around 30–40 days after planting.

Micronutrient deficiencies of zinc, manganese, sulphur and boron are appearing in certain high intensity cropping areas and should be rectified on the advice of experts. Use of zinc sulphate @ 25 kg/ha in zinc deficient soils is advised to rectify the defects.

Irrigation Schedule

Depending upon the water availability four to six irrigations are enough for wheat crop. These should be applied as per following schedule:

(I) Under optimum water availability:

(a) *Four irrigations*: Crown root initiation (CRI), *i.e.* 20 days after sowing, tiller completion, flowering and milk stages.

(b) *Five irrigations*: CRI, tiller completion, late jointing, flowering and milk stages.

(c) *Six irrigations*: CRI, tiller completion, late jointing, flowering, milk and dough stages.

(II) Under limited water availability:

(a) *One irrigation*: In between CRI and tillering stages

(b) *Two irrigations*: CRI and boot stages

(c) *Three irrigations*: CRI, boot and milk stages.

Weed Control

Phalaris minor, Avena ludoviciana, A. fatua, Lathyrus aphaca, Lolium temulentum, Poa annua, Asphodelus tenuifolius, Chenopodium album, Crisium arvense, Medicago denticulata, Cornopus didymus, Rumex dentatus, R. retroflexus, R. maritimus, Spergula arvensis, Vicia sativa, Angallis arvensis, Carthamus oxycantha, Convolvulus arvensis, Polypogon monspelnsis, Malwa neglecta, Melilotus indica, M. alba, Fumaria parviflora etc. are pre-dominant weeds associated with wheat crop. When these weeds grow in close proximally of wheat crop, the competition begins and the yield is reduced. Therefore, weed management is focused in cropping sequence. In rice-wheat cropping system, *Phalaris minor* is the major weed. If *P. minor* is the dominant weed Stomp 30 EC (Pendimethalin) 3300 ml/ha (1000 g a.i./ha) at 0–3 days after sowing in 500 litre of water/ha is applied as pre-emergence treatment. This herbicide controls both grasses and broad leaved weeds. For broad leaved weeds like *Chenopodium, Convolvulus* etc. 2, 4–D @ 500 g a.i./ha in 250–300 litres of water/ha or metasulfuron methyl @ 4 g a. i./ha in 250–300 litres of water/ha is applicable. Since weeds spread very fast, cultural practices are also used as preventive measures.

1.10 WHEAT DISEASES

Wheat plants are subject to a number of biotic and abiotic stresses in all stages of growth. Fungi, viruses, bacteria and nematodes interfere with normal functioning of physiological processes of plants and cause different

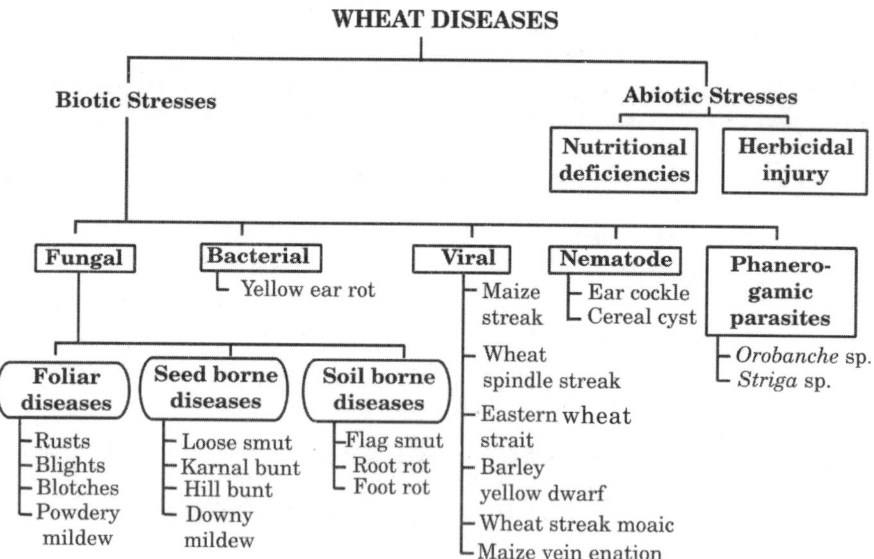

Figure 10: Major diseases of wheat prevalent in India

seed, soil and air borne diseases (Fig. 10). The potentially damaging diseases of wheat are three rusts, loose smut, Karnal bunt, foliar blights, powdery mildew whereas hill bunt is serious problem of hilly regions. Nematodes are significant in a particular region. Joshi *et al.*, (1978) have noted that viruses and bacteria are not very important pathogens of wheat. In certain wheat growing areas, the deficiency of nutrients like nitrogen, zinc, potash and sulphur has been found to cause physiological disorders in plants (Plates 1–8).

Though research on wheat diseases was initiated in the early twentieth century, intrusive work on this aspect was taken up after 1970 and information has been compiled and reviewed by many workers. In this context, diagnosis of wheat diseases, nature and extant of damage due to fungal, bacterial, viral and nematode diseases, and physiologic disorders; survey and surveillance in relation to distribution; dynamics of disease appearance; ecological factors governing disease development and epidemiology; variability in pathogens and strategic management planning in tackling the wheat health through integrated disease management approaches are introduced separately in different chapters of the book.

REFERENCES

Gill, K.S. 1979. *Research on Dwarf Wheat*. Indian Council of Agricultural Research, New Delhi, 180 p.

Gill, K.S , Indu Sharma and S.S. Aujla. 1993. *Karnal bunt and Wheat Production*., Punjab Agricultural University, Ludhiana, 153 p.

Hanchinal, R.R., N.B. Yenagi, G. Bhuvaneshwari and K.K. Math. 2005. *Grain Quality and Value addition of Emmer wheat,* University of Agricultural Sciences, Dharwad, 63 p.

Hanson, H., N.E. Borlaug, and R.G. Anderson. 1982. *Wheat in Third World*. Westview Press/Boulder, Colorado, USA, 174 p.

Jain, H.K. 1978. The Story of Wheat. **In**: *Souvenir* (Eds. M.S. Swaminathan, U.S. Rakesh, V.S. Mathur, J.L. Bhatt, S.K. Bhatia and D.S. Bajaj), pp. 21-24. Fifth Intern. Symposium, New Delhi.

Joshi, L.M., K.D. Srivastava, D.V. Singh, L.B. Goel and S. Nagarajan. 1978. *Annotated Compendium of Wheat Diseases in India*. ICAR, New Delhi, 332 p.

Joshi, L.M., D.V. Singh and K.D. Srivastava. 1986. Wheat and Wheat Diseases in India. **In**: *Problems and Progress of Wheat Pathology in South Asia* (Eds. L.M. Joshi, D.V. Singh and K.D. Srivastava), pp. 1-19. Malhotra Publishing House, New Delhi, 401 p.

Martin, J.M., W.H. Leonard and D.L. Stamp. 1976. *Principles of Field Crop Production*. McMillan Pub. Co., New York.

Mehra, K.L. 2000. History of crop cultivation in Pre-historic India. **In**: *Ancient and Medieval History of Indian Agriculture and its Relevance to Sustainable Agriculture in the 21st Century* pp. 11-27 (Eds. S.L. Chaudhary, G.S. Sharma and Y.L. Nene.) Proc. Summer School, Rajasthan College of Agriculture, Udaipur, 363 p.

Mishra, B., Jag Shoran, R. Chatrath, A.K. Sharma, R.K. Gupta, R.K. Sharma, J. Singh, J. Rane and Anuj Kumar. 2005. *Cost Effective and Sustainable Wheat Production Technologies*. Directorate of Wheat Research, Karnal Technical Bulletin No. 8, 36 p.

Pal, B.P. 1966. *Wheat*. Indian Council of Agricultural Research, New Delhi, 370 p.

Shaktawat, M.S. and Singhi, S.M. 2000. History of Cereals: Its significance for Achieving Sustainable production. **In**: *Ancient and Medieval History of Indian Agriculture and its Relevance to Sustainable Agriculture in 21st Century*. pp. 102-110. (Eds. S.L. Choudhary, G.S. Sharma and Y.L. nene). Proc. Summer School Rajasthan College of Agriculture, Udaipur, 363 p.

Tomar, S.M.S., Vinod and B. Singh 2004. Genetic resources and their utilization in wheat. **In**: *Wheat for Tropical Areas* . (Eds. K.A. Nayeem, M. Sivasamy and S. Nagarajan), pp. 3-9. Proc. Nat. Symp. Wheat Improvement for Tropical Areas, TNAU, Coimbatore, 300 p.

Vishnu-Mittre. 1971. Protohistoric records of Agriculture in India. J.C. Bose Endownmet Lecture. Trans. Bose Research Institute, Calcutta.

Vishnu-Mittre. 1981. Wild plants in Indian folk life- a historical perspective. **In**: *Glimpses of Indian Ethnobotany* (Ed. S.K. Jain), 37-58 pp. Oxford and IBH Publishing Co. Pvt. Ltd. New Delhi, India.

Wheeler, M. 1968. *The Indian Civilization*. Cambridge, U.K.

Zadoks, J.C., Chang, T.T. and Konzak, C.F. 1974. A. decimal code for the growth stages of cereals. *Weed Research* **14**: 415-421.

Diseases of Wheat

Stem rust

Leaf rust

Stripe rust

Spot blotch

PLATE – 1 Uredo-pustules of (A) Stem rust (*Puccinia graminis tritici*), (B) Leaf rust (*P. recondita*), (C) Stripe rust (*P. striiformis*) and lesions of (D) Spot blotch (*Drechslera sorokiniana*) on leaf blade and stem.

Diseases of Wheat

Tan spot

Alternaria leaf blight

Speckled leaf blotch

Glume blotch

PLATE – 2 Lesions of (A) Tan spot (*Drechslera tritici repentis*), (B) Alternaria leaf blight (*Alternaria triticina*), (C) Speckled leaf blotch (*Septoria tritici*), and (D) Glume blotch (*S. nodorum*) developed on leaf blade

Diseases of Wheat

Powdery mildew

Loose smut

Karnal bunt–infected spike

Karnal bunt–infected seeds

PLATE - 3 White mycelial growth of (A) Powdery mildew (*Erysiphe graminis tritici*) on leaf blade, (B) Loose smut (*Ustilago tritici*) and (C) Karnal bunt (*Tilletia indica*) infected spike, and (D) Wheat seeds showing Karnal bunt infection

Diseases of Wheat

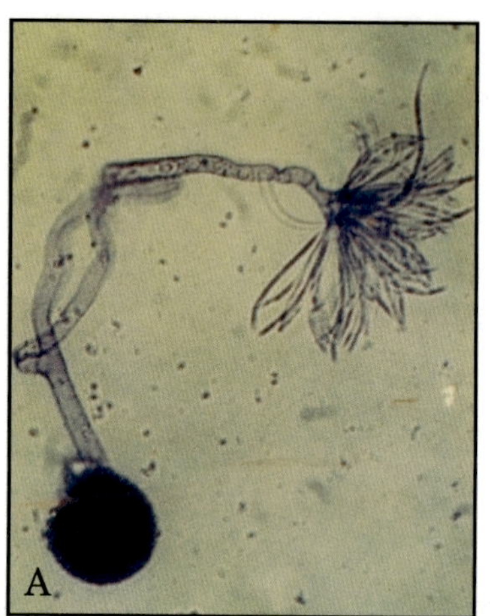

Teliospore of *Tilletia indica*

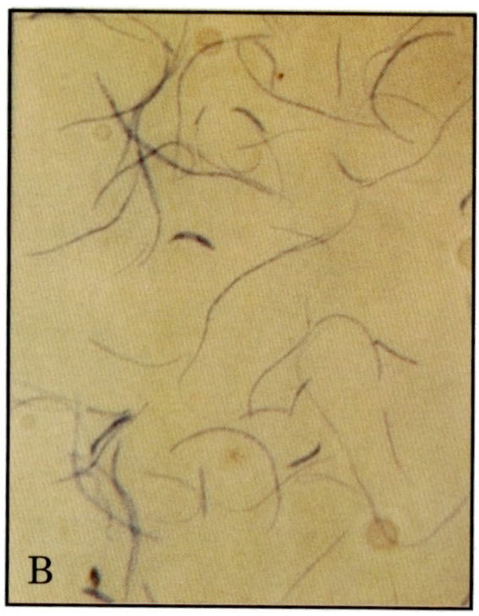

Sporidia of *Tilletia indica*

Hill bunt infected spikes and seed

Black point

PLATE - 4 (A) Germinated teliospore of Karnal bunt (*Tilletia indica*) showing promycelium and terminal sporidia (B) Secondary sporidia of Karnal bunt, (C) Hill bunt (*Tilletia caries/T. foetida*) infected spikes and grains, and (D) Wheat seeds with black point

Diseases of Wheat

Downy mildew

Flag smut

Barley yellow dwarf virus

Maize streak virus

PLATE - 5 Crazy top symptom of spike due to Downy mildew (*Sclerophthora macrospora*), (B) Black streaks of Flag smut (*Urocystis agropyri*) sori on leaf blade, (C) Stunting of plants and yellowing of leaves developed by barley yellow dwarf virus, and (D) Streaks of maize streak virus

Diseases of Wheat

Wheat spindle streak mosaic virus

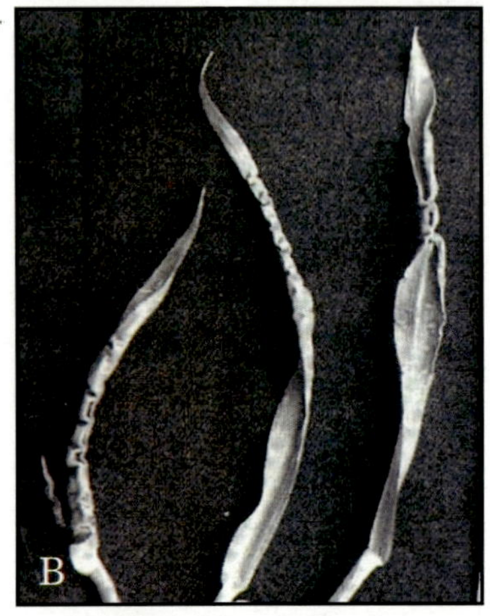

Ear cockle

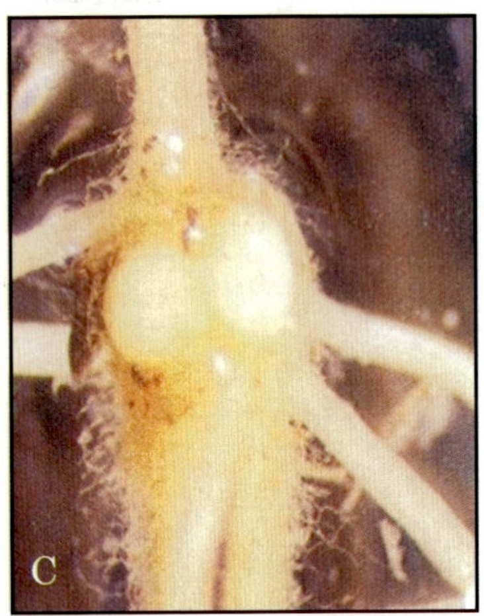

Cereal cyst nematode

Roots affected by cereal cyst nematode

PLATE - 6 (A) Chlorotic and necrotic dashes developed on leaf blade by wheat streak mosaic virus, (B) Twisted and crinkled leaves due to Ear cockle nematode (*Anguina tritici*), (C) Pearl like gravid females of *Heterodera avenae* (CCN – cereal cyst nematode), and (D) Roots of CCN susceptible Ⓢ and resistant Ⓡ plants

Diseases of Wheat

Yellow ear rot

Fusarium head scab

Orobanche sp.

Striga sp.

PLATE - 7 (A) Yellow ear rot (*Corynebacterium tritici*) – abortive spikes of wheat covered with yellow sticky bacterial exudate, (B) Pinkish mycelial growth on spike due to *Fusarium* species, and (C) *Orobanche* sp. and (D) *Striga* sp.

Mineral Deficiency Symptoms of Wheat

PLATE – 8 Deficiency symptoms produced by (A) Nitrogen (B) Phosphorus (C) Potassium (D) Sulphur (E) Magnesium (F) Iron (G) Manganese (H) Copper (I) Zinc

Chapter

Fungal Diseases

Wheat crop in India suffers from many insect-pests and diseases. Barring termites and grain-storage pests hardly any insect pest (cutworm, armyworm, leaf hopper, mite, etc.) has seriously threatened the crop. Contrary to insect pests, the crop right from the time of sowing till harvest is at risk from the attack of various diseases caused by diverse types of pathogens of fungal, bacterial, viral and nematode origin. Among these, fungal diseases of wheat are more important because they cause substantial reduction in yield as well as deterioration in quality of grain. Three rusts of wheat, spot blotch, Alternaria leaf blight, powdery mildew, Septoria blotches, smut, bunts and several other diseases associated with wheat need attention in maintaining the productivity curve of the crop. The fungal diseases are described in this chapter with respect to their symptoms, causal organism, nature and extent of damage, disease cycle, epidemiology and protection measures.

2.1 RUST DISEASES

Rusts are among the earliest known diseases which co-existed with wheat crop since ancient period. There has been great historical importance of rusts. Early records testify that Pliny, the Roman author in the 1st century A.D. described rust of wheat as the greatest pest of crops (Hanson *et al.*, 1982). Aristotle (384–322 B.C.) wrote of rust being produced by the "warm vapours" and mentioned the devastation of rust and years when rust epidemic occurred. Theophratus (371–286 B.C.) reported that rust was more severe on cereals than legumes (Roelfs *et al.*, 1992). There are Biblical allusions to wheat rusts. The earliest reference of this kind is the caution of Moses to the Israelies during their entry into Canaan (Deut 28.22) around 1360 B.C. Moses had warned the people of Israel that if they failed to observe the commandments of Jahova, "Lord would punish them with smut and rust." A similar reference of Bible in the prayer attributed to Solomon (I King 8:37; II Chron 6:28) offered at the dedication of the temple, is the testimony of the existence of rust around 950 B.C. The exhortation of astrologer Haggai (500 B.C.) to the people, "I smote you with smut and with rust and with hail in the labour of your hand, yet turned not to me", is another allusion of the prevalence of this disease in ancient period (Levine, 1919). Excavations in Israel have revealed uredospores of stem rust that have been dated at about 1300 B.C. (Kislev, 1982). Romans used to celebrate religious festival Robigala to appease the god Robigo or Robigus. "Stem Robigo, spare the herbage of cereals....withhold we pray thy roughening hand....", was part of the official prayer at Robigala ceremony as given by Ovid (43 B.C.-17 A.D.). It gives the impression that stem rust was serious disease in Italy during that time (Arthur, 1929).

Physiological specialization: The rust pathogen *Puccinia*, an obligate fungal parasite belongs to the division Eumycota, sub-division Basidiomycotina, class Teliomycetes, order Uredinales and family Pucciniaceae. The genus *Puccinia* was first named by P.A. Micheli (1729) and its relationship with wheat rust was established by Persoon in1794 who termed stem rust of wheat as *Puccinia graminis.* Subsequently, Schroeter (1879) observed specialization of parasitism in *Puccinia graminis* and Eriksson (1894) provided evidence of physiological specialization through cross inoculation experiments. Cross inoculation tests led to identification of five varieties in *Puccinia graminis* viz. *tritici* on wheat, *avenae* on oat, *secalis* on rye, *agrostidis* on *Agrostis* sp. and *poae* on *Poa* sp. and the term *formae specialis* (f. sp.) was coined. These varieties of *P. graminis* can be identified on the basis of their pathogenicity on specific host and on spore morphology. The *formae specialis tritici* of stem rust (*Puccinia graminis*) is able to attack wheat but does not infect oat, rye, *Poa* and *Agrostis*, though they are morphologically similar. Thus, the pathogen of stem rust of wheat is designated as *Puccinia graminis* Pers.f.sp. *tritici* Eriks. & Henn. Based on the same phenomenon leaf rust is named as *Puccinia recondita* Rob.ex Desm. f.sp.*tritici* and stripe rust as *Puccinia striiformis* Westend f.sp. *tritici.*

After the discovery of *formae specialis*, Stakman and Pimesiel (1917) further showed variability in the pathogen within *tritici* variety and demonstrated the existence of physiologic races. Further more rust races in turn could be divided into the ultimate sub-units called biotypes. A biotype is a population of individuals which are genetically identical. These entities are also named as biologic species, parasitic strains, physiologic forms, pathotypes etc. The races and biotypes are identified on standard differential hosts. Accordingly a set of differential varieties for identification of races was proposed for stem rust (Stakman and Levine, 1922), leaf rust (Johnston and Mains, 1932) and stripe rust of wheat (Gassner and Straib, 1932). In order to identify new races arising due to changes in varietal pattern few supplementary differentials were also added (Nayar *et al.,* 2002).

In India, Mehta (1940) initiated systematic race analysis of cereal rusts and subsequently others followed this system and identified races of three rusts. When an isolate of rust is subjected to inoculation on differential host varieties, it produces different infection types (host-pathogen response) as evident by development of uredo-pustule, (Joshi *et al., 1988).* These infection types (Fig. 11) are categorized and graded as 0 (immune), 0; (nearly immune), 1 (very resistant), 2 (moderately resistant), 3 (moderately susceptible), 4 (susceptible) and X (mesothetic).

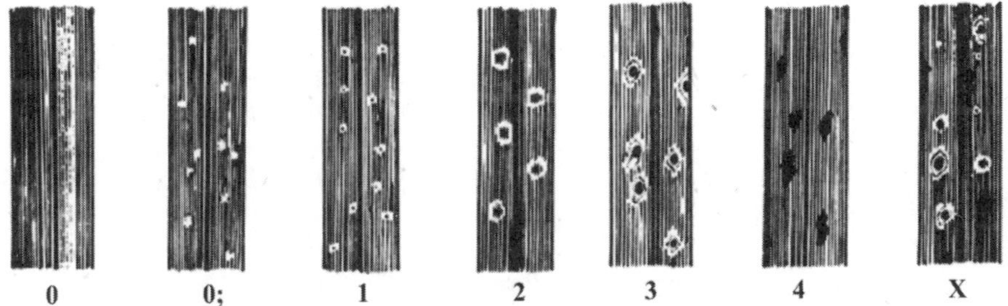

Figure 11: Infection types of stem rust (*Puccinia graminis tritici*) of wheat

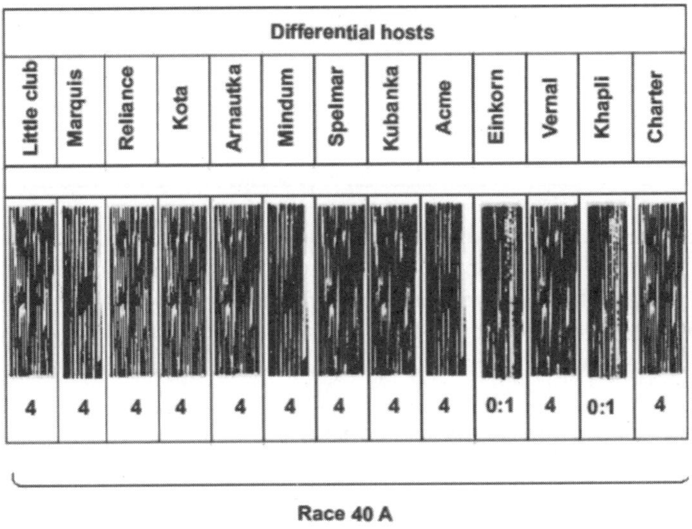

Differential hosts												
Little club	Marquis	Reliance	Kota	Arnautka	Mindum	Spelmar	Kubanka	Acme	Einkorn	Vernal	Khapli	Charter
4	4	4	4	4	4	4	4	4	0:1	4	0:1	4

Race 40 A

Figure 12: Designation of stem rust (*Puccinia graminis tritici*) race 40 A based on infection types.

Table 2. Differential sets 0, A and B constituted for the identification of races of *Puccinia* species infecting wheat in India

Set-0	Set-A	Set-B
	Stem rust (*Puccinia graminis tritici*)	
Sr 24	*Sr 13*	Marquis (*Sr76+*)
NI 5439	*Sr 9b*	Einkorn (*Sr 21)*
Sr 25	*Sr 11*	Kota (*Sr 28+)*
DWR 195	*Sr 28*	Reliance (*Sr 5+*)
HD 2189	*Sr 8b*	Charter (*Sr 11+*)
Lok 1	*Sr 9e*	Khapli (*Sr 7a, Sr 13, Sr14*)
HI 1077	*Sr 30*	
Barley local	*Sr 37*	
Agra local		
	Leaf Rust (*Puccinia recondita tritici*)	
IWP 94		
Kharchia mutant	*Lr 14a*	Loros (*Lr 2c*)
Raj 3765	*Lr 24*	Webster (*Lr 2a*)
PBW 343	*Lr 18*	Democrat (*Lr 3*)
UP 2338	*Lr 13*	Thew (*Lr 20*)
K 8804	*Lr 17*	Malakoff (*Lr 1*)
Raj 1555	*Lr 15*	Benno (*Lr 26*)
HD 2189	*Lr 10*	HP 1633 (*Lr 9+*)
Agra local	*Lr 19*	
	Stripe rust (*Puccinia striiformis*)	
WH 147	Chinese 166 (*Yr 1*)	Hybrid 46 (*Yr 4*)
Barley Local	Lee (*Yr 7*)	Heines VII (Yr 2+)
WH 416	Heines Kolben (*Yr 6*)	Compir (*Yr 8*)
HD 2329	Vilmorin 23 (*Yr 3*)	*T.spelta album*
HD 2667	Moro (*Yr 10*)	Tc6/*Lr 26 (Yr 9*)
PBW 343	Strubes Dickopf	Sonalika (*Yr 9*)
HS 240	Suwon 92 x Omar	Kalyansona (*Yr 2 Ks*)
Anza	Riebesel 47/51 (*Yr 9+*)	
A-9-30-1		

Observations pertaining to infection type of uredo-pustules are taken after expression of symptoms. The matching with the reactions recorded in the International Key for identification of races of different rusts confirms the designation of the physiologic race. For example, the isolation and designation of stem rust race 40 A is illustrated in Figure 12.

2.1.1 New Virulence Analysis System

Flor (1956) proposed the gene-for-gene theory and explained that for every virulence gene in the pathogen, there is a corresponding resistance gene in the host. It means that host resistance to a specific isolate of rust is a genetic trait. It was, therefore realized that the pathotypes identification system should be meaningful in deciding the avirulence/virulence structure of pathotypes. Consequently, the new systems using near isogenic lines for leaf rust (Nagarajan *et al.,* 1985), stem rust (Bahadur *et al.,* 1985) and stripe rust (Nagarajan *et al.,* 1985) were proposed in India. The new system works on three sets of differentials *i.e.* '0-set', 'A-set' and 'B-set' (Table 2).

Set-0 consists of a universal susceptible, a resistant line and present day cultivars. This set is not taken into consideration for nomenclature of the virulence. It acts as a watchdog in the detection of minor variability in the pathogen. The A-set consists near isogenic lines of Stem (*Sr*), Leaf (*Lr*), and Stripe (*Yr*) rusts. This set is useful in recognizing the effectiveness of genes and also quantifies virulence of the pathogen. In B-set, selected lines from old international differentials have been included. The old differentials in set-B are able to monitor the variability in the rust population effectively.

In the new virulence analysis system the race/pathotype is designated on the basis of binary notation system of Habgood (1970). Under this system the isogenic lines of A-set and entries of B-set have been arranged serially as per their strength to the different races of the pathogen and each has its decanary value which involves rising of a value to the base 2. By multiplying the decanary value with binary notation (2^0, 2^1, 2^2, 2^3, 2^4, 2^5, 2^6, 2^7) a decoded binary value is arrived (1, 2, 4, 8, 16, 32, 64, 128). When an isolate of the pathogen is tested on isogenic lines of A-set and differential hosts of B-set variable type of infection *i.e.* 0, 0; 1, 2, 3, 4 and X are developed. These are recorded and following binary coding system resistant (R) reactions (0, 0; 1, 2) are denoted as '0' and susceptible (S) reactions (3, 4, X) as '1'. By multiplying the binary notation of reactions with decanary value of isogenic line/differential host a number is arrived. For example the nomenclature of race 40A of *P. graminis tritici* is 62 G 29. The value 62 is on the basis of isogenic *Sr* lines in set-A where as 29 is the value on set-B. The value of set-A and set-B is separated by capital alphabet G for stem rust. (Table 3) In case of leaf rust alphabet R is used to denote the pathotypes of leaf rust and alphabet S is used for stripe rust.

Physiologic races and biotypes of stem, leaf and stripe rusts of wheat have been identified in India from time to time and pure cultures of the isolates are being maintained at Regional Station of Directorate of Wheat Research at Shimla in Himachal Pradesh (Table 4).

2.1.2 Evolution of Rust Races

In nature, the host and the pathogen co-exist in a state of balance. Breeding wheat resistant to the rust is the principal method used to control rusts. If a variety possesses desirable level of resistance, then it exerts strong

Table 3. New nomenclature of stem rust (*Puccinia graminis tritici*) race 40A

Decanary/ binary value	Set A							
	Sr 13	***Sr 9b***	***Sr 11***	***Sr 28***	***Sr 8b***	***Sr 9e***	***Sr 30***	***Sr 37***
Decanary value	2^0	2^1	2^2	2^3	2^4	2^5	2^6	2^7
Decoded value (a)	1	2	4	8	16	32	64	128
Rust reactions	R	S	S	S	S	S	R	R
Binary notation (b)	0	1	1	1	1	1	0	0
Decoded decanary value (a) × binary notation (b)	1×0 = 0	2×1 = 2	4×1 = 4	8×1 = 8	16×1 = 16	32×1 = 32	64×0 = 0	128×0 = 0
Total	0+2+4+8+16+32+0+0 = 62							

Table 3. *Contd.*

Decanary/ binary value	Set B					
	Marquis	**Einkorn**	**Kota**	**Reliance**	**Charter**	**Khapli**
Decanary value	2^0	2^1	2^2	2^3	2^4	2^5
Decoded value (a)	1	2	4	8	16	32
Rust reactions	S	R	S	S	S	R
Binary notation (b)	1	0	1	1	1	0
Decoded decanary value (a) × binary notation (b)	1×0 = 0	2×1 = 2	4×1 = 4	8×1 = 8	16×1 = 16	32×0 = 0
Total	1+0+4+8+16+0 = 29					
	Race 62 G 29					

Table 4. Physiological races and biotypes of stem rust, leaf rust and stripe rust recorded in India

Stem rust (*Puccinia graminis tritici*)		Leaf rust (*P. recondita tritici*)		Stripe rust (*P. striiformis*)	
Old name	**New name**	**Old name**	**New name**	**Old name**	**New name**
11	79G31	10	13R19	13	67S8
11A	203G15	11	0R8	14	66S0
14	16G2	12	5R5	14A	66S64
17	73G7	12-1	5R37	19	70S0-2
21	9G5	12-2	1R5	20	70S0
21-1	24G5	12-3	49R37	20A	70S64
21A-1	20G21	12-4	69R13	24	0S0-1
21A-2	75G5	12A	5R13	31	67S64
24	18G3	17	61R24	38	66S0-1
24A	5G19	20	5R27	38A	66S64-1
34	26G13	26		57	0S0
34-1	10G13	61		A	70S4
40	104G13	63	0R8-1	G	4S0
40A	62G29	70		G-1	4S0-3
40-1	62G29-1	77	45R31	I	38S102
42	19G35	77-1	109R63	K	47S102
42B	7G35	77-2	109R31-1	L	70S69
42B-1		77-3	125R55	M	1S0
42B-2		77-4	125R23-1	N	46S102
53		77-5	121R63-1	P	46S103
75		77-6	121R55-1	Q	5S0
117	37G3	77-7		T	47S103
117A	36G2	77A	109R31	U	102S100
117A-1	38G18	77A-1	109R23		46S119
117-1	166G2	77B			CI
117-2	33G3	104	17R23		CII
117-3	167G3	104-1	21R31-1		CIII
117-4	166G3	104-2	21R55	**Total races: 27**	
117-5	166G2-2	104-3	21R63		
117-6	37G19	104A	21R31		
122	7G11	104B	29R23		
184	53G1	106	0R9		
194		107	45R3		
222		107-1	45R35		
295	7G43	108	13R27		
Total races: 35		131			
		162	93R7		
		162A	93R15		
		162B			
		Total races: 39			

selection pressure on the pathogen. In order to survive, the pathogen has also its defense mechanism. When confronted by genetic resistance, the pathogen undergoes a change and it evolves into stronger/virulent forms or races to infect resistant host. Consequently, newer races or pathotypes continuously appear in nature. In wheat rust fungi five possible mechanisms, *viz.* hybridization, heterokaryosis, parasexuality, mutation and introduction from neighbouring countries operate for production of new races.

Hybridization: Hybridization as a mechanism of origin of new races through meiotic recombination is well documented in rust fungi. Craigie (1927) discovered the function of pycnia in the life cycle of *P. graminis tritici.* Based on Craigie's discovery it was shown that a diversity of wheat rust races results through recombination of factors for pathogenicity by hybridization on *Berberis* (alternate host). Obviously, new races are recorded in those regions where *Berberis* provides a common platform for sexual reproduction. When haploid basidiospores from different races of the fungus infect the same leaf of *Berberis* plant, monokaryon thalli with certain genetic characters are produced. The mycelium of the thalli grows intercellularly to give rise to pycnia or spermogonia. Hybridization is then accomplished by mating of resultant pycniospore of one race with receptive hyphae of another race to give to dikaryotic stage. It results in the formation of heterozygous characteristics. The heterokaryons that arise in the form of aeciospores are not capable to infect *Berberis* but they can initiate the life cycle of new diverse forms through repeating cycles of uredial stage.

Stakman *et al.* (1934) have shown that a number of physiologic races of *P. graminis tritici* are produced in areas like Pennysylvania and New York where *B. vulgaris* is common and functional. Stakman and Harrer (1958) isolated quite a few races from immediate neighborhood of *Berberis* bushes which were not found in USA and Mexico.

New recombinations can be obtained not only by hybridization but by selfing heterozygous races also. Newton *et al.* (1930) developed a method for crossing two rust races and could get a large number of recombinations along with new forms. In India, Mishra (1963) selfed stem rust races 21 and 42 and obtained altogether new races.

Heterokaryosis: In some countries like India, the alternate hosts, *Berberis* and *Thalictrum* are either absent or not functional; therefore, the chances of evolution of new races by hybridization are ruled out. Newton *et al.* (1955) regarded that in this situation heterokaryosis may perform prominent role in altering the pathogenicity of strains of wheat rusts by production of heterokaryons. The term heterokaryosis precisely describes the condition of a vegetative cell containing two or more genetically different nuclei. This condition arises on wheat from fungal hyphae of uredospores of two or more races by anastomosis. Anastomosis, as it is related to the establishment and development of heterokaryosis, involves fusion of hyphae, displacement of one or more nuclei into the fused cells.

In addition to anastomosis, heterokaryosis can also arise through gene mutation. The significance of heterokaryosis in the new pathogenic races has been demonstrated by Nelson (1956), Wilcoxson (1957) and Bridgman (1957) with *P. graminis tritici*, Vakili and Caldwell (1957) with *P. recondita* and Little and Manners (1969) with *P. striiformis.*

Parasexuality: Recombination of hereditary properties in the vegetative calls involving mitosis has been termed parasexualism. This process follows a sequence of steps. Heterokaryosis is the first step of parasexuality in which heterokaryotic mycelium of the fungus is produced through plasmogamy. This step is followed by fusion of unlike nuclei to give rise to heterozygous diploids. The next step is the multiplication and segregation of diploid nuclei by mitosis. As a result replicates of the diploid can develop from daughter nuclei possessing characters different from heterokaryons.

Possible role of parasexual cycle in *P. graminis tritici* and *P. recondita* has been indicated by Vakili and Caldwell (1957). Sharma and Prasada (1969) studied the phenomenon of production of new races by inoculating wheat cultivars with six pairs of races of *P. graminis tritici.* As a result of mixing races 15CM × 14 and 15CM × 17 fourteen races and biotypes were new and previously undescribed.

Mutation: It is another possibility of appearance of new races. Mutation is an abrupt change in genetic material of cell and occurs spontaneously in nature. Johnston (1930) recorded an aberrant race (race 27) of *P. recondita* arose by mutation. Mutation in rust fungi may bring about a change in pathogenicity or colour or both. In general rust mutants are less pathogenic but few races of more virulent mutants are also recorded. Watson (1957) recorded increased pathogenicity due to mutation in a race of *P. graminis tritici.* Colour mutants of races of *P. graminis tritici* and *P. recondita* have been recorded in India by Joshi and Kak (1955), Misra and Lele (1955), Sharma and Misra (1965) and Sharma (1967).

Race Introduction: Wind is an innocent carrier of rust uredospores with no regard for national borders. The light weight uredospores are produced in great number. The stem rust uredospores are produced @ 4×10^{12} uredospores /day/ha and that of leaf and stripe rust is said to be the order of 3.2×10^{13} and $0.6 - 2.0 \times 10^{13}$ respectively (Christensen, 1942). These spores can easily be transported over long distance by wind. This ability to spread on the wind, leave countries powerless to keep new races/ virulences/pathotypes from entering their territories. As an example of how efficiently races enter, it is speculated that stripe rust pathotypes 104

E 137 not prevalent in Australia prior to 1979, was introduced from Europe (Wellings and McIntosh, 1981). Another instance of rust advance is the dissemination of pathotypes 46 S 119 having virulence for *Yr9* in India via Ethopia, Turkey, Iran, Afghanistan and Pakistan.

Besides dispersal of uredospores by wind, human aided introduction of races/pathotypes by jet-age travel is also possible because people quickly travel between distant locations. Dubin (1984) speculated that pathotype 24 of stripe rust dispersed southward throughout the Anden area.

2.1.3 Rust Epidemics in India

Wheat rust epidemics have been recorded in India at different times and farmers have suffered from spectacular yield losses. Therefore, the rusts are listed as threatening diseases of national importance (Joshi *et al.,* 1978). It is apparent from the chronological account of rust epidemics (Nagarajan and Joshi, 1975) that famines due to rusts calamity were common during British period of East India Company. The earliest stem rust epidemic that occurred in the country was during 1786 A.D. in the district of Sagar (Central Province, now Madhya Pradesh). It was repeated during 1805, 1827, 1829 and many times thereafter. The British Army Major W.H. Sleeman noted in his diary that Narmada valley in Madhya Pradesh experienced a severe epidemic of rust in 1839 wherein the entire wheat crop was destroyed. In 1887 Col. K. Mackenzie also recorded an epidemic and mentioned that his clothes got red with uredospores while walking the fields of wheat in Central Province.

Epidemics of rusts have also occurred during India's post independence period. In 1946–47 an epidemic was recorded in the states of Bombay and Central Province. Bihar suffered from severe rust epidemic in 1956–57 (Prasada, 1960). In 1971–72 and 1972–73 the famous wheat variety Kalyansona was attacked by leaf and stripe rusts wherein nearly 0.8–1.5 million tonnes of wheat were lost. In 1978–79 the local wheat genotypes, Pissi Malvi and Pissi Red were destroyed in Narmada valley by stem rust epidemic causing 60–75% reduction in grain yield. Sonalika, a widely cultivated variety of wheat experienced leaf rust epidemic in entire Uttar Pradesh and Bihar and caused nearly one million ton reduction in wheat production (Joshi *et al,* 1975, 1980, 1984).

2.1.4 Nature and Extent of Damage

Wheat is subjected to attack of rust in all stages of its growth but the pathogen does not cause instant death of plants. Damage to plant depends on its growth relative to rust development. The rust infection that occurs at an early growth stage deprives the plants their normal development and vigour. The severe rust infection during flowering/heading stage is more detrimental. The impact of rust on wheat appears in different forms. There is reduction in number of tillers if plants are attacked by the rust before primodia formation. The rusted plants are stunted and they produce small spike in extreme situation. There is impairment of fertility of florets to reduce the number of seed set per head (Mains, 1930; Srivastava, 1974, 1984). Severe infection results in pre-mature ripening of plants. Thus, the major cause of loss in wheat yield is on account of production of shrivelled seeds with reduced size and grain weight (Pal, 1936; Gokhle and Patel, 1952; Gupta and Singh, 1981). The fungus interferes indirectly with normal growth of root and uptake of nutrients. All these factors reduce yield, flour milling quality and food value also.

The rust infection with in host tissues interrupts plants physiological processes. The rate of transpiration increases with the rupture of epidermis of diseased plant, malfunctioning of stomata and increased cell wall permeability. An increase in evaporation of water from rusted plants particularly at seed formation stage, results in shrivelling of grains. Increased water loss can seriously damage the plants under drought conditions. The respiration in infected plants greatly increases. Under the influence of rapid respiration carbohydrates are degraded to release carbon dioxide and energy. The photosynthetic area of wheat plant is reduced with onset of sporulation. It results in destruction of chlorophyll and reduction of green leaf area needed for reception of sunlight. Obviously the combined effect is reduced food for the production of seed.

2.1.5 Life Cycle

Anton de Bary (1886) resolved heteroecism in the rust and recorded that the life cycle of. *P. graminis* is completed on two different hosts. Wheat serves as the main host of the fungus and *Berberis* is the alternate host. Errikson and Henning in1896 further elaborated that three species of rust have as many as five spore stages (Chester, 1946). Pycnial or spermogonial stage (stage 0) and aecidial stage (stage I) occur on alternate hosts (*Berberis/ Thalictrum*) for completion of life cycle of the pathogen. The other three stages *i.e.* uredial (stage II), telial (stage III) and basidiosporal (stage IV) occur on wheat.

The uredospores are dikaryotic, meaning they have two nuclei (n+n). These spores are disseminated by the wind. They germinate on wheat plant by producing a germ tube. In case of stem and leaf rust appressoria over the stomata are formed but infection by the stripe rust takes place without formation of appressorium. The appressorium sends a small infection peg in the space beneath epidermis and forms a sub-stomatal vesicle. These vesicles then produce hyphae that grow between the host cells. The fungus gets its nourishment from protoplasm of the host cells through haustoria. After developing in the host tissue for about a week, the fungus begins to produce uredospores. The uredospores propogate the fungus asexually through repeated cycles and the uredial stage is reffered to as the repeating stage. Normally, the uredospores complete one cycle in 9 to 10 days under optimum conditions of temperature but at low temperatures incubation period is very much prolonged. Uredospore production is followed by teliospore development within uredia in separate telial sori at the end of the crop season. These spores are black, binucleate, 2-celled and have thick smooth walls. The teliospores are the resting spores and they remain attached to the host material. A period of freezing weather is required before the teliospores of stem and leaf rust fungi can germinate. Prior to germination, the two nuclei in each cell fuse to produce a true diploid condition (2n). The diploid nucleus formed in teliospores undergoes meiosis and germinates into four-celled promycelium (basidium) with uninuceate haploid sporidia (basidiospore). The basidiospores are of two different strains, namely + and – each being self incompatible. These sporidia are carried by wind, get deposited on alternate host and infect young leaves, producing yellow orange pycnia on upper leaf surface. Within the pycnium uninucleate, hyaline pycniospores (spermatium) are produced which are functional sex-gametes. In addition to pycniospres the pycnia also produce receptive or flexous hyphae. Pycnia arise by rupturing the epidermis of the leaf within 3–4 days. Receptive hyphae extend beyond the opening of pycnia. Insects, rain water and other agents transfer pycniospores from one pycnium onto the receptive hyphae of another pycnium. The pycniospore contents along with nucleus pass into the receptive hyphae through a pore by dissolving the wall at the place of contact. The nucleus divides into the compatible receptive hyphae. The extra nucleus migrates into an adjacent cell where the process is repeated and eventually the mycelium of pycnium becomes dikaryotized. The dikaryotic mycelium thus formed thereafter proliferates and forms aecial cups rupturing the epidermis on lower surface of leaves of alternate host. From the base of the aecium numerous rows of aeciospores are produced in chains. The dikaryotic spores are yellow, hexagonal, fairly thick walled and echinulate, having six germ pores. Aeciospores are incapable of infecting *Berberis/Thalictrum*. These get blown off by the wind to wheat where they produce infections that in turn produce urediospore, and the life cycle is completed (Fig. 13).

2.1.6 Perpetuation and Dissemination of Wheat Rusts in India

The rusts of wheat are polymorphic in nature as they produce five spore stages for completion of life cycle. The role of these spores is considered important in the perpetuation of the disease. Two stages of the rust fungus (*i.e.* uredial and telial) are produced on wheat and other two stages (*i.e.*

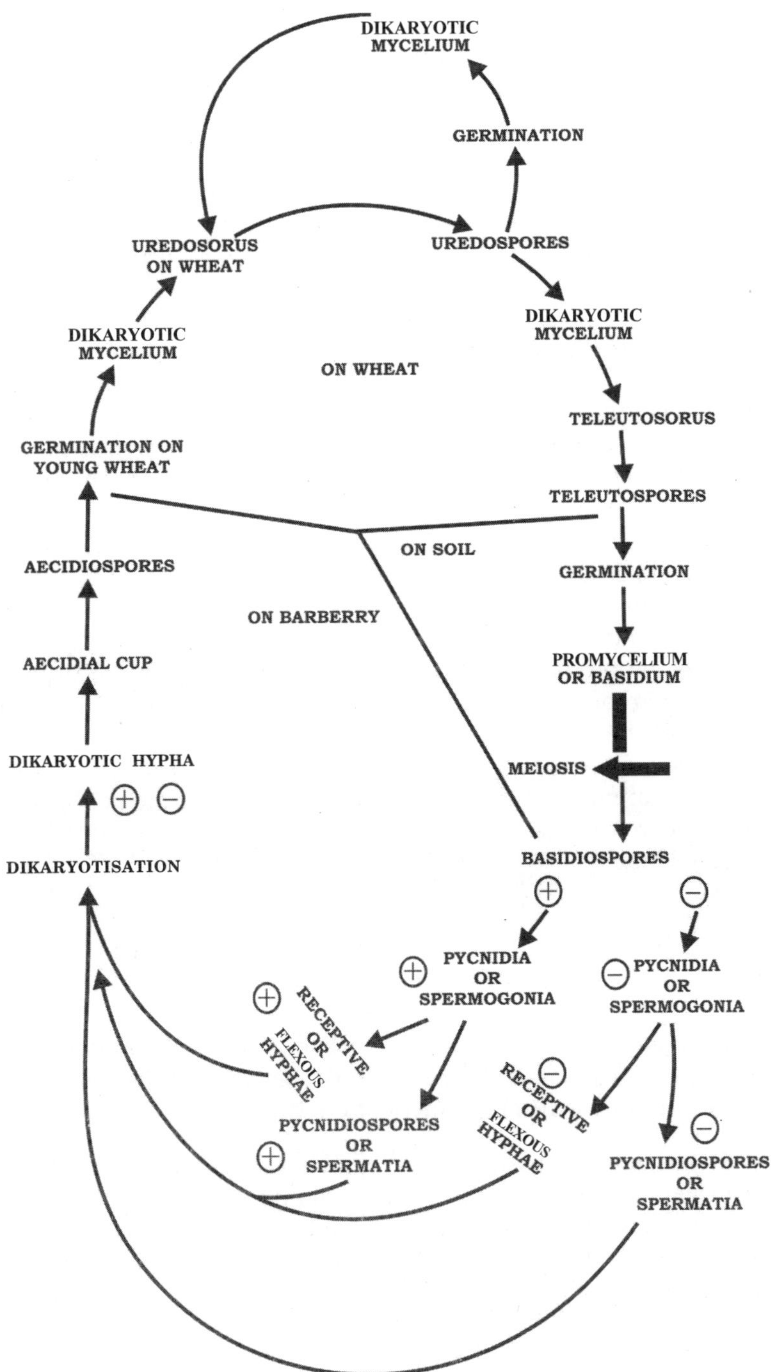

Figure 13: Generalized life cycle of the wheat rust fungi

pycnial and aecial) are produced on *Berberis* or *Thalictrum*. While it has been shown in USA and European countries that alternate hosts are the main source of inoculum for the onset of the disease on wheat but in India they do not appear to play any role of in perpetuation. Although Barclay (1887) in India had recorded an aecial stage on *Berberis aristata* in the neighbourhood of Shimla hills and he thought that this stage could be connected with *Puccinia graminis* but he did not confirm its genetic relationship. E.J. Butler, a pioneer in plant pathology in India raised doubts on the role of alternate host in the perpetuation of stem rust. Butler (1918) felt that the fungus has probably lost its aecial stage in India and the aecial stage described by Barclay may not be connected with stem rust of wheat. Butler's investigations were followed by the classical work of K.C. Mehta on cereal rusts at Agra college, Agra and Rust Research Laboratory, Shimla from 1922–1950. Mehta (1940, 1952) clarified that the aecial stages occurring on different *Berberis* spp. in the hills have no relationship with *P. graminis* and they are of no consequence in the perpetuation of stem rust in the country. Further studies confirmed that aecial stages existing in Shimla hills belong to *Puccinia graminis agropyrii* (Prasada, 1947), *P. poae nemoralis* (Joshi and Payak, 1963), *P. brachypodii* (Payak, 1965). According to Mehta (1940) there is scarcity of germinable teleutospores, and if at all viable spores germinate and produce basidiospores at higher altitudes, they do not infect barberries as their leaves are no longer tender and vulnerable to infection. So far no case of any outbreak of stem rust from barberries is on record. *B. vulgaris,* the alternate host of stem rust of wheat does not occur in the plains of India. Secondly, teleutospores formed on wheat crop also does not survive the intense heat during May-June following harvest of the crop. Under these conditions there is no local source of infection in the plains.

Thalictrum species (Jackson and Mains, 1921) are non-functional as far as perpetuation of leaf rust is concerned. The earlier recorded aecial stage of *Thalictrum javanicum* in Shimla hills was connected with *Puccinia persistens* of *Agropyron semicostatum* (Prasada, 1946). Other alternate hosts viz. *Isopyrum fumarioides* (Chester, 1946) *Clematis* spp. (de Oliveira and Samborski, 1966) are also not functional in India.

In the absence of any local source of infection, the problem of annual recurrence of three rusts in the plains was investigated by Mehta (1952). He noticed that dissemination of rusts that occurs in India, results from the planting and harvesting period of wheat crop in hills and plains and prevailing climate of these regions. Mehta visualized that the wheat crop in Nilgiri and Pulney hills in South is cultivated round the year. The rusts oversummer in their uredial stage in the cool climate of hills on wheat crop, self sown plants and ratoon tillers to act as source of inoculum. Similarily, the Himalayan chain covering India and Nepal has several valleys and gorges where wheat is grown during April-May and harvested in September-October. Sivalik ranges, Hindukush Mountain and North Western Frontier Province have been considered to be active foci of infection for dissemination of rust uredospores to Indo-Gangetic plain. Although several grasses as collateral host, such as *Bromus coloratus, B. carinatus, B. mollis, B. japonicus, B. patulus, Hordeum disticheom, H. murinum, H. stenostachys, Lolium perenne, Brachypodium sylvaticum, Hilaria jamesii, Aegilops squarrosa, A. ventricosa* and *A. trigecilis* have also been recorded to be susceptible to wheat rusts (Prasada, 1951, Vasudeva *et al.*, 1979) but they are not considered significantly important in the annual recurrence of rusts in the plains. Hence, the summer wheat crop of Nilgiris and Pulney hills in South and Himalayan ranges in North carries infection of rusts. These places serve as the primary foci of rust inoculum of three *Puccinia* species.

The fungus which multiplies in the form of uredospores on summer hill crop proceeds downwards through foot hills to infect early sown wheat crop. Once the amount of uredospores is produced in successive generations at regular interval, the wind borne uredospores are disseminated from foot hills to the plains in the main wheat crop season during winter months. After the spores are deposited on host, local environmental conditions greatly influence the subsequent infection and development processes. Since the wheat growing parts of India differ environmentally, particularly in temperature and moisture regime, the epidemic growth of different rusts varies from region to region. Mehta made outstanding contributions on epidemiology of wheat rusts in India and published classical scientific monograph entitled "Further studies on cereal rusts in India". Hardly any work on wheat rust epidemiology appears to have been done after 1940. However, in 1967, when India ushered an era of Mexican dwarf wheat, the Indian Agricultural Research Institute, New Delhi initiated a National Wheat Disease Survey and Surveillance Programme which rejuvenated the interest in the epidemiological studies of wheat rusts in the country (Joshi *et al.*,1971). Nagarajan and Singh (1973) divulged that in addition to wind, the inoculum of stem and leaf rusts that is available in the South Indian hills is carried further by tropical cyclones and is deposited along with rain over wheat crop in central-peninsular India. Some of these epidemiological aspects are further elaborated in the following pages.

Epidemiology of stem rust: The earlier held view of Mehta (1952) that both Himalayas and South Indian hills act as the primary foci of infection of *P. graminis tritici*, was re-investigated. Joshi *et al.*, (1974) revealed that stem rust uredospores spread primarily from Nilgiri and Pulney hills and Himalayas play a minor role in recurrence of epidemics of the disease in northern India. It has now experimentally established that the sporulation of stem rust over naturally infected fields in southern hills is faster under warm weather. The cumulate value of rust uredospores in air is several times more in South than in North in the month of November. Hence, many places in southern-peninsular India get early outbreaks of stem rust in December and January. In contrast, low temperature prevailing in Himalayan region during winter months hampers the sporulation and multiplication of *P. graminis tritici.* Therefore, stem rust in Himalayas remains dormant with almost no sporulation. Resultantly, the disease does not appear on wheat until late in season. Moreover, variation in temperature profile of southern and northern regions during crop season also appears to influence the incubation period of stem rust. In northern region of India, the duration of incubation period of the rust is longer. The pathogen under cool climate of Shimla takes 40–50 days to sporulate in winter months. As a result of longer period of incubation the pathogen is unable to multiply quickly and very little amount of inoculum is available in northern hilly region. On the contrary, *P. graminis tritici* takes 8–10 days for sporulation in Nilgiris and the multiplication of inoculum is faster through repeated cycles of uredospores. So these conditions suggest that the Himalayan region as primary foci of infection contributes little, if at all to stem rust epidemic in the main wheat belt and in all possibilities South Indian hills are considered the chief foci of stem rust infection (Joshi, 1976).

In India, uredospores of *P. graminis tritici* move from South to North and get deposited with rain drops. It is observed that the cyclonic disturbances that occur during October/November in the Bay of Bengal cross the Coromandal coast and then recurve inwards towards central-peninsular region of the country. The lasting weather pattern is invariably accompanied by heavy rain. Nagarajan and Singh (1974) examined the rain samples collected in a rain sampler designed by Roelf *et al.* (1970) along with satellite television cloud photographs (STCPs) and trajectories and proved that under the influence of cyclonic wind circulations the uredospores of *P. graminis tritici* present in the higher elevation of Nilgiris are lifted to conviction currents to 700 mb level and are transported to hundreds of kilometers towards North. The accompanying rain washes down the uredospores on 30 days old crop, traveling a distance of more than 700 km and the persistent damp weather favours infection of stem rust (Fig. 14).

This 700 mb synoptic weather situation is a fundamental requirement for the initiation of stem rust foci in peninsular and central wheat zones very early and its spread later to main wheat belt of the country. Based on this observation, three synoptic rules called the Indian Stem Rust Rules (ISR) have been formulated (Fig. 15):

ISR 1: A storm of depression should be formed either in the Bay of Bengal or in the Arabian Sea demarcated by longitude 65–81° East and 8–15° North.

ISR 2: A persistent high pressure cell over the southern part of central India (not far from the Nilgiris) must be present.

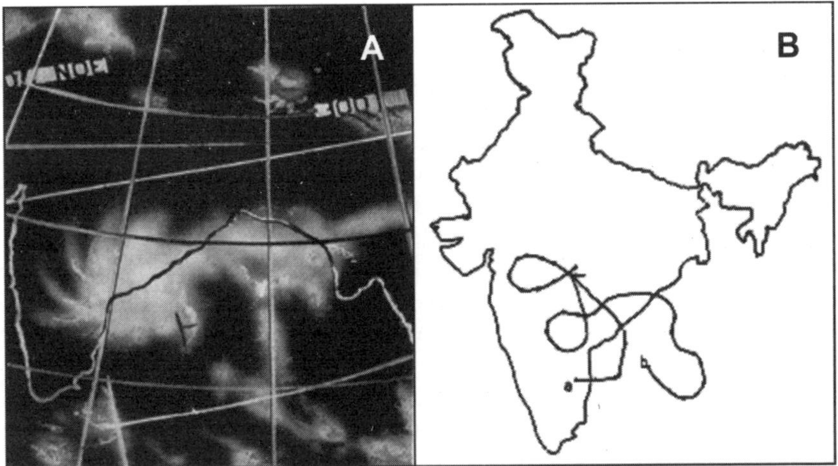

Figure 14: (A) Satellite photo of cloud cover over Indian sub-continent and (B) back-ward wind trajectories for 700 mb and 850 mb levels for Wardha.

ISR 3: Appearance of a deep trough extending upto southern India and caused by onward movement of western disturbance and associated rainfall over central India.

The weather pattern at 700 mb level can be monitored by scanning the weather satellite images and from drawing of corresponding backward wind trajectories for the transport of uredospores of *P. graminis tritici*. Using these tools it is possible to predict the appearance of stem rust in peninsular and central zones a month ahead of the field scouts observations. If the conditions of ISR 1 occur, then viable uredospores are deposited all over the central India where secondary disease foci appear. Since ISR 1 invariably occurs during November, only this weather situation brings in stem rust inoculum in adequate quantities to trigger and sustain the epidemic. The weather conditions stipulated in ISR 2 and ISR 3 occur after the month of

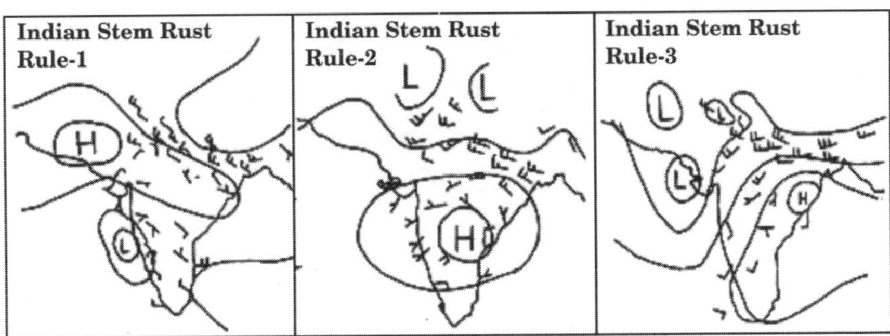

Figure 15: The 'Indian Stem Rust Rules'

December and therefore, only traces of infection appear. Since wheat crop matures in the country by early March, and due to the absence of conditions set in ISR 2 and ISR 3, weather is never able to promote stem rust epidemic in North Indian plains (Nagarajan and Muralidharan, 1995). This prediction system was proved to be robust and the annual travel route of the *P. graminis tritici* that became clear was designated as the "*Puccinia* path" of India by Nagarajan and Joshi (1980). They considered the whole Indian subcontinent to be a single epidemiological zone for stem rust which can be split into four sub zones (Fig. 16). Sub-zone 1, by virtue of its proximity to primary foci of infection, gets inoculum by means of katabatic wind. In sub-zone 2, the inoculum arrives either by wind currents or with rain around November while in sub-zone 3 the inoculum is transported by cyclonic disturbances and deposited by rains during October/November. This zone can serve as the main secondary focus for sub-zone 4 which remains free from stem rust in the absence of an appropriate tropical cyclone.

Epidemiology of leaf rust: Mehta (1940, 1952) suggested that leaf rust of wheat spreads both from South and North Indian hills. This view is supported by recent studies. It has been demonstrated that the first build up of leaf rust like stem rust takes place in the plains of Karnataka in South India, generally in the last week of December. Dissemination of leaf rust from southern foci to central-peninsular India is like that of stem rust and appearance of the disease can be predicted on the basis of Indian Stem Rust Rules. At the same time the infection also establishes in Bihar in North-eastern region of the country. The rust population from southern

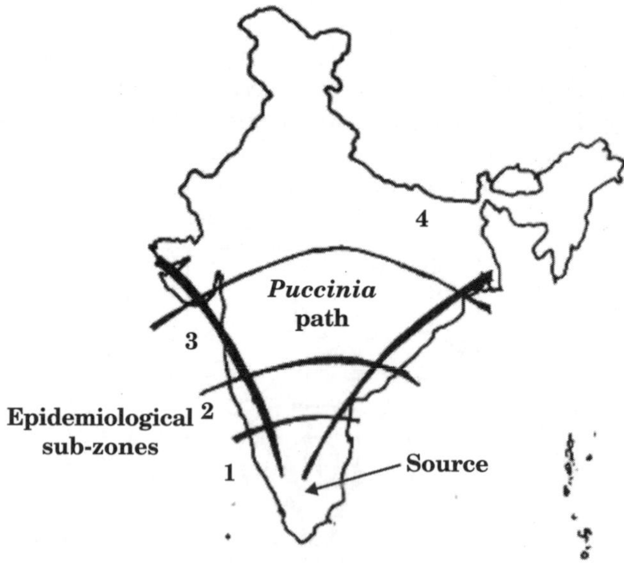

Figure 16: The 'Puccinia Path' of India and sub-epidemiological zones.

region moves northwards towards Maharashtra and Madhya Pradesh and population moves from northern region towards south. Finally both the populations, moving in opposite directions merge into each other (Joshi *et al.,* 1974).

Recent investigations have revealed that inoculum of *P. recondita* available in central Nepal becomes wind borne and arrive in the foot hills and adjoining plains of Bihar sometime during Christmas. The uredospores of leaf rust land on the wheat seedlings and the dew condensation on leaves favours infection and establishment of primary foci in North-eastern region. Such foci have been found up to 270 km from the nearest foot hills into parts of North east India (Joshi *et al.,* 1977). On the other hand few isolated pockets of infection along the foot hills of Jammu and Kashmir, Punjab, Himachal Pradesh and western Uttar Pradesh also appear but they remain less active because of cool climate. In comparison to North-western region the weather of North-eastern sector is warmer which enables the leaf rust to multiply many times more. Therefore, by early February leaf rust is well established in the eastern region of Indo Gangetic plains and then the inoculum moves from this area towards North-western direction. By the end of February with rise in temperature, the less active pockets of infection along the foot hills of Uttranchal, Himachal Pradesh, Punjab and Jammu region of J&K also become active and spread slowly towards North-east. Finally, the actively spreading population of uredospores from North-east gets mixed with the population from North-west. Thereafter, the two migrating populations are undistinguishable (Fig. 17).

Nagarajan *et al.* (1978) explored that the uredospores of leaf rust from northern region are also carried away by western disturbances. This observation suggests that if more western disturbances accompanied by frequent rains occur in North India, there is apparently good chance for the spread and built up of leaf rust in North-western region to lead to an epidemic.

Epidemiology of stripe rust: Stripe rust pathogen (*P. striiformis*) prefers low temperature for infection and symptom expression so it survives in uredial form at several locations in the Himalayan ranges in the absence of alternate host (Mehta, 1940). There are very few foci of infection which are capable of harbouring the stripe rust all the year round but in general the pathogen oversummers in the inner valley of Himalayas within a range of 2200–2500 m a.s.l. or above, shifting up and down from one place to another as the environment changes (Joshi *et al.,* 1977). By December/ January when the crop is about a month old the rust appears along the foot hills of northern hill zone/North-western plain zone along the mouth of rivers such as Tawi, Ravi, Beas, Sutluz and Yamuna where temperatures are low and dew is abundant (Nagarajan *et al.,* 2006). As in the leaf rust, in case of stripe rust also the katabatic winds carry uredospores of the pathogen

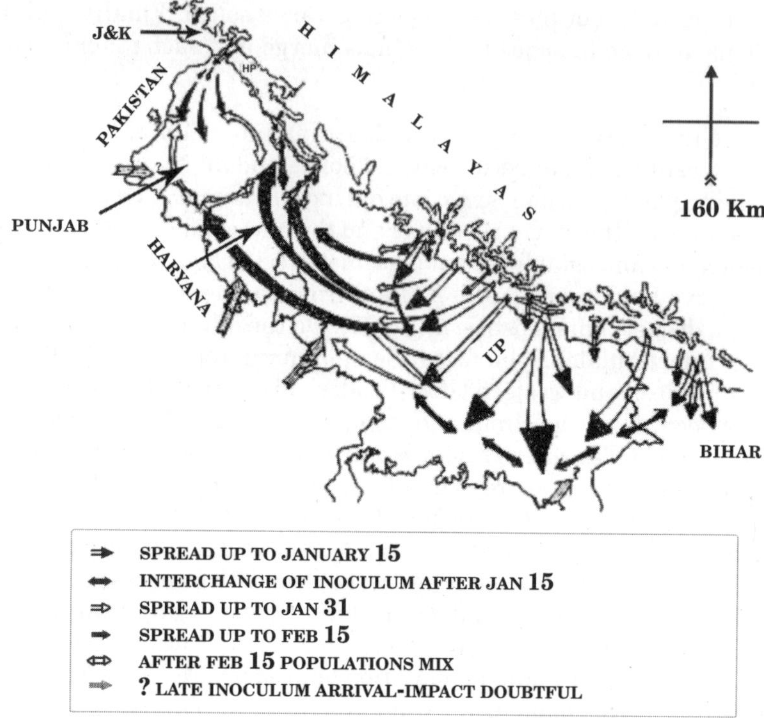

Figure 17: Spread of leaf rust (*Puccinia recondita*) in North-western India.

to the foot hills. From this region, the stripe rust spreads southwards and becomes well established in the northern part of the country by the end of February.

The disease can survive in Nilgiri and Pulney hills but it cannot spread even to the foot hills of Nlgiris due to unfavourable warm weather. So far there is no evidence to suggest that stripe rust inoculum from Nilgiri hills is being introduced in the northern wheat belt (Joshi *et al.,* 1984). Hence, this disease is essentially a major problem of North and North-western region of the country.

2.1.7 Approaches to Rust Management

Avoiding rust epidemics in the country is a complex challenge, given that agronomically superior varieties with good level of rust resistance become victim of the boom and bust cycle. When high yielding disease resistant varieties are released for cultivation they increase in acreage and occupy large areas. This phase of increase is called the boom. Simultaneously, there is also pathogen response whenever there is an increase in the level of host resistance. This response is due to selection pressure exerted by the host and when most of the rust races are unable to infect resistant host, the pathogen in response to it undergoes a change and evolves new virulent races. These races get selected and their frequency increases gradually over years, and an epidemic is resulted (Bahadur, 1986). This part is termed as the bust or the burst. If the gene involved is a weak gene, the boom and bust cycle is fast. Therefore, one has to delay or avert an unforeseen bust cycle to avoid an epidemic. Delaying of bust cycle can be made through various approaches as discussed herein-

Varietal Mixture: The idea of varietal mixture as an epidemiological strategy was emanated from Cambridge, U.K. (Wolfe and Barrett, 1980). This approach implies that precaution should be taken with regard to the genetic diversity in host. In case of pure line cultivar all plants in the field are identical, once infection gets established their further spread is unabated. If high yielding variety mixed together with resistant components is grown as a composite population then resistant plants would buffer against the pathogen thereby reducing the epidemic progress of the disease. The reduction in the rust intensity is possible on account of decrease in the percentage of susceptible ones and secondly, due to impediment and retardation of spread of inoculum and infections created by resistant plants (Basant Ram *et al.*, 1989).

Multiline Approach: The multiline approach that was originally proposed by Jensen (1952) is another means of generating diversity at field level. B›rlaug (1953) gave the multiline concept as an approach of long term effectiveness of host resistance. In this strategy several isogenic lines of similar phenotypic characters but differing with regards to specific gene for resistance, are physically mixed to constitute a multiline. By introducing diverse genes in a population of isogenic lines, the planning is to lower the rate of multiplication of the pathogen (r) and reduce the load of inoculum (X_0) in the micro-environment of the host. Thus overall behaviour is like horizontal resistance which makes multilines more enduring (Marshal *et al.*, 1972). The mechanism by which the multiline cultivar buffers against pathogen has been explained by Frey *et al.* (1975). In multiline cultivar, a number of genes for vertical resistance operate and each gene protects the cultivar against particular race of the pathogen. The spores from initial spore shower, if land on the resistant components they are trapped out of reproduction. In this way, both initial inoculum (X_0) and rate of infection (r) are bound to reduce in multiline population. The effective inoculum in each generation therefore, is reduced. The process of reduction in inoculum load is similar to sanitation and it results in time delay (t) in the build up of rust epidemic. This strategy has, therefore, epidemiological advantage in rust management. If multiline cultivars are deployed in North-eastern region (NEPZ) of India, it is believed that the initial inoculum of leaf rust introduced from Central Nepal will be reduced and multiplied at slow rate thereby minimising the massive spread of leaf rust inoculum from NEPZ to North-western region (Sivastava *et al.*, 1984).

KSLM-3, Bithoor and MLKS-II multiline varieties of Kalyansona were developed in India for cultivation (Aujla and Sharma, 1986) but due to certain limitations like agronomic conservativeness, large scale seed production, labour involvement, additional cost in seed production etc., these varieties have not been preferred (Kulshrestha, 1986).

Multigenic Varieties: Race-specific resistance is widely used in breeding programmes for the control of rusts. This type of resistance is based on a single major gene (or a combination of major genes). It is preferred because incorporation of race-specific gene into improved genotype through hybridization is simple. However, it is acknowledged that the use of race specific genes leads to boom and bust production cycles where genotypes succumb in short time to new virulent race of the pathogen. Therefore, Watson and Singh (1952) proposed that instead of single gene several genes for resistance, preferably early unused ones should be combined together in a desirable agronomic background to get durable resistance against rusts. Gill *et al.*, (1994) explained that well made combinations of several minor genes create evolutionary difficulties for the pathogen and thereby render the multigenic varieties more durable. This type of non race-specific resistance is fairly adequate against all the races of the pathogen for a number of years over a range of environment.

The conventional breeding technique following the pedigree and backcross method is used for pyramiding resistance genes for rust resistance in wheat. Although several genes can be incorporated by simple backcross breeding programme but this method is constrained by the influence of environment on the disease expression so the genes fail to express their trait consistently. Moreover, the necessity of adequate inoculum load of the pathogen is also crucial to establish the disease for any conventional methodology. Therefore, molecular markers linked to resistance genes have been identified. These markers are fragments of DNA and can be genetically located. Besides RFLP (restriction fragment length polymorphism) several other markers such as STS (sequence-tagged site), RAPD (randomly amplified polymorphic DNA), SCAR (sequence characterized amplified region), AFLP (amplified fragment length polymorphism), SSR (single sequence repeats or micro satellite) etc. have come into existence for their application in breeding programmes (Gupta *et al.*, 1999). Utilization of these markers show promise for marker assisted selection (MAS) of resistance genes and their pyramiding in one background. Das *et al.* (2007) have developed SCAR markers which can be used for pyramiding of the *Sr31* gene with other rust resistance genes in marker assisted selection.

Breeding programmes world over are designed to bring in two or more than two alien *Sr, Lr* and *Yr* genes in desirable agronomic background for direct use as multigenic varieties. McIntosh *et al.* (2003) have tagged a few resistance genes for rusts through molecular markers. In India NCL, Pune; PAU, Ludhiana; UAS, Dharwad and Agarkar Research Institute, Pune have attempted pyramiding of *Lr15* and *Lr34* genes in the background of drought tolerant wheat NI 5439 using various ISSR and SSR markers (Gupta *et al.*, 2007). SCAR and RAPD markers have been employed to pyramid the gene *Lr24* and *Lr28* together into well adopted and widely grown wheats, HD 2329 and PBW 343 at IARI, New Delhi and PAU, Ludhiana (Prabhu, 2004, Prabhu *et al.,* 2004). SCAR markers linked to *Lr9* have been used to combine *Lr9* + *Lr24* and *Lr9* + *Lr28* in one genetic background. Pyramids of rust resistance gene *Lr24*, *Lr28* and *Lr37* in popular Indian wheat HD 2733, HD 2687, PBW 343, HUW 234, Lok1 and WH 147 have been generated recently. The incorporation of *Sr24* with *Sr26*, *Sr27* and *Sr36* in Indian wheat cultivars is complete and molecularly confirmed (Anon, 2007).

Gene Cycling: The gene cycling assumes gene for gene relationship between host and the pathogen. When a new variety is released, a resistant allele in the host is effective and frequencies of the related virulence gene are low. As the frequencies of the related virulence gene increase in the pathogen, the effectivity of the resistance gene in the host is generally lost and the variety becomes susceptible. In order to deal with this problem Person (1967) suggested the utilization of resistance genes on rotational basis over a number of years. Thus gene cycling strategy aims at cyclic shift in the resistance genes used for breeding for disease resistance. In this approach the genes which have lost their effectiveness are withdrawn temporarily and are re-introduced when appropriate changes in the genetic content of the pathogen have taken place. For example, if a wheat variety having resistance gene *Lr13* becomes susceptible to leaf rust, then such a variety as a basic principle of gene cycling should be withdrawn from cultivation along with discontinuation of the use of related *Lr13* gene in the breeding programme for sometime. Such a discontinuity shall minimise the chances of evolution of complex virulences. This strategy facilitates gradual drop in the frequency of virulences matching corresponding resistance gene. Consequently, after a gap of few years the same variety again tends to behave as resistant and can be re-introduced for cultivation. However, Wolfe (1988) pointed out that except in rare circumstances, re-cycling a variety that has become susceptible is unlikely to be successful unless the resistance is hybridized into a new variety with a genetic background selected in the presence of matching virulence. In this context, Tomar and Menon (2001) developed near isogenic wheat varieties of popular Indian bread wheat cultivars like C 306, HD 2402, HD 2285, HS 240, HD 2329, Kalyansona, Sonalika, UP 262, WL 711, Lok1, WH 147, PBW 343 etc. by introgressing effective rust resistance genes. Many of these lines developed through such programme have become cultivars in India in recent times (Nayeem and Sivasamy, 2004).

Gene Deployment: The epidemiology of rust has an important bearing on their management in India. As explained earlier, Nagarajan and Joshi (1980) recognized that Puccinia Path consisting of four sub-epidemiological zones exists in India, therefore they explored the possibility of obstructing the path through deployment of horizontal resistance, slow rusting varieties in South and North Indian hills and vertical resistance against prevalent pathotypes in other zones to minimize the threatening rust epidemics or pendemics. Horizontal resistance appears to be useful at foci of rust infection because it is evenly spread against all races of the pathogen (VanderPlank, 1968) while in slow rusting rust develop slowly and never reaches a high degree of severity (Caldwell, 1968; Wilcoxson, 1981). As a result damage is slight. The stability of slow rusting is an important question. There is now ample evidence that minor genes for resistance can affect rust development at various stages, for example: receptivity, length of latent period, pustule size and spore production. Each gene has a relatively small effect but when several of them are combined, satisfactory resistance can result (Knott, 1988). Kulkarni and Chopra (1980, 1982) have noticed environment as the cause of differential interaction between slow rusting cultivars and pathogenic races. In comparison to horizontal resistance, the deployment of vertical resistance which is effective against some races seems to be useful in Zones 2, 3 & 4. Nagarajan *et al.* (1984) and Bahadur and Nagarajan (1985) suggested that gene combination *Lr2a* + *Lr15* and *Sr9b* + *Sr36* in Zone 2; combination *Lr15* + *Lr20* and *Sr11* + *Sr13* in Zone 3 and *Lr9* + *Lr10* and *Sr6* + *Sr8* in Zone 4 would be useful for deployment along the Puccinia Path. Recently, a revised strategy for managing stem and leaf rusts of wheat in six different ecological areas of the country has been proposed (Bahadur *et al., 1994)*. In areas 1 and 6, use of horizontal resistance/slow rusting varieties has been re-advocated as recommended earlier (Sharma and Pal, 1993). In addition to earlier recommended combinations, an addition of *Lr13* and *Sr2* in area 2; *Lr19* and *Lr23* and *Sr2* and *Sr26* in area 3; *Sr2* and *Sr31* and *Lr34* and *Lr26* in area 4 and *Lr24* and *Sr2* and *Sr24* in area 5 is considered to be useful in increasing the productive life of wheat cultivars (Fig. 18). However Bhardwaj *et al.* (2005) recorded that *Lr19* resistance in wheat has become susceptible to *P. recondita* in India.

Disease Prediction and Chemical Control: Disease prediction is of fundamental importance for environmentally safe, efficient and economical use of fungicides for the control of diseases (Roelfs, 1988). Bio-climatological and computer based prediction models are developed to anticipate probable time of the occurrence of the disease for deciding the period of chemical application. In India, the ground survey data and information collected through rain sampler, wind trajectories, satellite cloud

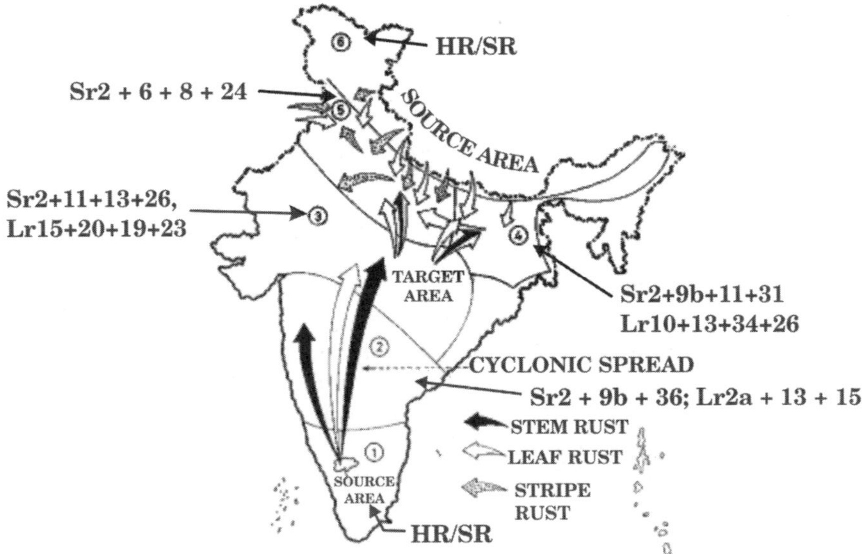

Figure 18: 'Puccinia Path' and deployment of *Sr* and *Lr* gene combinations for rust resistance.

photography etc. have been used for developing bio-climatic models as well as linear prediction equations (Nagarajan, 1977). A bio-climatic model was developed to predict stem rust appearance in central India based on stem rust rules (Nagarajan and Singh, 1976). On the basis of bio-climatic model developed for stem rust, some successful predictions have been made on the probability of rust appearance in central-peninsular India during the years of cyclones (Joshi *et al.,* 1984). Like stem rust some common features on the development of the leaf rust in North India occurring every season, have also provided the basics of bio-climatic prediction model in North-western India (Nagarajan *et al.,* 1979). This model indicates that the following criteria must be satisfied if leaf rust is to appear:

1. At least 5–6 infection sites should be observed at a maximum distance of 25 ± 5 km around 15–20 January in Uttar Pradesh and North Bihar.
2. The number of rainy days between January to middle of April over North-western India should be at least twice the normal number.
3. The weekly mean maximum temperature during March to middle of April should be within 1°C of the normal temperature.

It has been assumed that if the susceptible cultivars in the entire Indo-Gangetic plain are cultivated (although in reality they are not) and 1st and 2nd criteria are satisfied then a severe epidemic will tend to occur. If they are partly satisfied isolated outbreaks of the disease may probably occur.

The bio-climatic model can give an idea of probable dates of rust appearance but further development of the disease will depend on environmental factors prevailing after the establishment of infection. In this direction, Nagarajan and Joshi (1978) developed a following mathematical model for 7-day forecast of stem rust severity. The linear equation outlines as under:

$$Y = 29.3733 + 1.820\,X_1 + 1.7735\,X_2 + 0.2516\,X_3$$

In this linear equation Y represents the predicted disease severity and X_1, X_2, and X_3 are previous disease severity, weekly mean of minimum temperature and maximum relative humidity, respectively. Srivastava *et al.* (1985) also evolved a functional equation for weekly prediction of leaf rust. These prediction models are useful in estimating the severity of stem and leaf rust in advance, thereby giving enough time to decide the use of systemic fungicides like triadimefon, oxycarboxin, triarimol, fenapanil and propiconazole.

2.1.8 Stem Rust

Stem rust of wheat, also known as black rust, though prevalent all over the country, can assume damaging proportions in central and South Indian region of the country. Stem rust is basically a disease of warmer climate and is generally prevalent in Nilgiris, foot hills of Nigiris and Pulney hills, Karnataka, Maharashtra and Madhya Pradesh. Sometimes, it also appears in eastern Uttar Pradesh, Bihar and West Bengal. The uredospores of stem rust from central zone to northen region never arrive in time so the disease appears late in this part and usually does not cause much damage.

The yield losses implicated by stem rust have been enormous. The epidemic of the disease in 1946–47 in Madhya Pradesh and Maharashtra destroyed over two million tons of wheat grain (Asthana, 1948). In 1956–57 rust created almost famine conditions in West Bengal, Bihar and eastern parts of Uttar Pradesh (Prasada, 1965). An isolated but severe epidemic of stem and leaf rusts was recorded on wheat in Jalore district of Rajasthan in 1973. It is estimated that nearly one-third of the entire crop (approximately 8000 hectares) was completely destroyed and not even harvested in this area. In 1978–79 a large area of the Narmada valley in Madhya Pradesh was hit badly by stem and leaf rusts causing 60–75 percent losses to local wheat varieties Pissi and Malvi local (Joshi *et al.,* 1980).

Symptoms: Stem rust appears on all the above-ground parts of the plant. Disease symptoms most commonly develop on the stem and leaf sheaths but leaf blade and spikes may also become infected. The first symptom of the rust is characterized by the appearance of small chlorotic spots or flecks on both sides of affected parts of the plant. These flecks after passing through incubation period give rise to dark brown, large elongated pustules or uredosori. In the beginning, the pustules are scattered but later they may coalesce. Millions of uredospores are produced in pustules that rupture, throwing up large fragments of the epidermis and exposing masses of reddish brown spores. Remanants of the epidermis are visible on the margins of the pustule, giving it a ragged appearance (Plate 1A). As the infection advances, the uredospores are gradually replaced by black teliospores in the same pustule and telia are formed. They are conspicuous, linear, oblong, dark brown to black and often merging with one another to cause linear patches of black lesions.

Causal Organism: Persoon (1794) made the detailed study of causal agent of stem rust of wheat and termed the fungus on barberry as *Aecidium berberidis* and the form on wheat as *Puccinia graminis* in 1794. Eriksson (1894) demonstrated physiological specialization in this species and described strains as specialized forms (*formae specialis*). The pathogen is therefore named as *Puccinia graminis* Pers. f.sp. *tritici* Eriks & Henn.

P. graminis tritici is macrocyclic, heteroecious fungus. Two stages of the fungus *i.e.* uredial and telial stages are produced on wheat and other two stages, *viz.* pycnial and aecial are developed on alternate hosts like *Berberis* and *Mahonia. B. vulgaris* is the most common alternate host of *P. graminis tritici.* Since under Indian conditions alternate hosts are not functional, the uredospores, therefore, play major role in the survival and perpetuation of the fungus.

The uredospores are oval, single celled with long pedicel, minutely echinulated having 4 equatorial germpores and measuring 25–30 × 17–20 microns (Fig. 19). When the uredospores are lodged on the host surface, they start germination under suitable climatic conditions. The germ tubes emerge from uredospores through germpores. Sometimes more than one germ tube is produced. In such a case only the longer one becomes infective,

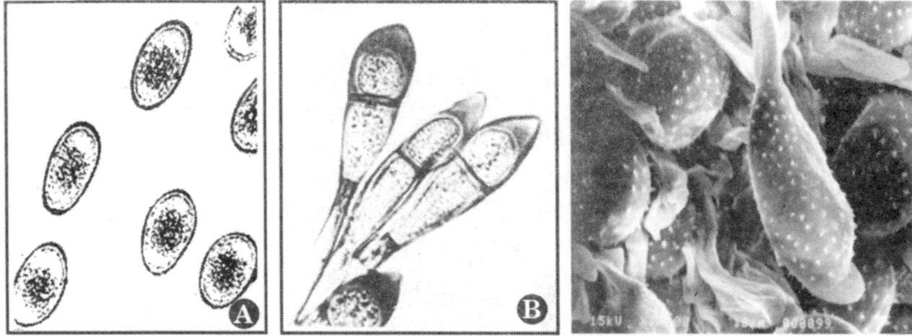

Figure 19: Stem rust: (A) Uredospores (B) Teliospores/Teleutospores (C) Scanning electron microscope photograph of uredospores of *Puccinia graminis tritici*

the others wither away. In order to facilitate penetration, the tip of germ tube swells and produce an elongated appresorium from which infection pegs are produced. Then entry of fungus in the host is through stomatal opening. The infection peg emerging from appresorium enters the sub-stomatal cavity where a sub stomatal vesicle is formed, from which hyphae strands arise. These hyphae spread intercellularly, sending haustoria into cells to absorb nutrients. When fully established, the fungus produces the uredosori which erupt and release powdery mass of brick red coloured uredospores.

The production of uredospores is favoured by suitable temperature and availability of sufficient moisture. The spore germination initiates when dew for 2 hours or more is available on the host surface. Temperature ranging from >5–30°C (optimum 15–24°C) at this stage promotes the germination process. During germ tube growth and appresorium formation, the germinating spores require dew for 10 hours and optimum temperature varying from 16–27°C. The penetration of infection pegs through stomata is rapid under film of free water. Light intensity above 3.2×10^3 lux hampers germination of uredospores, which ceases at 11×10^3 lux (Table 5).

Slow drying in dark maintains viability of appresoria (Hogg *et al.,* 1961). Prabhu and Wallin (1971) established that stem rust spore production from a single pustule has a relationship with temperature. They noticed that the production of uredospores increases with rise in temperature.

When uredospore production ceases at maturity of the plants, telia arise from the same mycelium at high temperature. The teliospores which

Table 5. Influence of environmental factors on the germination of uredospores and the infection process of stem rust (*Puccinia graminis tritici*) of wheat

Process	**Temperature (°C)**	**Moisture**	**Light**
Germination	Minimum > 5	Necessary dew period for 2 or more hours required	Hamper above 3.2×10^3 lux and ceases at 11×10^3 lux
	Optimum 15–24 Maximum 30		
Growth of germtube	Optimum 15–20	Optimal with 10 hours dew	As above
Appressorium formation	Optimum 20	Dew necessary	Darkness maintain viability
Penetration	Optimum 20	Not necessary	At least 5.5×10^3 lux
Growth of uredo-mycelium	Optimum 20	Essential	Favours host pathogen interaction

develop in the same uredia are dark brown, two celled and some what wedged shape. They retain a portion of the pedicel or stalk, have thick walls and measure 40–60 × 18–26 microns. The apical cell of teliospore is rounded or slightly pointed (Fig. 19). Each cell of teliospore has one pore, the upper being at apex and that of the lower being at the side just below the septum. The teliospores frequently erupt through the epidermis, persist in the plant tissue during off season and serve as first stage in sexual cycle of the fungus. They germinate after a long resting period and expose to freezing temperatures. On germination, a long, thin walled, hyaline four celled promycelium (basidium) is produced. Each cell produces a sterigma bearing a thin walled, hyaline roundish sporidium (basidiospore). The basidiospores germinate by germ tube and infect alternate host to produce flask shaped pycnia on the upper surface. Inside the pycnia numerous pycniosporophores are formed, mixed with sterile hyphae strands called paraphyses. At the tip of each pycniosporophere small unicellular, spherical, thin walled and hyaline pycniospore (spermatia) develop. These spores and the mycelium from which they develop are all monocaryotic. The pycniospores are the agents which bring about diploidization of the monokaryotic mycelium. The dikaryotic mycelium then produces aecial cups on the lower surface of the host. They contain chains of aeciospores produced on short stalks developing from the base of the cup. The aeciospores are yellow, roundish or angular, echinulate, fairly thick walled with 6 germpores, measuring 24–26 microns in diameter. They germinate readily in water and are capable of infecting wheat and certain grasses to repeat uredial stage.

Over 35 physiological races of *P. gramins tritici* have been recorded in India. They are 11, 11A, 14, 17, 21, 21-1, 21-A-1, 21A-2, 24, 24A, 34, 34-1, 40, 40A, 40-1, 42, 42B, 42B-1, 42B-2, 53, 75, 117A, 117A-1, 117-1, 117-2, 117-3, 117-4, 117-5, 117-6, 122, 184, 194, 222 and 295. At present race 40A (62 G 29) occur in the most virulent form in certain parts of the country (Bhardwaj *et al.,* 2006).

Disease Cycle: *P. gramins tritici* is an obligate parasite, being unable to complete its life cycle in the absence of living plant host. It has developed a very complex life cycle involving five spore types alternating between two host species *i.e.* wheat and *Berberis / Mahonia.* In a number of temperate countries *Berberis* species serve as the alternate host and help to perpetuate the disease from one season to the next. Teleutospores of the rust are produced in autumn before harvest and remain dormant till the spring. In the early spring, the teleutospores germinate to produce four basidiosores. These are carried by wind to infect young leaves of *Berberis*, producing pycnia and pycniospores. The pycniospores are functional sex gametes, which after fertilization give rise to aeciospores in aecial cups. The aeciospores re-infect the young wheat plants and produce uredospores in uredia. Uredospores are capable to repeat reproduction asexually (Fig. 20).

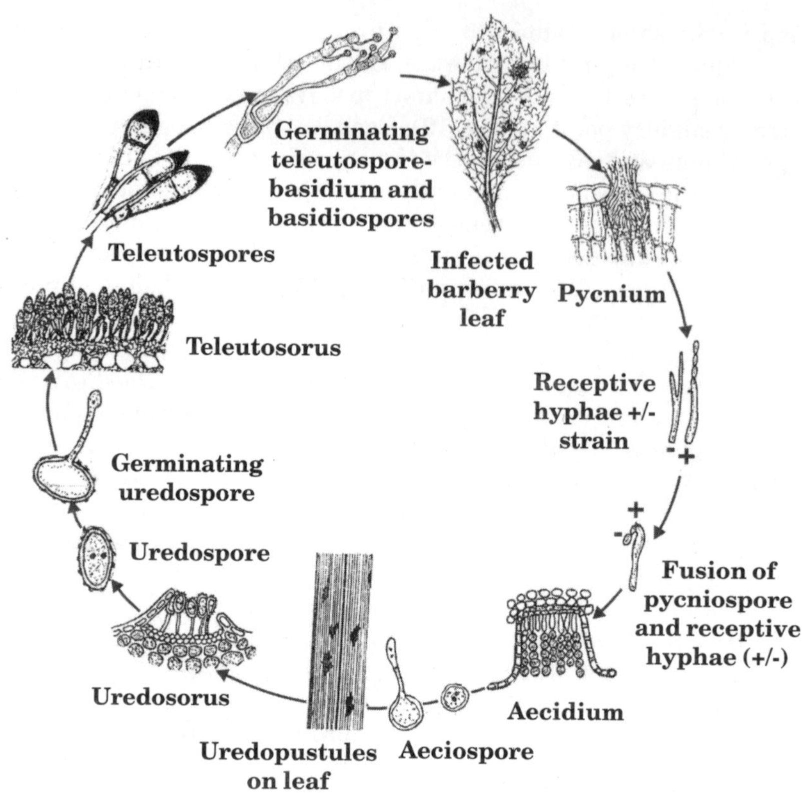

Figure 20: Disease cycle of stem rust (*Puccinia graminis tritici*) of wheat

Under Indian conditions where alternate hosts are not functional, no pycnial and aecial stages of the fungus have been discovered. The teleutospores are not capable of immediate germination under hot and dry conditions prevailing in summer months. The basidiospores are not formed to infect any known plant species (Mehta, 1940). Therefore, survival of the rust in the Indian sub-continent is through uredospores produced on collateral hosts at high altitudes on the hills. In hilly areas, besides summer wheat crop and self sown wheat plants, some graminaceous hosts (grasses) are found throughout the year, which allows the rust to survive in the area in the form of uredospores. The rust infections initially may also arise from inoculum introduced from some distant source to begin fresh life on regular wheat crop.

2.1.9 Leaf Rust

Leaf rust of wheat, also named as brown rust is the most important rust, widely distributed and most frequently occurring disease of wheat in the

Indian sub-continent. It is the most common rust prevalent in North-western and North-eastern, central and peninsular region of the country. In South India, it is found on the crop grown both in the Nilgiris and Pulney hills and in the plains including Tamil Nadu, Karnataka and Andhra Pradesh.

It is economically most important disease because of its capacity to cause epidemics in almost all parts of the country. In 1972–73 leaf rust appeared in epidemic form severely affecting extensively grown wheat cultivar Kalyansona in Punjab, Haryana and western Uttar Pradesh. The crop suffered a loss of nearly 1.5 million tonnes of wheat (Joshi, 1975). The "Sonalika epidemic" of leaf rust which swept over the entire Uttar Pradesh and part of Bihar in 1980 caused a loss of one million tonnes (Joshi *et al.*, 1985).

Symptoms: Disease appears as small, circular, brown to orange brown pustules scattered mainly on upper surface of leaf blade and sometime on leaf sheath, peduncles, internodes and ear heads also. The pustules break through the epidermis, but do not cause loose epidermal tissue at the margins as is the characteristic of stem rust uredia (Plate 1B). The telia develop during the later stage of plant development in uredo-sori on leaf sheaths and on both surfaces of leaf blades. They are small, oval or linear and dull black. The teliospores remain in the leaf tissues and are covered by the epidermis.

Causal Organism: In 1815 the leaf rust pathogen was called *Uredo rubigo-vera*. Eriksson and Henning (1894) described the causal organism of both wheat and rye leaf rust as *Puccinia dispersa*. Eriksson (1899) separated the wheat and rye leaf rust fungi and the causal agent of wheat leaf rust became *P. triticina*. Savile (1984) stated that *P. triticina* should be the binomial for wheat rust and *P. recondita* for rye leaf rust. The current binomical used by most pathologists is *P. recondita* (Cummins and Caldwell, 1956) and *P. recondita* f. sp. *tritici* is used by most workers (Roelfs *et al.*, 1992).

Puccinia recondita Rob. ex. Desm. f. sp. *tritici* Eriks. & Henn. (*P. triticina* = *P. rubigo-vera*), is heteroecious fungus. The uredial and telial stages appear on wheat and collateral hosts and the pycnial and aecial stages on alternate host, *Thalictrum* (*T. javanicum*, *T. flavum*)and *Isopyrum* (*I. fumarioides*) species. The pycniospores and aeciospores are not produced in India, so the survival and perpetual of the fungus is through uredospores.

The uredospores of *P. recondita* are brown, spherical, minutely echinulated with 7–10 germpores, and usually measure 20–28 micron in diameter (Fig. 21). Spores initiate germination with in 30 minutes after contact with free water at temperatures varying from 15–25°C. The germ tube grows along the moist leaf surface until it reaches a stoma and forms

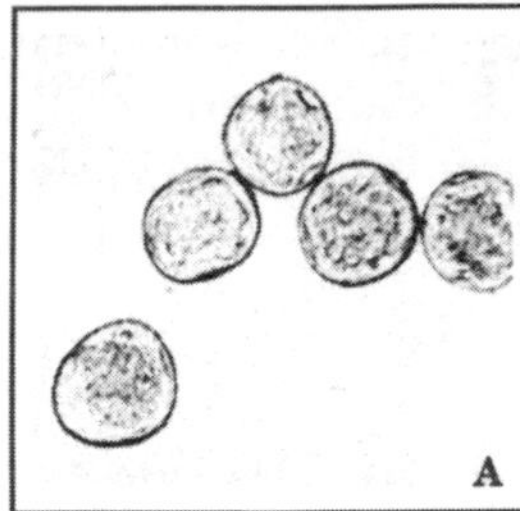

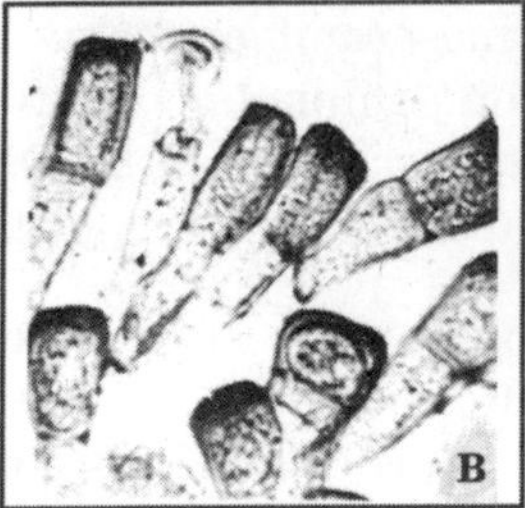

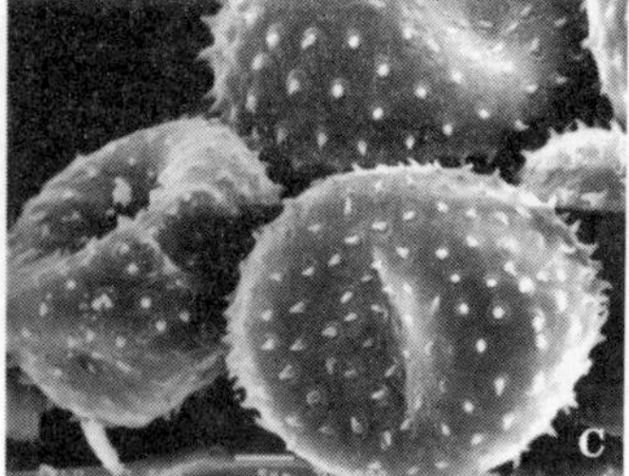

Figure 21: Leaf rust: (A) Uredospores (B) Teliospores (c) Scanning electron microscope photograph of uredospores of *Puccinia recondita tritici*

an appressorium. The appressorium formation and development of penetration peg and substomatal vesicle take place at 20°C temperature and high humidity (Table 6).

Telia usually formed on under surface of the leaf, are deeply embedded in host tissues. The teliospores are dark brown; two celled with thick walls, and are rounded or flattened at the apex, arranged in groups separated by paraphyses. Thirty nine races of the fungus, namely 10, 11, 12, 12–1, 12–2, 12–3, 12–4, 12A, 17, 20, 26, 61, 63, 70, 77, 77–1, 77–2, 77–3, 77–4, 77–5, 77–6, 77–7, 77A, 77A-1, 77B, 104, 104–1, 104–2, 104–3, 104A, 104B, 106, 107, 107-1, 108, 131, 162, 162A and162B have been recorded in the country from time to time.

Disease Cycle: Leaf rust of wheat is macrocyclic rust. In the absence of alternate hosts the fungus perpetuates in the form of uredial stage. The uredospores like that of stem rust, survives on wheat, self sown plants and collateral hosts (grasses) in South Indian hills and Himalayas through out

Table 6. Influence of environmental factors on the germination of uredospores and the infection process of leaf rust (*Puccinia recondita*) of wheat

Process	Temperature (°C)	Moisture	Light
Germination	Minimum 2–3 Optimum 8–28 Maximum 32	Essential	Retards germination
Growth of germtube	Optimum 15–20	High humidity essential	As above
Appressorium formation	Optimum 18–25	Free water required	Favoured by darkness
Penetration	Optimum 20	Not necessary	No effect
Growth of uredo-mycelium	Optimum 20	Essential	Favours host pathogen interaction

the year. The rust inoculum in the form of uredospores spreads from hilly region into plains where they cause infection on regular wheat crop. These spores are repeatedly reproduced asexually on wheat plants during crop season. It is because of this reason, the uredospores are also called repeating spores. In this way the pathogen completes its disease cycle with the uredospores only.

2.1.10 Stripe Rust

Stripe rust of wheat, also called as yellow rust, prefers low temperature for infection and symptom expression. Therefore, the disease occurs in cooler wheat tracts of North-western region of the country, Himalayan ranges and adjoining foot hills. Sometimes it also appears in Bihar, eastern Uttar Pradesh and even in central India (Joshi, 1978). The disease is totally absent from southern India except for higher elevation of Nilgiri and Pulney hills but it cannot spread even to the foothills of the Nilgiris obviously due to warmer weather. In South Indian hills the incidence of stripe rust has been recorded on durum wheats and barley varieties but the situation has changed since 1972 and there are reports of stripe rust appearance on aestivum wheats in these hills.

In general, stripe rust is more destructive than stem and leaf rusts. In the epidemic years it may implicit considerable losses due to destruction of foliage and sometimes by sterility of spikelets and shrivelling of grains. In 1971–72 the disease appeared in an epidemic form along with leaf rust and severely attacked Kalyansona variety loosing approximately 0.8 million tonnes of wheat (Joshi *et al.*, 1985).

Symptoms: The uredo-pustules develop in the form of narrow, yellow, linear stripes mainly on leaves and spikelets (Plate 1C). When earheads are infected, the uredia appear on the inner surface of the glumes and lemmas, occasionally invading the developing grains. The pathogen on infecting leaf blade remains partially systemic and the stripes keep on enlarging. Later on these stripes turn black when teliosori are formed. The teliosori are also arranged in long stripes and covered by epidermis. They are dull black in colour.

Causal Organism: Stripe rust was first described by Gadd and Bjerkander in 1777 on rye. Schmidtt designated the pathogen as *Uredo glumarum* in 1827. But Westerndrop named the stripe rust pathogen of rye as *Puccinia striaeformis* in 1854. Eriksson and Henning (1896) proposed the name as *P. glumarum* [(Schm.) Eriks. & Henn.]. However, Hylander *et al.* (1953) and Cummins and Stevenson (1956) chose the name *P. striiformis* West. which is currently in use.

P. striiformis is a heteroecious, with uredial and telial stages occurring on wheat. The uredospores are yellow to orange in colour, more or less spherical to oval with some angular walls and measure 20–30 microns in diameter and possess 6–16 germpores. However, it is difficult to identify the leaf and stripe rust fungi by uredospores characters if they are observed separately. The teleutospores are dark brown to black and are two celled with thick walls, measure 35–63 × 12–24 microns and are interspersed with brown unicellular paraphyses. They are more or less similar in size and shape to those of *P. recondita*, except that the cap or crown is flattened, not rounded.

P. striiformis prefers cooler environmental conditions for its normal growth. The uredospores initiate germination at 9–13°C with free water. The germtube growth takes place at 10–15°C. An optimal temperature 8–13°C favours formation of appressorium and sub-stomatal vesicle under humid conditions (Table 7).

Disease Cycle: Stripe rust can exist independent of an alternate host. The pathogen over-summers in the inner valleys of Himalayas on wheat and volunteer plants in uredial stage under cool climate. If temperatures are very low, infection results in extended latency. Sporulating uredeospores can survive to a temperature of minus 4°C and incipient infections can prevail as long as the host leaf survives. When favourable temperature returns, the sori of rust bust into uredia with abundant uredospores and the inoculum moves towards foothills of northern India. By December/ January when the crop is about a month old, the inoculum then spreads by katabatic winds to sub-mountainous parts of North western plain zone. Primary infection foci occur in this region along the mouth of rivers, *viz.* Tavi, Ravi, Beas, Satluj and Yamuna. Later on the katabatic winds carry the uredospores to adjoining plains and cause infection. The pathogen on infecting a leaf remains partially systemic with in that leaf and thereafter,

Table 7. Influence of environmental factors on the germination of uredospores and the infection process of stripe rust (*Puccinia striiformis*) of wheat

Process	Temperature (°C)	Moisture	Light
Germination	Minimum 0 Optimum 7–15 Maximum 23–26	Necessary	Varies with temperature, favourable at > 15°C
Growth of germtube	Optimum 10–15	Free water required	As above
Penetration	Optimum 8–13	Not necessary	No effect
Growth of uredo-mycelium	Optimum 12–15	Favours host pathogen interaction	No effect

several uredo cycles on infected plants are completed resulting in high disease severity. At maturity of the crop, the temperature begins to rise and it is not congenial for uredospore production. The spread of the rust is checked above 25°C. With the ceasing of sporulation the telial stage develops. The teliospores apparently serve no function in the absence of alternate host and the recurrence of the disease takes place through air borne uredospores from primary foci of infection.

Physilogical races of *P. striiformis* recorded in India are 13, 14, 14A, 19, 20, 20A, 24, 31, 38, 38A, 57, A, G, G-1, I, K, L, M, N, P, Q, T, U, C I, C II, C III and 46 S 119.

2.1.11 Control Measures

The epidemic growth of rusts is influenced by four factors, *viz.* availability of virulent race of the pathogen, susceptible host, favourable weather and suitable time. This interaction relationship results in its representation as a 'disease pyramid'. Keeping this in view it can be possible to delay or avert the growth of the pathogen if these factors are somehow manipulated. In this context, various management approaches depending upon the nature of rust epidemic, have been recommended as discussed earlier. However, the major emphasis for control of rusts is on combination of cultural practices with host resistance and use of chemicals.

Cultural Control: Cultivation of wheat with other non-cereal crops is an old Indian practice to ensure some output in addition to wheat. This practice of mixed cropping of wheat with pulses and oil seed crops like gram, pea, lentil, linseed, mustard and rapeseed appears to be useful. In mixed cropping, the uredospores liberated from infected wheat plants are distributed over adjacent non-wheat plants. In this process the spores that land on plants of companion crop are unable to reproduce. As a result the numbers of successful infection sites are decreased and rust incidence is reduced.

High dose of nitrogenous fertilizers to the soil makes wheat plants more susceptible to rust infection. Application of potash at higher rate has the opposite effect (Singh, 1978). In general, the rate of application of NPK is 120:60:40, if a variety is susceptible then excessive use of fertilizers, especially the nitrogenous fertilizers should be avoided with reduction in the proportion of nitrogen in NPK ratio.

Adjustments in date of sowing are advised to circumvent attack of rusts by avoiding the peak period of infection. However, the knowledge of the conditions favouring appearance of disease and vulnerable age of the plants at which they are susceptible to infection is essential to put this concept into practice. It has been noticed that stem rust severity increases with progressive delay in sowing of wheat crop in northern region of the country. For example, the prevailing environmental conditions in Delhi become favourable for incidence of stem rust in March or early April, if the crop is caught in the milk or dough stage then not only the severity of the disease is increased but substantial yield losses are also caused (Patil, 1972; Joshi and Pathak, 1975). In this situation the normal sowing of early maturing varieties with adult plant resistance is likely to be useful in containing the rust under control.

Resistance: Growing resistant varieties appears to be most practical approach to the rust problem but virulent races of the pathogen appear quiet frequently and varieties become susceptible. Therefore, need for lasting resistance is always felt. In recent times, alien sources have been utilized for rust resistance. Cytological observations of the presence of 1B/1R carrying *Sr31/Lr26/Yr9* indicate that several wheat varieties in India carry 1B/1R translocation (Nayar *et al.*, 1993; Anon, 2001). 1B/1R substitution from rye to wheat has not only contributed for rust resistance but also revolutionized the production. Stem rust resistance gene *Sr26* is known to confer resistance for the last many years (Luig, 1983). Wheat varieties carrying *Sr24* produce low infection types to most of the stem rust pathotypes. Resistance gene *Sr2*, in addition to other unknown minor genes derived from cultivars Hope and H44 provides durable resistance. For example, variety Sonalika released in1960 in the Indian sub-continent has remained resistant, carry *Sr2* gene. *Sr5* is known to produce resistant type of infection to *P. graminis tritici* pathotypes. Many wheat varieties throughout the world have resistance gene *Sr31* which was introgressed into wheat on translocated chromosomal fragment from rye. This gene has provided highly effective resistance for many years. Among 320 lines of Indian wheat, 56 lines have been found to possess resistance gene *Sr31*. *Sr2* has been identified in 22 lines. Of these DL 803-3, DWR 195, GW 190, HUW 318, MACS 2496 carrying *Sr2* + *Sr31* are suitable for cultivation in stem rust prone areas in central and peninsular India. Wheat varieties GW 273, HW 1085, JW 515, NIAW 34, and RAJ 3675 having *Sr5* are recommended for central zone (Prashar *et al.*, 2004)

In 1999, a new variant of *P. graminis tritici Ug99* on gene *Sr31* was picked up from Uganda in Africa and subsequently it was detected from Kenya, Ethopia, Yemen and Sudan (Singh *et al.,* 2008). It is feared that the spores of *Ug99* may migrate and threaten wheat production in the Middle East and central Asia. The value of *Sr31* is still rated high in Indian Wheat programme as this gene shows enhanced resistance to stem rust in combination with *Agropyron elongatum* derived gene *Sr24* and *Sr25* in the background of many bread wheat varieties (Tomar *et al.*, 2004). Therefore, use of stem rust resistance genes *Sr24, Sr25* and *Sr26* in Indian wheat cultivars is recommended as a pre-emptive measure to face the challenge of *Ug99* (DWR, 2006). Some Indian wheat varieties like DBW17 and PBW 550 are reported resistant to this race.

The exploitation of genes for leaf rust resistance has led to the development of several wheat varieties carrying genes *Lr9, Lr10, Lr13, Lr14, Lr23, Lr24, Lr26* and *Lr28*. The diversification of varieties and their strategic deployment in different agro-climatic zones has been instrumental in arresting the spread of rust to major wheat growing areas of the country. *Lr3* in mid thirties, *Lr23* in early seventies and *Lr26* in eighties were used in breeding for leaf rust resistance. However, races with matching virulences for these resistance genes were selected during 1980-2004. In view of these reports, search for new resistance genes should continue and marker assisted selection be exploited for improving leaf rust resistance. Saini (2002) has identified two new genes *Lr48* and *Lr49* which are available for incorporation. According to Singh (2004) and Parasher *et al.* (2008) resistance genes *Lr34* and *Lr46* and other minor genes hold promise for durable resistance to leaf rust. *Lr34* gene is linked to *Yr18*. The presence of *Lr34* can be indicated by the presence of leaf tip necrosis in adult plants. Combination of these genes results in adequate resistance levels in most environments. Similarly slow rusting *Yr29* is completely linked to gene *Lr46* which confers moderate resistance to leaf rust (William *et al.,* 2003). Pyramiding of *Lr34* has been attempted through molecular assisted selection (Rao *et al.,* 2007). The genetic basis of seedling and adult plant resistance has been analysed in wheat lines CS 2A/2M 4/2 and CS 2A/2M/318 by Bansal *et al.* (2008) which showed that leaf rust resistance is conferred by *Lr 28* gene.

In mid 1960's, Kalyansona and Sonalika, the most popular wheat cultivars in India exhibited high degree of resistance to stripe rust, presumably due to the presence of *Yr2* gene. The extensive cultivation of Kalyansona exposed the resistance to new virulence *Yr2 (Ks)* in 1971–72, causing stripe rust epidemic in the Indo-Gangetic plains. Subsequently, Kalyansona was replaced by Mexican genotypes having chromosomal translocation 1B/1R that carries *Yr9*. A new virulence of *P. striiformis* matching *Yr9*, which was first recorded from Eastern African highlands in 1991, has been picked up from Iran, Afghanistan, Pakistan, Nepal, China and India (Singh, 2004). In recent years, some Veery # 5 selections like Pak 81, Pirsabak 85, Rohtas 90, Annapurna I and PBW 343 have shown susceptibility to *Yr9* virulence in Indian sub-continent (Nagarajan *et al.,* 2006). Therefore, the breakdown of *Yr9* underlines the need for cultivars having durable resistance genes such as *Yr18* (Singh, 1992; McIntosh, 1992). According to Tomar *et al.* (2004), the use of APR genes with few other genes like *Yr5*, *Yr10* and *Yr15* shall be useful for achieving long term rust control. Two independent dominant genes of wheat variety HP 1731 responsible for adult plant resistance have been isolated (Datta *et al.,* 2007). This type of single gene of APR can be useful in breeding program for leaf rust resistance. Nayar (2008) has also reported some resistance genes effective in India against three rusts of wheat (Table 8).

The resistance sources in wheat against rusts have been identified through multilocation tests (Sharma *et al.,* 2002). The genetic stocks FLW 10, FLW 16, FLW 17, FLW 18 and FLW 24 carrying *Yr5, Yr10* and *Yr32* are being persued at DWR, Karnal.

Table 8. Rust resistance gene effective in India

Rust	Gene
Stem rust (*P. graminis tritici*)	*Sr25, Sr26, Sr27, Sr31, Sr32, Sr39*
Leaf rust (*P. recondita*)	*Lr24, Lr25, Lr28, Lr32, Lr39, Lr45*
Stripe rust (*P. striiformis*)	*Yr10, Yr11, Yr12, Yr13, Yr14, Yr16*

Hyperparasitism: Certain fungal strains, viz. *Trichothecium roseum, Penicillium notatum, Darluca flium* and *Fusarium roseum* act as hyperparasites/mycoparasites of uredospores of *Puccinia* species attacking wheat but their practical utility as biocontrol agent has not been ascertained (Joshi *et al*., 1978).

Chemical Control: Several chemicals have been tested and recommended for controlling wheat rusts. Some of the earliest studies in this direction were with sulphur and hydrophobic colloidal sulphur and sulpha drugs (Gattani, 1953; Grover and Joshi, 1962). Although dusting wheat with sulphur is effective but fields must be dusted 5–10 times during the season, and thus such an option is not economically feasible. Dithiocarbamate fungicides with zinc and manganese bases have been tried. Among these compounds, 3 to 4 foliar sprays of zineb or mancozeb @ 2 to 3 kg/ha at 10–15 days interval have proved effective in controlling rusts (Grewal and Dharamvir, 1959). Systemic fungicides have shown great promise for rust control with oxicarboxin (@ 0.1%) and triadimefon (@ 0.1%), benodanil (@ 2 kg/ha). Some other compounds like dichlobutrazol, piperazine, triforine, fenapanil and Sicarol also effectively reduce rust severity. An appropriate spray with propiconazole (Tilt) @ 0.1 percent is very effective against three rusts of wheat (Singh and Rai, 1978). However, the success of chemical control will depend, to a great extent on disease forecasting system.

REFERENCES

Anonymous. 2007. *Annual Report*, Indian Agricultural Research Institute, New Delhi.

Arthur. J.C. 1929. *The Plant Rusts* (*Uredinales*). John Wiley and Sons, New York. 446 p.

Asthana, R.P. 1948. Wheat rusts and their control. *Mag. Agric. Coll. Nagpur* **22**: 136-143.

Aujla, S.S. and Indu Sharma. 1986. Multiline approach in wheat improvement. **In**: *Problems and Progress of Wheat Pathology in South Asia.* (Eds. L.M. Joshi, D.V. Singh and K.D. Srivastava), pp. 345-373. Malhotra Publishing House, New Delhi, 401 p.

Bahadur, P. 1986. Physiologic specialization in wheat rusts. **In**: *Problems and Progress of Wheat Pathology in South Asia.* (Eds. L.M. Joshi, D.V. Singh and K.D. Srivastava), pp. 69-91. Malhotra Publishing House, New Delhi, 401 p.

Bahadur, P. and S. Nagarajan. 1985. Gene deployment for management of stem rust of wheat *Puccinia graminis* (Pers.) f. sp. *tritici* (Erikss. & Henn.) in India. *Indian J. Agric. Sci.* **55**: 219-222.

Bahadur, P., S. Nagarajan and S.K. Nayar. 1985. A proposed system of virulence designation in India. II. *Puccinia graminis* f. sp. *tritici. Proc. Indian Acad. Sci.* (Plant Sci.) **95**: 29-33.

Bahadur, P., D.V. Singh and K.D. Srivastava. 1994. Management of wheat rusts-A revised strategy for gene deployment. *Indian Phytopath.* **47**: 41-47.

Barclay, A. 1887. A descriptive list of Uredinales occuring in the neighbourhood of Simla (Western Himalayas). *J. Asiatic. Soc. Bengal* **56**: 350-375.

Basant Ram, A.S. Radhu and S. Singh. 1989. Development of rusts and powdery mildew in mixtures of wheat varieties. *Cereal Res. Commu.* **17**: 195-201.

Bhardwaj, S.C., M. Prashar and S.B. Singh. 2006. Physiologic specialization of *Puccinia graminis tritici* on wheat (*Triticum aestivum*) in India during 2002-2004. *Indian J. Agric. Sci.* **76**: 386-388.

Bhardwaj, S.C., M. Prashar, M. Kumar, S.K. Jain and D. Datta (2005). *Lr19* resistance in wheat becomes susceptible to *Puccinia triticina* in India. *Plant Disease* **89**: 1360

Borlaug, N.E. 1958. The use of multilineal or composite varieties to control airborne epidemic diseases of self-pollinated crop plants. *Proc. Int. Wheat Genet. Symp.* Winnepeg, Canada, 12-27 pp.

Bridgmon, G.H. 1957. The production of new races of *Puccinia graminis tritici* by hyphal fusion on wheat. *Phytopathology* **47**: 517.

Butler, E.J. 1918. *Fungi and Disease in Plant*. Thacker & Spinck and Co. Calcutta, 547 p.

Caldwell, R.M. 1968. Breeding for general and or specific plant disease resistance. pp. 263-272. **In**: *Proc. 3rd Int. Wheat Genet. Symp*. (Eds. K.W. Finlay and K.W. Shepherd), Canberra, Australia.

Chester, K.S. 1946. *The Cereal Rusts, the Nature and Prevention of the Cereal Rusts as Examplified in the Leaf Rust of Wheat.* Chronica Botanica Co. Weltham, Mass, 269 p.

Christensen, J.J. 1942. *Long term dissemination of plant pathogens. Aerobiology No. 17* (Eds. F.R. Moullon), American Association for the Advancement of Science, Washington, D.C., 78 p.

Craigie, J.H. 1927. Experiments on sex in rust fungi. *Nature* **120**: 116.

Cummins, G.B. and R.M. Caldwell. 1956. The validity of binomials in the leaf rust complex of cereal and grasses. *Phytopathology* **46**: 81-82.

Das, B.K., A. Saini, S.G.Bhagwat and N. Jawali. 2006. Development of SCAR markers for identification of stem rust resistance gene *Sr31* in the homozygous or heterozygous condition in bread wheat. *Plant Breading* **125**: 544-549.

Datta, D., M. Prashar, S.C. Bhardwaj and S. Kumar. 2007. Genetic analysis of adult plant leaf rust resistance in three bread wheat cultivars. *Euphytica* **154**: 75-82.

De Bary, A. 1866. Morphologie and Physiologie der Pilze, Flechten and Myxomyceeten, W. Eugelmann, Leipzing, 316 p.

D'Oliveira, B. and D.J. Samborski. 1966. Aecial stage of *Puccinia recondita* on Ranunculaceae and Boraginaceae in Portugal. *Proc. Cereal Rusts Conference* 1964, 133-150 pp.

Dubin, J.H. 1984. *Puccinia striiformis* f. sp. *hordei*: cause of barley yellow rust epidemic in South America. *Phytopathology* **74**: 820.

Eriksson, J. 1894. Ueber die specialisier ung des parasitismus bei den getreider ostpilzen. Berlin Deut. *Bot. Ges.* **12**: 292-331.

Eriksson, J. and E. Henning. 1896. Die Getreideroste. Lhre Geschichte und Natur sowie Massregein gegen dieselben. P.A. Norstedt and Soner. Stockholm, 463 p.

Eriksson. J. 1899. Nouvelles etudes sur la rouille brune des cereales. *Ann. Sci. Nat. Bot. Ser.* **8,9**: 241-287.

Flor, H.H. 1956. The complementary genic systems in flax and flax rust. *Adv. Genet.* **8**: 29-54.

Frey, K.J., J.A. Browning and M.D. Simons. 1975. Multiline cultivars of autogamous crop plants. *SABRAO Jour.* **7**: 113-123.

Gassner, G. and W. Straib. 1932. Die Bestimmung der biologischen Rassen des Weizengeebrostes. *Puccinia glumarum* f. sp. *tritici* (Schmidt). Erikss & Henn. *Arb. Biol. Reichsausst. fur Land and Forstivistsch.* **20**: 141-163.

Gattani, M.L. 1952. Hydrophobic colloidal sulphur spray for wheat rust control. *Proc. 39th Indian Science Congr.* Part III, 307 p.

Gill, K.S., K. Bains and M.M. Verma. 1974. Genetic and breeding work to control wheat rusts in India. *Indian J. Genet.* **34A**: 540-550,

Gokhale, V.P. and M.K. Patel. 1952. Effect of stem rust particularly on grain weight of some improved Indian wheats. *Indian Phytopath.* **5**: 89-103.

Grewal, J.S. and Dharmvir. 1959. Efficacy of different fungicides II. Field trials in relation to wheat rusts. *Indian Phytopath.* **12**: 168-171.

Grover, R.K. and L.M. Joshi. 1962. Effect of some sulpha drugs on wheat stem rust. *Indian Phytopath.* **15**: 264-268.

Gupta, P.K., R.K. Varshney, P.C. Sharma and B. Ramesh. 1999. Molecular markers and their application in wheat breeding. *Plant Breed.* **118**: 369-390.

Gupta, R.P. and A. Singh. 1981. Field evaluation to tolerance in wheat to brown rust. *Indian Phytopath.* **34**: 300-303.

Gupta, V., V.S.P. Rao and R.G. Saini. 2007. Improving rust resistance in wheat suitable for the marginal rainfed and semi-arid zone of central and peninsular India through molecular markers. ISCB Project Report, National Chemical Lab., Pune (India).

Habgood, R.M. 1970. Designation of physiological races of plant pathogens. *Nature* (London) **227**: 1268-1269.

Hanson, H., N.E. Borlaug and R.G. Anderson. 1982. *Wheat in the Third world.* Westview Press/Boulder, Colorado, USA 174 p.

Hogg, W.H., C.E. Hounam, A.K. Malik and J.C. Zadoks. 1969. Meteorological factors affecting the epidemiology of wheat rusts. *Tech. Note* 99, Geneva, WMO, 143 p.

Jackson, H.S. and E.B. Main. 1921. Aecial stage of the orange leaf rust of wheat, *Puccinia triticina* Erikss. *J. Agric. Res.* **22**: 151-172.

Jensen, N.F. 1952. Intravarietal diversification in oat breeding. *Agron. J.* **44**: 30-34.

Johnston, C.D. and E.B. Mains. 1932. Studies of physiologic specialization in *Puccinia triticina*, *U.S. Dept. Agric. Tech. Bull.* 313, 1-22 pp.

Johnson, T. and M. Newton. 1946. Specialization, hybridization and mutation in cereal rusts. *Bot. Rev.* **12**: 337-392.

Joshi, L.M. 1975. Surveys on wheat rusts in India: The rust situation since 1972. *Cereal Rust Bull.* **3**: 7-9.

Joshi, L.M. 1976. Recent contributions towards epidemiology of wheat rusts in India.- Presidential Address of the Indian Phytopathological Society. *Indian Phytopath.* **29**: 1-16.

Joshi, L.M. 1978. Dissemination of wheat rusts in the Indian sub-continent. *Bot. Progress* **1**: 1-5.

Joshi, L.M. and D. Kak. 1955. Mutation in *Puccinia graminis tritici* (Pers.) Erikss & Henn. physiologic race 194. *Indian Phytopath.* **8**: 95-96.

Joshi, L.M. and S. Nagarajan. 1978. Regional deployment of *Lr* genes for brown rust management. **In**: *Genetics and Wheat Improvement* (Ed. Gupta, A.K.). Oxford & IBH Publishing Co, New Delhi, 268 p.

Joshi, L.M. and K.D. Pathak. 1975. Loss in yield due to brown rust on late sown wheat crop. *Indian J. Agron.* **20**: 79-81.

Joshi, L.M. and M.M. Payak. 1963. A *Berberis* aecidium in Lahaul valley western Himalaya. *Mycologia* **55**: 247-250.

Joshi, L.M., S. Nagarajan and K.D. Srivastava. 1977. Epidemiology of brown and yellow rusts of wheat in North India. I. Place and time of appearance and spread. *Phytopath. Z.* **90**: 116-122.

Joshi, L.M., E.E. Saari and S.D. Gera. 1971. Epidemiological aspects of *Puccinia graminis* var. *tritici* in India. *Proc. Indian Nat. Sci. Acad.* **37B**: 449-453.

Joshi, L.M., E.E. Saari, S.D. Gera and S. Nagarajan 1974. Survey and epidemiology of wheat rust in India. **In**: *Current Trends in Plant Pathology* (Eds. S.P. Raychaudhuri and J.P. Verma), 150-159 pp.

Joshi, L.M., D.V. Singh and K.D. Srivastava. 1984. Fluctuation in incidence of rusts and other wheat diseases during past decade and strategies for their containment. 23rd All India Wheat Res. Workers Workshop held at Kanpur, Aug. 1984.

Joshi, L.M., D.V. Singh and K.D. Srivastva. 1988. *Manual of Wheat Disease*. Malhotra Publishing House, New Delhi, 75 p.

Joshi, L.M., K.D. Srivastava and D.V. Singh. 1984. Wheat Diseases News Letter **17**: 1-20, IARI, New Delhi.

Joshi, L.M., K.D. Srivastava and D.V. Singh 1985. Monitoring of wheat rusts in the Indian subcontinent. *Proc. Indian Acad. Sci. (Plant Sci.)* **94**: 387-407.

Joshi, L.M., K.D. Srivastava, D.V. Singh and S. Nagarajan. 1978. *Annotated Compendium of Wheat Diseases in India*. ICAR, New Delhi, 332 p.

Joshi, L.M., K.D. Srivastava and K. Ramanujam. 1975. An analysis of brown rust epidemic of 1971-72 and 1972-73. *Indian Phytopath.* **28**: 138.

Joshi, L.M., K.D. Srivastava, D.V. Singh and K. Ramanujam 1980. Wheat rust epidemics in India since 1970. *Cereal Rusts Bull.* **8**: 7-21.

Kislev, M.E. 1982. Stem rust of wheat 3300 years old found in Israel. *Science* **216**: 993-994.

Knott, D.R. 1988. Using polygenic resistance to breed for stem rust resistance in wheat. **In**: *Breeding strategies for resistance to the rusts of wheat*. (Eds. N.W. Simmonds and S. Rajaram), pp. 39-47. Mexico. D.F. CIMMYT, 151 p.

Kulkarni, R.N. and V.L. Chopra. 1980. Slow rusting resistance: Its components, nature and inheritance. *Pflanzenh Pflanzen.* **87**: 562-573.

Kulkarni, R.N. and V.L. Chopra. 1982. Environment as a cause of differential interaction between host cultivars and pathogenic races. *Phytopathology* **72**: 1384-1386.

Kulshrestha, V.P. 1986. Genetic aspects of rust resistance. **In**: *Problems and Progress of Wheat Pathology in South Asia* (Eds. L.M. Joshi, D.V. Singh and K.D. Srivastava), pp. 320-332. Malhotra Publishig House, New Delhi, 401 p.

Levine, M.N. 1919. The epidemiology of cereal rusts in general and black rust in particular. USDA (Mimeograph.)

Little, R. and J.G. Manners. 1969. Somatic recombinations in yellow rust in wheat (*Puccinia striiformis*): The production and possible origin of two new physiologic races. *Trans. Brit. Mycol. Soc.* **53**: 251-258.

Luig, N.H. 1983. *A survey of virulence genes in wheat stem rust Puccinia graminis f.* sp. *tritici.* Parey, Berlin, Hamburg. 199 p.

Mains, E.B. 1930. Effect of leaf rust (*Puccinia triticina* Erikss.) on yield of wheat. *J. agric. Res.* **40**: 417-446.

Marshal, D.R. and A.H.D. Brown. 1972. Stability of performance of mixtures and multilines. *Euphytica* **22**: 405-412.

Mc Intosh, R.A. 1992. Close genetic linkage of genes conferring adult-plant resistance to leaf rust and stripe rust in wheat. *Plant Pathol.* **41**: 523-527.

Mc Intosh, R.A., C.R. Wellings and R.F. Park. 1995. *Wheat rusts: An Atlas of Resistance Genes*. CSIRO Publications, Australia. 190 p.

Mc Intosh, R.A., Y. Yamazaki, K.M. Devos, J. Dubcovsky, W.J. Rogers and R. Appels. 2003. Catalogue of gene symbols for wheat. *Proc. 10th Int. Wheat Genet. Symp.* (Eds. N.E. Ponga, M. Romano, E.A. Ponga and G. Galterio). Vol. **4** pp. 1-34 and associated CD-Rom. (S.I.M.I. via N. Nisco 3/A-00179 Rome, Italy.

Mehta, K.C. 1940. *Further Studies on Cereal Rusts in India.* Sci. Monogr. No. 14, Imperial Council Agric. Res., 224 pp.

Mehta, K.C. 1952. *Further Studies on Cereal Rusts in India*. Sci. Monogr. No. 18, Indian Council Agric. Res., 165 p.

Misra, D.P. and V.C. Lele. 1955. Mutation of *Puccinia graminis tritici* (Pers.) Erikss. & Henn., physiologic race 15-C. *Indian Phytopath.* **8**: 79-81.

Nagarajan, S. and H. Singh 1973. Satellite television photography as a possible tool to forecast plant disease spread. *Curr. Sci.* **42**: 273-274.

Nagarajan, S. and H. Singh 1974. Satellite Television Cloud Phytopathology- a new method to study wheat rust dissemination. *Indian J. Genet.* **34A**: 486-490.

Nagarajan, S. and H. Singh. 1976. Preliminary studies on forecasting of wheat stem rust appearance. *Agric. Meteorl.* **7**: 281-289.

Nagarajan, S. and L.M. Joshi. 1975. Historic account of wheat rust epidemics in India and thier significance. *Cereal Rusts Bull.* **3**: 25-33.

Nagarajan, S. and L.M. Joshi. 1977. Sources of primary inoculum of wheat stem rust in India. *Plant Dis. Reptr.* **61**: 454-457.

Nagarajan, S. and L.M. Joshi. 1980. Further investigations on predicting wheat rusts appearance in central and peninsular India. *Phytopath. Z.* **98**: 84-94.

Nagarajan, S. and L.M. Joshi. 1985. Epidemiology in Indian Subcontinent. **In**: *The Cereal Rusts* (Eds. A.P. Roelfs and W.R. Bushnell), 371-402, Orlando, Florida, Academic Press.

Nagarajan, S. and K. Muralidharan. 1995. *Dynamics of Plant Disease*. Allied Publishers Ltd., New Delhi, 247 p.

Nagarajan, S., P. Bahadur and S.K. Nayar. 1984. Contemplating the brown rust resistance (*Lr*) genes to mitigate the spread of *Puccinia recondita* f. sp. *tritici*. *Indian Phytopath.* **37**: 490-497.

Nagarajan, S., L.M. Joshi, K.D. Srivastava and D.V. Singh. 1979. Epidemiology of brown and yellow rusts of wheat in North India. IV. Disease management recommendations. *Cereal Rusts Bull.* **7**: 15-20.

Nagarajan, S., S.K. Nayar and P. Bahadur. 1983. The proposed brown rust of wheat (*Puccinia recondita* f. sp. *tritici*) virulence analysis system. *Curr. Sci.* **52**: 413-416.

Nagarajan, S., S.K. Nayar and P. Bahadur. 1985. The proposed system of virulence analysis III. *Puccinia striiformis* West. *Kavak* **13**: 33-36.

Nagarajan, S., S.K. Nayar and J. Kumar. 2006. Checking the *Puccinia* species that reduce the productivity of wheat (*Triticum spp.)* in India – A review. *Proc. Indian Nat. Sci. Acad.* **72**: 239-247.

Nayar, S.K., S.C. Bhardwaj and S.K. Jain. 2002. Fungal Disease of Wheat and Barley: Rusts. **In**: *Diseases of Field Crops* (Eds. V.K. Gupta and Y.S. Paul), pp 1-39. Indus Publishing Co., New Delhi, 464 p.

Nayar, S.K., M. Parashar, J. Kumar and S.C. Bhardwaj. 1993. Adult plant race specific low infection type leaf rust resistance in three wheats carrying *Lr. 26*. *Cereal Rusts and Powdery Mildews Bulletin* **21**: 27-30.

Nayeem, K.A. and M. Sivasamy. 2004. Transfer of alien genes for rust resistance through back cross method. **In**: *Wheat for Tropical Areas* (Eds. K.A. Nayeem, M. Sivasamy and S. Nagarajan), pp 18-32. IARI Regional Station, Wellington, Nilgiris (Tamil Nadu), 300 p.

Nelson, R.R., R.D. Wilcoxson and J.J. Christensen. 1955. Heterokaryosis as a basic for variation in *Puccinia graminis* var. *tritici*. *Phytopathology* **45**: 639-643.

Newton, M., T. Johnson and A.M. Brown. 1930. A primary study on hybridization of physiologic forms of *Puccinia graminis tritici*. *Sci. Agric.* **10**: 721-731.

Pal, B.P. 1936. Effect of brown rust attack on wheat. *Indian J. Agri. Sci.* **6**: 127-128.

Payak, M.M. 1965. *Berberis* as the aecial host of *Puccinia brachypodii* in Simla hills, (India). *Phytopath. Z.* **52**: 49-54.

Person, C. 1967. Genetic aspects of parasitism. *Can. J. Botany.* **45**: 1193-1204.

Persoon, C.H. 1794. Neuer Versuch einer sptematischen Eintheilung der Sehwamme Romer's neues Mag. *Bot.* **1**: 63-128.

Prabhu, A.S. and J.R. Wallin. 1971. Cyclical variations in spore production by *Puccinia graminis tritici. Proc. Indian Nat. Sci. Acad.* **37B***:* 454-459.

Prabhu, K.V. 2004. Marker assisted selection for leaf rust resistance in wheat. 2004. **In**: *Wheat for Tropical Areas* (Eds. K.A. Nayeem, M. Sivasamy and S. Nagarajan), pp. 67-79. IARI, Regional Station, Wellington, Nilgiris (Tamil Nadu), 300 p.

Prabhu, K.V., S.K. Gupta, A. Charpe and S. Koul. 2004. SCAR marker tagged to the alien leaf rust resistance gene *Lr19* uniquely marks the *Agropyron elongatum* derived gene *Lr24* in wheat. *Plant Breed.* **123**: 417-420.

Prasada, R. 1960. Fight the wheat rusts. *Indian Phytopath.* **13**: 1-15.

Prasada, R. 1946. The uredo stage of aecidium found on *Thalictrum* in Simla hills. *Curr. Sci.* **15**: 254-255.

Prasada, R. 1947. Discovery of the uredo stage connected with the aecidia so commonly found on species of *Berberis* in the Simla hills. *Indian J. Agric. Sci.* **17**: 137-151.

Prasada, R. 1948. Studies on the formation and germination of teliospores of rust. *Indian Phytopath.* **1**: 119-126.

Prasada, R. 1951. Rusts on wild grasses. *Curr. Sci.* **20**: 243.

Prasada, R. 1965. Our elusive fungal foes. Presidential address: Section of Agricultural Sciences, *Indian Sci. Congr. Assoc.*

Prashar, M., S.C. Bhardwaj and D. Datta. 2004. Wheat rust- challenges ahead. **In**: *Wheat for Tropical Areas* (Eds. K.A. Nayeem, M. Sivasamy and S. Nagarajan), IARI, Regional Station, Wellington, Nilgiris (Tamil Nadu), 300 p.

Priyadarshini, S. 2008. How real is the wheat stem rust. Nature (India) doi: 10. 1038/nindia 2008. 182: published online, 1-3 pp.

Roelfs, A.P. 1988. Resistance to leaf and stem rusts in wheat. In: *Breeding Strategies for Resistance to Rusts in Wheat* (Eds. N.W. Simonds and S. Rajaram), pp. 10-22, CIMMYT, Mexico.

Roelfs, A.P., J.B. Rowell and R.W. Romig. 1970. Sampler for monitoring cereal rust uredospores in rain. *Phytopathology* **60**: 187-188.

Roelfs, A.P., R.P. Singh and E.E. Saari. 1992. *Rust Diseases of Wheat: Concepts and Methods of Disease Management*, Mexico. D.F., CIMMYT, Mexico, 81 p.

Saini, R.G., M. Kaur, B. Singh, S. Sharma, G.S. Nanda, S.K. Nayar, A.K. Gupta and S. Nagarajan. 2002. Genes *Lr48* and *Lr49* for hypersensitive adult plant leaf rust resistance in wheat (*Triticum aestivum* L.). *Euphytica* **124**: 365-370.

Savile, D.B.O. 1984. Taxonomy of the cereal rust fungi. pp. 79-112. **In**: *The cereal Rusts* Vol. 1 (Eds. W.R. Bushnell and A.P. Roelfs), Academic Press, Orlandso.

Schroeter, J. 1879. Entuicklungs geschichte einiger Rost pilze. *Beitr. Biol. Pfl.* **3**: 51-93.

Sharma, A.K., D.P. Singh, J. Kumar, M.S. Saharan, K.S. Babu, A.K. Singh and S. Nagarajan. 2002. Diseases and Insect Pest Resistant Genotypes of Wheat and Triticale. *Res. Bull. No. 15*, Directorate of Wheat Research, Karnal.

Sharma, S.K. and S. Pal. 1993. Slow rusting in wheat varieties to leaf rust. *Cereal Rusts and Powdery Mildews Bull.* **21**: 31-34.

Sharma, S.K. and D.P. Mishra. 1965. Mutation in race 107 of brown rust of wheat-*Puccinia recondita. Indian Phytopath.* **18**: 363-366.

Sharma, S.K. and R. Prasada. 1967. Origin of physiologic races in wheat rusts through mutations. *Proc. First Int. Symp. Plant Pathology,* IARI, New Delhi, 4-5 pp.

Sharma, S.K. and R. Prasada. 1969. Production of new races of *Puccinia graminis* var. *tritici* from mixture of races on wheat seedlings. *Australian J. Agric. Res.* **20**: 981-985.

Singh, A. and R.C. Rai. 1986. Chemical control of wheat diseases in India. 1986. **In**: *Problems and Progress of Wheat Pathology in South Asia* (Eds. L.M. Joshi, D.V. Singh and K.D. Srivastava), pp. 374-394. Malhotra Publishing House, New Delhi, 401 p.

Singh, R.P. 1992. Genetic association of leaf rust resistance gene *Lr34* with adult plant resistance to stripe rust in bread wheat. *Phytopathology* **82**: 835-838.

Singh, R.P. 1992a. Association between gene *Lr34* for leaf rust resistance and leaf tip necrosis in wheat. *Crop Sci.* **32**: 874-878.

Singh, R.P., H.M. William, J.H. Espino and G. Rosewarne. 2004. Wheat Rust in Asia: Meeting the challenges with old and new technologies. *Proc. 4th Int. Crop Sci. Cong.*, Brisbane, Australia, 26 Sept. 10 Oct. 2004.

Srivastava, K.D., L.M. Joshi and R.K. Malhotra. 1974. Effect of brown rust on yield components of wheat variety Lal Bahadur. *Indian Phytopath.* **27**: 286-290.

Srivastava, K.D., L.M. Joshi and S. Nagarajan 1984. An assessment of leaf rust in multiline population of wheat. **In**: *Progress in Microbial Ecology* (Eds. K.G. Mukherjee, V.P. Agnihotri and R.P. Singh), pp. 179-193. Print House (India), Lucknow.

Srivastava, K.D., L.M. Joshi and S. Nagarajan 1985. A linear equation for predicting severity of leaf rust of wheat. *Indian Phytopath.* **38**: 116-120.

Srivastava, K.D., L.M. Joshi and D.V. Singh 1984. Losses in yield components of certain cultivars due to leaf rust of wheat. *Indian Phytopath.* **27**: 286-290.

Stakman, E.C. and J.G. Harrar. 1957. *Principles of Plant Pathology*, Ronald Press, New York: 581 p.

Stakman, E.C. and M.N. Levine 1922. Biologic from of *Puccinia graminis* on *Triticum* spp. *Minn. Agric. Exp. Station Tech. Bull* 8.

Stakman, E.C. and F.J. Piemeisel. 1917. Biologic forms of *Puccinia graminis* in cereals and grasses. *J. Agric. Res.* **10**: 429-502.

Stakman, E.C., M.N. Levine and R. U. Cotter. 1930. Origin of physiologic forms of *Puccinia graminis* through hybridization and mutation. *Sci. Agril.* **10**: 707-720.

Tomar, S.M.S. and M.K. Menon. 2001. *Genes for Resistance to Rust and Powdery mildew in Wheat*, IARI, New Delhi, 152 p.

Tomar, S.M.S., Vinod and B. Singh. 2004. *Distant Hybridization in Wheat*, IARI, New Delhi, 160 p.

Vakilli, N.G. and R.M. Caldwell 1957. Recombination of spore colour and pathogenicity between uredial clones of *Puccinia recondita* f. sp. *tritici*. *Phytopathology* **47**: 536.

Vasudeva, R.S., L.M. Joshi and V.C. Lele. 1953. Susceptibility of some grasses to cereal rusts. *Indian Phytopath.* **6**: 39-46.

Watson, I.A. 1957. Mutation for increased pathogenicity in *Puccinia graminis* var. *tritici*. *Phytopathology* **47**: 507-509.

Wolfe, M.S. 1985. The current status and prospects of multiline cultivars and variety mixtures for disease resistance. *Ann. Rev. Phytopathol.* **23**: 251-273.

Wolfe, M.S. and J.A. Barret. 1980. Can we lead the pathogen astray? *Plant Disease* 64: 148-155.

Watson, I.A. and D. Singh. 1952. The future of rust resistant wheat in Australia. *J. Aust. Inst. Agric. Sci.* **18**: 190-197.

Wellings, C.R. and R.A. Mc Intosh, 1981. Stripe rust- a new challenge to the wheat industry. *Agr. Gaz.*, N.S.W. **92**: 2-4.

Wilcoxson, R.D. 1986. Slow rusting of cereals. **In**: *Problems and Progress of Wheat Pathology in South Asia* (Eds. L.M. Joshi, D.V. Singh and K.D. Srivasatava), pp. 333-344. Malhotra Publishing House, New Delhi, 401 p.

Wilcoxson, R.D., J.F. Tuite and S. Tucker. 1957. Anastomosis of germ tubes in *Puccinia graminis*. *Phytopathology* **47**: 537.

2.2 LEAF BLIGHTS AND BLOTCHES

In India, foliar blight on wheat crop is considered a complex syndrome due to involvement of several pathogens. Amongst them *Drechslera sorokiniana* (Sacc.) Subram. & Jain; *D. tritici repentis* (Died) Shoemaker; *D. tetramera* (Mc Kinney) Subram. & Jain; *D. halodes* (Drech.) Subram. & Jain var. *tritici* Mitra; *D. bicolor* Mitra; *D. catenaria* (Drech.) Ito; *D. rostrata* (Drech.) Richardson and Fraser; *D. nodulosa* (Berk. & Mc Kinney) Subram & Jain; *Helminthosporium atypicum* Deshpande and Deshpande; *Alternaria triticina* Prasada and Prabhu; *A. alternata* (Fr.) Keissler; *A. tenuissima* (Kunze ex. Pers.) Wiltshire; *A. triticola* Vasant Rao; *Dilophospora alopecuri* (Fr.) Fr; *Leptosphaerulina trifolii* (Rostr.) Petr; *Chaetomium dolichotrichum* Ames.; *Cladosporium herbarum* (Pers.) Link ex. S.F. Gay and *Pyricularia oryzae* Sacc. are associated with foliar blight complex. Monitoring of the disease by National Centre on Foliar Blight (Faizabad, U.P.) indicated that *D. sorokiniana* affects the wheat crop more occasionally followed by *A. triticina* (DWR, 1995) and other pathogens are of minor importance (Joshi *et al.*, 1978).

The leaf spots/blights/blotches, though reported from almost all parts of the country, are economically important in eastern region.

2.2.1 Spot Blotch

Spot blotch/leaf blight caused by *Drechslera sorokiniana* (Sacc.) Subram. & Jain (Syn. *Helminthosporium sativum* P.B.K., *Bipolaris sorokiniana* Sacc.) is a serious disease of warm and humid environments as found in wheat growing areas in South Africa, Zambia, Tanzania, Medagaskar, Brazil, Mexico, Argentina, Paraguay, Bolivia, Indonesia, Thailand, Philippines and China (Hetzler *et al.*, 1991; Mehta, 1993; Ginkel and Rajaram, 1998). In the Indian sub-continent warm humid conditions occur in Bangladesh, *Tarai* region of Nepal and North-eastern India, where the disease is widely prevalent (Singh and Srivastava, 1997). Helminthosporium leaf blight was first reported by Sorokin in 1890 from Ussuria (Russia) and in India, it was first noticed by Mohy in 1914 from Pusa, Bihar. Later on McRae (1924) observed that the pathogen appears commonly as leaf parasite in the vicinity of Pusa. Spot blotch occurs in Bihar, West Bengal, eastern Uttar Pradesh and Orissa (Chattopadhyay and Chakrabarti, 1968; Narain *et al.*, 1973; Singh *et al.*, 1998). Of late, the disease has been extended to North-western region comprising of Punjab, Haryana, Delhi and western Uttar Pradesh under rice-wheat cropping system (Mahto *et al.*, 2002) and also in peninsular areas of Maharashtra and Karnataka.

The importance of Spot blotch is expressed in terms of yield losses which are variable but are very significant. Losses upto 85% were estimated in Zamibia (Raemaekers, 1988) and 40% in Philippines (Lapis, 1985). In Brazil the disease caused 100% yield loss in some fields (Hetzler *et al.*, 1991). In China the yield loss has commonly been 20 and 30% or above 75 percent (Chang and Yousan, 1998, Xiao Zhimin *et al.*, 1998). Duveiller *et al.* (1998) recorded 49–90% losses in Mexico. In farmers' field in Bangladesh the average loss due to spot blotch was assessed to be 15% (Alam *et al.*, 1998). Yield loss assessment conducted in Nepal showed 24–27% damage on highly susceptible varieties (Mahto, 1999).

Nema and Joshi (1971) observed that yield loss due to spot blotch is variety dependent. They observed that on variety Sonora 64, the loss incurred was 3% while S 227 suffered a loss of about 20% in India. The loss in grain weight increased with the increase in disease intensity (Karwasra *et al.*, 1998). It was 45.6% in infection grades 7, 8 and 9 and 19.4, 28.7 and 34.4 percent respectively in infection grade 4, 5 and 6. According to Singh *et al.* (1998) leaf blight caused 21.7% loss in variety HP 1633 with 9.8% reduction in 1000-grain weight. Recently, Singh (2003) estimated yield losses under artificial epiphytotic field conditions at Delhi which varied from 6.18 to 36.67 percent.

Symptoms: Spot blotch is basically a foliar disease, although the associated fungus attacks all the plant parts including roots, sub-crown, awans, glumes and seeds. The disease manifests itself in different types of symptoms, *viz.* seed rot, seedling blight, root rot, foot rot and head blight (Narain *et al.*, 1973; Bidari, 1974). However, Mitra (1930) considered that foot rot and seedling blight phases of the disease are not as important as leaf blight phase. Leaf infection coincides with wet and warm weather via airborne conidia of fungus. The first symptom starts appearing with a number of small, oval to oblong, chlorotic young spots on the lower leaves of the plants at seedling stage. Soon these lesions turn light brown having reddish brown centre surrounded by yellow margin (Plate 1D). Gradually, with the advancement of plant growth the fungus forms larger, irregular, necrotic blotches on upper leaves which are surrounded by chlorotic margin (Nema and Joshi, 1971). In severe cases the necrotic blotches covers 50–80 percent area of the leaf blade and leaves dry prematurely, resultantly whole field looks blighted. The drying of the leaf starts from the tip downwards or from margin of the leaves.

The symptoms are also commonly observed on leaf sheath, nodes, internodes and ear heads including the awns. Brownish black discolouration appears at lower nodes. Diseased nodes take on a greyish black, velvety appearance due to dense sporulation. Diseased plants may become stunted and chlorotic (Mehta, 1993). Under severe infection seeds are heavily affected and shrivelled. The grains show light brown to blackish discolouration around the germination point of the seed causing black point.

Causal Organism: The pathogen was first reported by Sorokin in 1890 from infected spikes of wheat and it was identified by Saccardo as *Helminthosporium sorokinianum* Sacc. But Drechsler (1923) believed that the organism which Saccardo identified was synonymous to *H. sativum*. No attempt was made to segregate the graminicolous *Helminthsporium* species until Nisikado (1928) erected two sub-genera *i.e. Cylindro Helminthosporium* having cylindrical conidia that germinates from any cell as represented by *H. triticii* and *Eu Helminthosporium* which has fusoid or curved conidia and germinates from the end cells as represented by *H. sativum*. To accommodate members of sub-genus *Eu Helminthosporium* Shoemaker (1959) erected a new genus *Bipolaris* for *Helminthosporium* species. However, Subramanium and Jain (1966) argued that accommodation of graminicolous *Helminthosporium* species in two genera- *Drechslera* and *Bipolaris* was not warranted and all species previously disposed as *Drechslera* and *Bipolaris* should be described under genera *Drechslera*. Therefore, different nomenclature assigned to spot blotch pathogen are *Helminthosporium sativum*, *Drechslera sorokiniana* and *Bipolaris sorokiniana*. The pathogen is a Demataceous Hyphomyceteous fungus.

D. sorokiniana grows well at 25°C on culture media having pH 6.5 (Rehman *et al.*, 1996). In axenic culture the fungus produces light grey colonies which are composed of hyphae interwoven as a loose cottony mass. These colonies turn black to olivaceous black. Conidiophores which are simple or sometimes branched emerge solitary or in fascicles of 2–4, occasionally more from the stomata or between epidermal cells of the host. They are brown to dark olivaceous in colour, 220 µm long, 6–10 µm, 2–10 septate, septa 10.92–43.68 µm apart with well defined geniculations. Scar is present marking the points of attachment on conidia. The conidia develop terminally and are laterally attached with conidiophore. They are variable in shape, straight or slightly curved, fusoid, elliptical, 3–12 septate, 60–100 × 8–13 µm in size, brown to dark olivaceous in colour having thick peripheral wall and a conspicuous small hilum situated with in the contour of rounded basal end. The spores are somewhat widest near the middle and tapering towards both the abruptly rounded ends (Fig. 22). Scanning electron microscopy of the conidia shows thickenings at the places of pseudosepta and scar at the point of attachment with conidiophore (Das *et al.*, 1999). The germination of conidium is both unipolar and bipolar. Sometimes interseptal germination is also observed (Mishra, 1973).

The perfect stage of the fungus was observed in the laboratory on natural medium in the presence of opposite mating types, and was first described as *Ophiobolus sativus*. It was later renamed *Cochliobolus sativus* (Ito and Kuribayashi) Drechsler ex. Dastur (Dastur, 1942). Under natural conditions, the perfect stage has not been noticed to occur anywhere except in Zambia (Raemaekers, 1988). Fungus produces pseudothecia which are

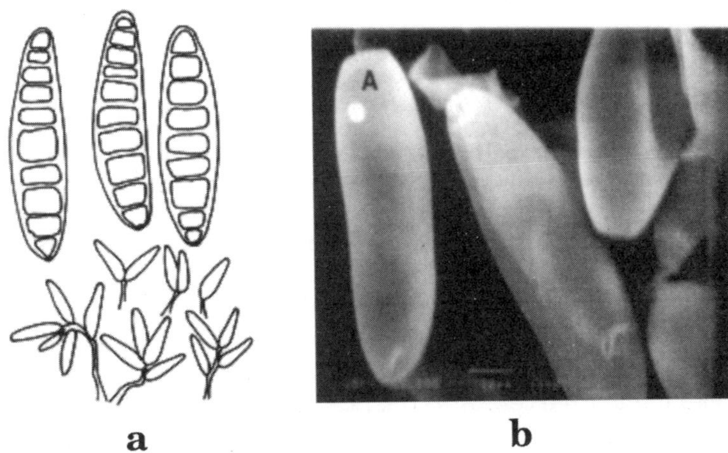

Figure 22: Conidia of *Drechslera sorokiniana* (a) and (b) ultra-structural features of conidium.

dark brown to black, flask shaped upto 530 µm broad with a cylindrical ostiolar beak 80–110 µm long. Asci are cylindrical to clavate and contain 1–8 ascospores. Ascospores are binucleate, hyaline, light brown, 6–13 septate and filiform. Pseudoparaphyses are visible in the asci and these are hyaline, filiform and branched.

Christensen (1922) attempted identification of races of *D. sorokiniana* as early as in 1922 and detected at least 37 races. Mehta (1981) tested isolates of the fungus on a set of 13 differential hosts and recorded the existence of 32 races in Brazil. In India, Bidari and Govindu (1975) and Mishra *et al.* (1981) found that leaf isolates of the fungus differ in their parasitic capability. Significant differences in incubation period, lesion length and sporulation were observed among 9 isolates by Akram and Singh (2000). Inoculation with individual isolates of *D. sorokiniana* on susceptible cultivar Arnez showed different infection index (Singh *et al.*, 2004). Although it is evident from these reports that isolates of *D. sorokiniana* differ in their pathogenic capability but at the same time it is also visualized that each progeny from single conidium of the fungus may differ in pathogenicity. Kumar *et al.* (2002) expressed the view that anatomosis between hyphae that stem from different conidia may result in somatic hybridization and the emergence of new fungal variant. Moreover, isolates loose their pathogenic capability after a period of time and reaction pattern is altered. Constant mutation and saltation are other problems which make race identification still more difficult (Singh and Srivastava, 1997). The term 'race' therefore, has been questioned.

D. sorokiniana produces non-specific toxins. The presence of phytotoxic substances in culture filtrates of the fungus were detected by Ludwig (1957).

Later, De Mayo (1961) isolated sesquiterpenoid toxin related to helminthosporal and pre-helminthosporal. Indian isolates of the fungus were also found to produce highly toxic substance in culture which inhibited the shoot and root growth of wheat seedlings (Das and Srivastava, 1969, 1971). In addition to helminthosporal and pre-helminthosporal toxins, another compound sorokinianin was isolated which proved to be the product of the sesquiterpenoid and tri-carboxylic acid pathways (Nakajama *et al.*, 1994, 1998). Of these, pre-helminthosporal is the most active and is produced in abundance (Olbe *et al.*, 1995). The bioassay with toxin develops symptoms like water soaking, chlorosis and necrosis of the leaf. Histo-pathological studies indicates that toxin inoculated susceptible wheat genotypes show alterations such as degradation of cytoplasm, shrinkage of cell and separation of plasmalemma from cell wall (Das, 2003; Jahani *et al.*, 2006).

Disease Cycle and Epidemiology: Spot blotch pathogen, *D. sorokiniana* is transmitted through contaminated or infected seed, infected crop residue, volunteer plants and free dormant conidia in soil. The conidia in the soil can remain viable upto 1–10 cm layer for 1–3 years (Reis *et al.*, 1998). In the rice-wheat rotation system the spores can survive in the field and may increase the load of primary inoculum in the subsequent wheat crop (Pandian, 2004). However, the investigation shows that the pathogen in India, perpetuates principally on infected seed and cause local infection. Shahner (1981) also attributed that the disease transmission rate is higher through seed than other source of inoculum.

The fungus which is present as mycelium within the pericarp of the seed starts growing soon after sowing and externally infects the emerging coleoptile and sometimes the first leaf. Once inoculum is introduced by the seed, disease develops on above ground plant parts where pathogen sporulates in the presence of direct sunlight. The conidia are wind dispersed or rain splashed to new infection sites for secondary infection cycles (Fig. 23). The conidia germinate in contact with host within four hours and produce penetration pegs and appressoria like structures before penetration (Aggarwal *et al.*, 2002). The appressorial formation is frequent at the juncture of the epidermal cell wall. The germtube can penetrate directly through epidermal cells or stomata. Inside the host, the mycelium multiplies both inter and intracellularly. Hyphae from initially infected cells then enter adjacent mesophyll cells and bring about histopathological changes within host. There is degradation of cytoplasm due to the effect of toxin produced by the pathogen in its necrotrophic phase. Jagdeesh and Nema (1978) observed that cellulolytic and pectinolytic enzymes have role in disruption of infected cells. Cytoplasm becomes granular 24 hours after pathogenesis and host nuclei and other orgnelles degenerate (Aggarwal *et al.*, 2008). The invaded cells get shrunken and water soaked symptoms appear on leaf blade and leaf sheath. Disease generally progresses upward in the crop

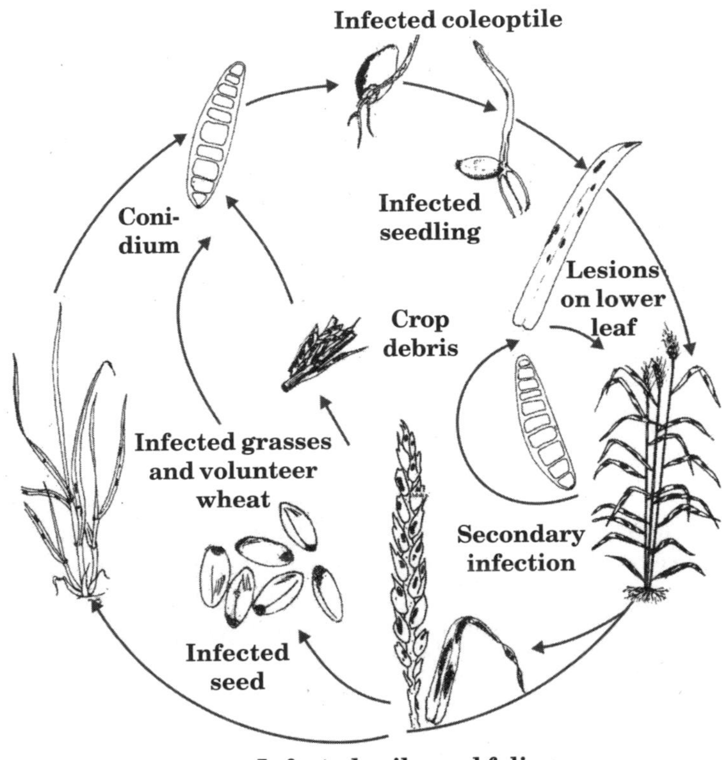

Figure 23: Disease cycle of spot blotch (*Drechslera sorokiniana*) of wheat

canopy, culminating in infection of the ear and finally the seed. The fungus gains access to the seed via an infected glume, lemma or palea and survives at least 2 years within the pericarp of an infected seed. During harvesting and threshing of diseased plants, conidia of the fungus contaminate the seeds. Survival in soil is accomplished by the ability of the fungus to colonize diseased wheat straw (Nair, 1962). High levels of moisture and addition of nitrogen, manganese, boron and zinc cause disappearance of the fungus from the straw (Subramanium, 1962).

The environmental factors particularly temperature and humidity influence the disease intensity, sporulation and spore deposition (Mitra and Bose, 1935). Although infection can occur over a wide range of temperatures, the optimum temperature around 28°C and relative humidity 92 percent favour the maximum development of foliar blight (Nema and Joshi, 1971; Saari, 1985). Frequent rains and prolonged wetting of the foliage increase production of conidia and infection efficiency. A minimum of 18 hours leaf wetness with a mean temperature greater than 18°C is required for disease development in susceptible wheat.

D. sorokiniana in addition to wheat has a large number of grasses that co-exist in an area (Nagarajan and Kumar, 1998). Lapis (1985) reported that various grasses and broad leaf weeds such as *Commelina diffusa, Chloris barbate, Dactyloctentium aegyptium, Eleusine indica, Cyperus difformis, C. fimbicatus, Imperata cylindrica, Cynodon dactylon, Paspalum conjugatum, Leptochloa chinensis, Brachiaria distachya, B. mutica* and *Echinocloa colona* grow year round in Philippines and harbour the pathogen. The role of weeds in the perpetuation of disease in Indian sub-continent is still not clear, but the association of *Phalaris minor* and *Launia splenifolia* with *D. sorokiniana* was noticed in India by Singh *et al.* (1995). Pandian (2004) recorded natural infection of the fungus on *Avena fatua* and could also infect *Allium canadense* under artificial inoculation.

Management

Resistance: Inheritance studies on resistance to spot blotch indicate that both monogenic (Srivastava *et al.*, 1971; Adlakha *et al.*, 1984) and polygenic (Velazquez, 1994) resistance are operative. Dubin and Rajaram (1996) reported that resistance is under polygenic control. Recently, Joshi *et al.* (2004) used both qualitative and quantitative approach of genetics of resistance and proved that spot blotch resistance is conditioned by only three additive genes. It seems that information about the underlying genetic mechanism of resistance is still scant.

The inheritance of leaf tip necrosis (*Ltn*) in relation to spot blotch was tested by Singh *et al.* (1992) which proved to be monogenic. Under Indian scenario *Ltn* is reported to be associated with moderate resistance to spot blotch (Joshi *et al.*, 2004). Singh and Rajaram (1992) suggested that *Ltn* can be exploited as a marker for selection for *Lr34/Yr18/Bdv1* genes to achieve moderate level of multiple resistances to leaf rust, stripe rust, barley yellow dwarf virus and spot blotch.

Wheat genotypes show bio-histochemical defense response against *D. sorokiniana*. Das *et al.* (1999) noticed that resistant genotypes like Pusa T3336 (Lok Bharti x DARF) contain C-homologues wax particles over the leaf epidermis and around guard cells of stomata, which are related to resistance against spot blotch pathogen. Callose deposition on the epidermal cell wall and lignin around the mesophyll tissues also contribute to the resistance response of host (Aggarwal *et al.*, 2008). Role of defense related enzymes –peroxidase and estrase in resistance to spot blotch has also been defined (Das *et al.*, 2003) and such biochemical parameters have their utility in differentiating the susceptible and resistant genotypes of wheat.

Intensive efforts are underway for identification of sources of resistance against *D. sorokiniana*. These sources are broadly grouped into three categories, viz. Latin American, Chinese and wild relatives and alien species (Ginkel and Rajaram, 1988). Among Latin American sources BH 1146, CNT 1, CNT 8 etc. have the best resistance to spot blotch (Mehta, 1985). Chinese sources are Suzhae # 8, Yangmai, Longmai and Shanghai. Alien sources include *Thynosporium curvifolium* and *Aegilops squarrosa*. At CIMMYT, transfer of resistance genes from alien species (*T. curvifolium, Elymus curvifolius* and *Triticum tauschii*) to wheat germplasm has also been achieved (Mujeeb Kazi *et al.*, 1996). By utilizing some of the resistance genes from alien materials, promising genotypes such as Sabuf, Chyria 1, Cugap, Ocepar 7 and Mayoor were developed at CIMMYT, Mexico (Sharma *et al.*, 1997). The information presented by Directorate of Wheat Research, Karnal on resistance suggests that foliar blight can be kept under control with the adoption of disease tolerant wheat varieties like VL738, VL804, HS277, PBW34, PBW343, HD2189, HD2687, HD2733, HW2004, HW2045, DDK1009, DWR14, HUW234, HUW533 and NW1012 (Singh, 2008).

Cultural Control: Manipulation of agronomic practices is considered useful in IPM system. Use of healthy seeds collected from disease free crop to some extant reduces the chances of transmission through contaminated or infected seeds.

Tandon (1985) observed that the late sown crop in North-western region of the country experiences rainfall during February and March with rise in temperature, and such a warm and humid weather favours spot blotch incidence aggressively. Singh (2003) recorded that the disease in late sown crop progresses at higher rate than normal sowing in North-eastern region. Late planting of wheat should be avoided so that crop does not coincide with hot and humid period. Srivastava *et al.* (2004) emphasised that wheat under very late sowing conditions beyond 15th December attains physiological aging and at this stage the warm weather is not favourable for spot blotch development. Although the disease severity is comparatively less in very late sown crop but yield gain is low.

It has been experimented that dense vegetative growth of dwarf wheat varieties provides a distinct microclimate congenial for the build up of the disease through profuse tillering (Mehta and Gaudenico, 1991). Singh *et al.* (1998) noted that wider spaced row (23 cms) and lower seed rate (100 kg/ha) reduces foliar blight intensity. Pandian (2004) also proved experimentally that wider row spacing keeps disease severity of spot blotch at low level by reducing excessive vegetative growth and humid conditions. Clearing and ploughing of the stubble, grass weeds and volunteer plants reduce inoculum as in case of crop rotation (Diehl *et al.*, 1982).

Excessive use of NPK is associated with severe spot blotch development. Maity (2002) evaluated the effect of different doses and combinations of

NPK on foliar blight severity and found that disease development was severe with 120 N/ha. and minimum with NPK ratio 60:80:80. Singh (2003) also recorded that nitrogen application @ 60 kg/ha is the optimum dose for getting normal yield with low severity for foliar blight. But foliar blight decreased when higher dose of potash (K=70 kg/ha) was added to plots. Keeping this in view Srivastava *et al.*, (2004) suggested that application of NPK in the ratio of 60:70:70 may be restricted in high disease incidence area.

Chemical Control: As the disease is initially seed borne, treatment with fungicides like captan, mancozeb, maneb, thiram, pentachloronitrobenzene (PCNB), carboxin, iprodione, triadimefon, propiconazole and tebuconazole controls the pathogen in seed (Mehta, 1993; Srivastava *et al.*, 2004; Singh, 2008). The foliar blight phase of the disease is managed effectively with sprays of mancozeb and propiconazole (Lapis, 1985; Singh and Rai, 1986), triadimefon and tebuconazole (Srivastava and Tiwari, 2002; Singh *et al.*, 2003) when the crop is sprayed with appropriate fungicide soon after the appearance of first sign of the disease followed by second application 2–3 weeks later if necessary.

Biological Control: Microbial control of the disease was attempted by Mandal (1996) using the isolates of *Chaetomium globosum, Trichothesium roseum, Trichoderma reesei, T. hamatum* and *Talaromyces flavus*. Aggarwal *et al.* (1996) used these biocontrol agents to study ultra structure of hyphal interference between antagonisis and *D. sorokiniana*. They found suppression of mycelial growth of the host pathogen by antagonistic fungi through penetration of the hyphae by pegs, tight coiling of hyphae, creation of holes and lysis of mycelia. Biswas (2000) determined the antagonistic potential of *C. globosum* and recorded high inhibitory effect on mycelial growth and conidial germination. Scanning electron microscopy of inoculated wheat leaves sprayed with 0.2% antifungal compound showed distortion in conidial wall and disordered mycelial growth of the pathogen, succeeding in reduced spot blotching. *C. globosum* is reported to produce antifungal metabolites such as cochliodinol, mollicellin-G, chaetomin etc. which are correlated with antagonism to *D. sorokiniana* (Biswas, 2002). Biochemical characterization of *C. globosum* has also shown the production of β 1, 3-glucanase and xylanase (Ahammed *et al.*, 2008). Aggarwal *et al.* (2004) evaluated different strains of the biocontrol agent under *in vitro* and *in vivo* and indicated antibiosis as the principal mechanism of antagonism. Pre-inoculation spray of *C. globosum* is also found to induce resistance in wheat against *D. sorokiniana* (Biswas *et al.*, 2002, 2003). Nutritional requirements for mass multiplication of *C. globosum* were standardised and a bio-formulation effective against spot blotch was developed for its application in the field (Selvakumar *et al.*, 2001; Ahammed *et al.*, 2005).

2.2.2 Other Pathogenic *Drechslera* species

Drechslera tritici repentis

McRae (1922) recorded *D. tritici repentis* Died (Shoem.) on wheat at Pusa in 1922 and found that this pathogen was more common than *D. sorokiniana*. Later, Mitra (1931) described *D. tritici repentis* causing tan spot on wheat leaves and leaf sheath. Although this pathogen has been reported from cooler regions of India but it is never considered a major pathogen. According to Singh *et al.* (1998) *D. sorokiniana* and *D. tiritici repentis* are associated with distinguishable blight spots, sometimes combined on a single leaf. It is a principal foliar blight pathogen in the *Tarai* of Nepal (Mahto and Bimb, 1996).

Symptoms: The symptoms of the disease appear as minute light brown spots surrounded by yellowish hallo, measuring 1 × 0.5 mm. The spots enlarge longitudinally along the leaf vein and are oval or fusiform measuring 10 × 2.5 mm. Fully matured spots develop a uniform dried straw colour or greyish brown colour. The margin of the mature spots sometimes is encircled by yellowish zone (Plate 2A). The spots are never surrounded by darker brown margin as in case of *D. sorokiniana* (Misra, 1973). Misra and Singh (1972) collected three isolates of the fungus, namely Htr-1, Htr-2 and Htr-3 which exhibited difference in size of the spots and other characters as well as pathogenicity.

Causal Organism: *Drechslera tritici repentis* (Died). Shoem. (Syn. *Helminthosporium tritici repentis* grows well on potato dextrose agar medium at 20–30°C, but no growth is observed at 35°C (Mishra, 1973). Circular, dirty grey coloured colonies having regular margin appear on

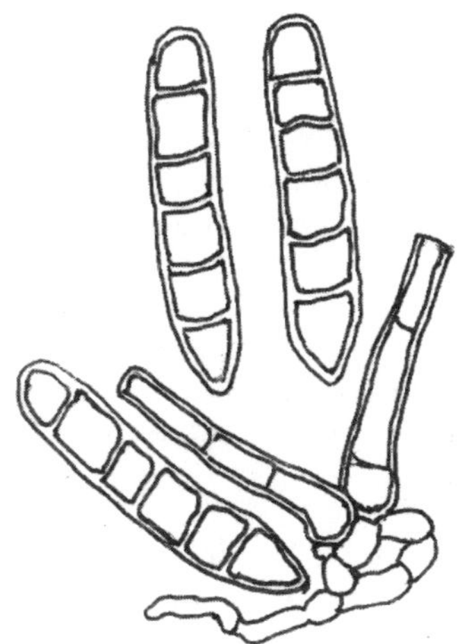

Figure 24: Conidia of *Drechslera tritici repentis*.

culture medium. Aerial mycelium bears numerous conidiophores and conidia. Conidiophores emerge solitary or in fascicles of 2–4 from the stomata or between epidermal cells. They are simple, olive to olivaceous in colour, 3–10 septate, measuring 80–400 × 6–9 µm and bear single conidia (Fig. 24). The conidia are cylindrical with a rounded apex and a characteristic conical basal cell with an inconspicuous scar (Zillinsky, 1983).

The perfect stage of the pathogen, *Pyrenophora tritici repentis* (Died.) Drechs. reported from outside India, occurs on weathered wheat stubbles and straw. The pseudothecia are black, erumpent and 0.2–0.35 mm in diameter with dark spines surrounding the short beak. The ascospores are oval, yellowish-brown and have 3 transverse and 1–2 longitudinal septa. They measure 45–70 × 18–28 µm and have a gelatinous covering (Horsford, 1972).

Disease Cycle and Epidemiology: Infected seeds, grasses and infected wheat stubble are important source of primary inoculum. The mycelium present within the pericarp of the seed infects emerging coleoptile and develops lesions. As the primary lesions enlarge, conidia are produced under humid conditions. Conidia are then wind borne to cause infection on lower leaves. Conidia surviving on the infested stubbles and infected grasses are also wind dispersed over long distances and responsible for secondary infection cycles (Fig. 25). The optimum temperature ranging from 20-28°C, frequent rains and prolonged wetting of foliage favours disease development.

Control: Seed dressing fungicides, *viz.*, guzatine, imazalil, triadimenol, bitertanol, carboxin and carbendazim give good control of the fungus. A number of foliar fungicides including protectant mancozeb and systemic fungicide propiconazole provide acceptable control (Mehta, 1993).

Drechslera tetramera

A new blight disease of wheat caused by *D. tetramera* (McKinney) Subram. & Jain was reported from Maharashtra state for the first time in 1972 (Kalekar and Kulkarni, 1972). The disease is not widely prevalent in the country.

Small, yellow lesions of the disease are first visible on leaf blade. In advance cases several lesions coalesce and cover extensive area of the green tissues. In seedling stage symptoms of foot rot are also seen (Kalekar and Kulkarni, 1972; Misra, 1973).

The fungus grows luxuriantly and sporulates profusely on PDA at 30°C. Olive-grey, circular colonies with zonations are produced on culture medium. Simple or compound, dark olivaceous to brown conidiophores, measuring 59.29–187.20 × 4.68–6.24 µm, 3–10 septate emerge solitary or in fascicle of

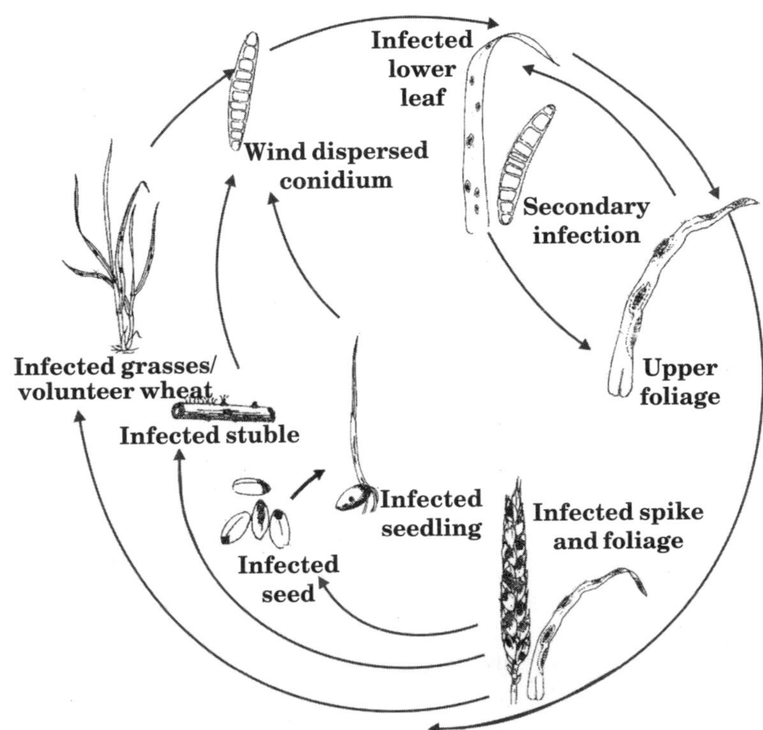

Figure 25: Disease cycle of tan spot (*Drechslera tritici repentis*) of wheat

2–5 through the stomatal opening or between epidermal cells. Conidia are usually 4-celled, symmetrical in shape, tapering towards the rounded ends, dark olivaceous to dark brown in colour, measuring 18.72–34.32 × 6.24–9.36 µm and borne in cluster of 2–30 or rarely more (Fig. 26). The pathogen is considered to be weak and is not economically important.

Some fungicides *i.e.* Agrosan, Captan, Cuman and Aureofungin are reported effective for checking the growth and sporulation of the fungus *in*

Figure 26: Conidia of *Dreschslera tetramera*

vitro (Kalekar and Patil, 1973). Varietal resistance to *D. tetramera* was recorded in some old wheat varieties (Kalekar, 1973).

Drechslera catenaria

Misra and Singh (1971) recorded *D. catenaria* (Drech.) Ito (Syn = *Helminthosporium catenarium*) as a new species of *Drechslera*, causing foliar blight in wheat. The disease appears as minute light brown spots measuring 1 × 0.5 mm on leaves of nearly two months old plants. The spots are clearly visible against light, later these enlarge in size and measure 8 × 2.5 mm and give a straw coloured to yellowish brown appearance. In severe cases the affected leaves are discernibly bent at tips and wither away.

The fungus grows well on PDA and wheat extract agar media at 28°C. Aerial mycelium is scanty and pale purplish in colour. Colonies are circular and saltation and zonation absent. Conidiophores arising from mycelium are simple, light olive-brown, measuring 68.64–349.44 × 7.80–9.36 µm, 4–13 septate (Fig. 27). The point of attachment of conidia is marked by scar at the apices of strongly geniculated conidiophore. Conidia are sub-hyaline when young and yellowish at maturity, measuring 43.68–134.6 × 9.36–15.60 µm, 2–9 septate, sometimes the septa associated with straight constrictions, produced in chains of 2–3 or rarely more by apical and lateral proliferation. Conidia are of two types, shorter and longer ones.

Wheat varieties HP 584, EA 210-1 and E 5533 were found resistant (Misra, 1973). Treatment of seed with captan or thiram and spraying of Dithane Z-78 as routine control measure are beneficial.

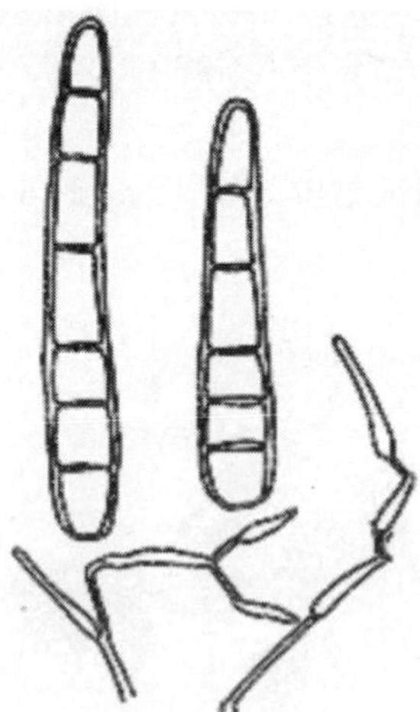

Figure 27: Conidia of *Dreshslera catenaria*

Drechslera specifera

Wheat blight incited by *Drechslera specifera* [(Syn. *Helminthosporium speciferum* (Bains) Nicob. (Fig. 28)] was first observed on Mexican wheat variety S 227 in 1967 (Singh and Singh, 1968). Conidia arising on solitary conidiophores are straight, oblong or cylindrical rounded at the ends. This disease became more serious in 1968 on several varieties. Disease index yield loss relationship revealed that in susceptible variety S 227, the loss was 72% and 63% in Lerma Rojo (Singh and Singh, 1971).

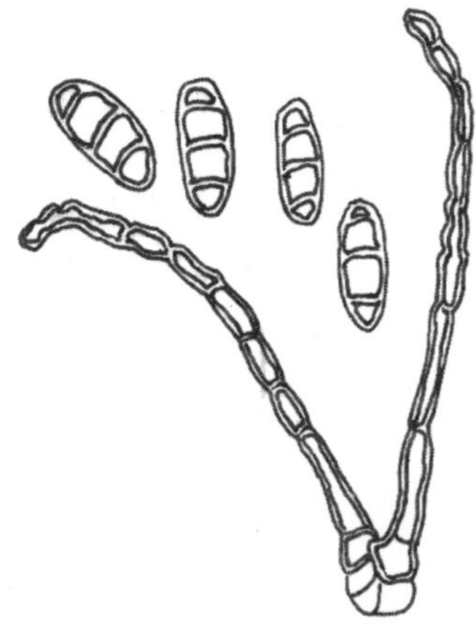

Figure 28: Conidia of *Drechslera specifera*

The disease manifests itself as yellowing of the leaves starting from leaf tip and progresses downward towards the base. Finally the affected leaves turn necrotic and die off. The disease plants are stunted and devoid of tillers. The grains become shrivelled and develop the black point symptoms.

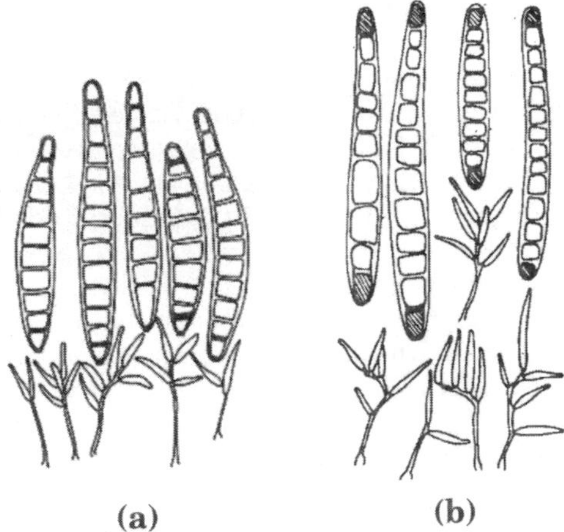

Figure 29: Conidia of (a) *Drechslera rostrata* and (b) *Drechslera bicolor*

Singh and Singh (1971) reported that *D. specifera* has wide host range in *Kharif* season and can survive all the year round. It can also survive in plant debris in off season, if any, due to its faculty of growing at wider temperature range of 10–45°C with an optimum at 30°C.

Forty three wheat varieties were listed as resistant to the pathogen. Dithane Z-78 as foliar spray @ 0.2 % at 15 days interval was found to protect plants from leaf blight phase of the disease.

Some other records of *Drechslera* species are: *D. gramineum* from Sagar in Madhya Pradesh (Grover, 1957); *D. nodulosa* var. *tritici* from Maharashtra and Gujarat (Patel *et al.*, 1953); *D. rostrata* from Bihar and Maharashtra (Misra and Singh, 1971; Kalekar and Patil, 1973) and *D. bicolor* (Fig. 29) from Bihar (Mitra, 1931). However, analysis of large number of diseased leaf samples indicates that *D. sorokiniana* is the main causal agent of foliar blight in India (Meena Kumari, 1985; Singh *et al.*, 1998) and other pathogenic species are insignificant.

REFERENCES

Adlakha, K.L., R.D. Wilcoxson, and S.P. Raychaudhuri. 1984. Resistance of wheat to leaf spot caused by *Bipolaris sorokiniana*. *Plant Dis.* **68**: 320-321.

Aggarwal, R., A.K. Tiwari, K.D. Srivastava and D.V. Singh. 2004. Role of antibiosis in the biological control of spot blotch (*Cochliobolus sativus*) of wheat by *Chaetomium globosum*. *Mycopathologia* **157**: 369-377.

Aggarwal, R., K.D. Srivastava and D.V. Singh. 1996. Mechanism of combat between plant pathogens and antagonists, p 7. Proc. Annual Meeting, Indian Phytopathological Society held at PAU. Ludhiana 14-16 Feb., 1996.

Aggarwal, R., Soma Das, D.V. Singh and K.D. Srivastava. 2002. SEM studies on spore morphology and infection process of spot blotch pathogen in wheat. *Indian Phytopath.* **35**: 197-199.

Aggarwal, R., Soma Das, J. Mehdi and D.V. Singh. 2008. Histopathology of spot blotch [*Bipolaris sorokiniana* (telemorph: *Cochliobolus sativus*)] infection in wheat. *Acta Phytopathologica* **43**: 23-30.

Ahamed, S.K., R. Aggarwal and D.V. Singh 2005. Optimizing nutritional conditions for mass multiplication of *Chaetomium globosum* an efficient biocontrol agent of *Bipolaris sorokiniana* (Sacc.) Shoemaker. *Indian J. Pl. Protect.* **33**: 90-93.

Ahammed, S.K., R. Aggarwal, Sneh and H.C. Kapoor 2008. Production, partial purification and characterization of extracellular xylanase from *Chaetomium globosum*. *J. Plant Biochem. Biotechnol.* **17**: 95-98.

Akram, M. and A. Singh 2000. Studies on tolerance to *Helmithosporium sativum*, incitant of spot blotch of wheat. *Nat. Symp. on Role of Resistance in Intensive Agriculture*, held at DMR, Karnal 15-17, Feb. 2000.

Alam, K.B., S.P. Banu and M.A. Saheed 1998. Occurrence and significance of spot blotch in Bangladesh. **In**: *Helminthosporium Blight of Wheat: Spot Blotch and Tan Spot* (Eds. E. Duveiller and associates), pp. 63-66, CIMMYT, Mexico, D.E.

Bidari, V.B. 1974. Studies on blight of wheat caused by *Helminthosporium sativum* Pam., King and Bakke in Mysore State. *Mysore J. Agric. Sci.* **8**: 475-477.

Bidari, V.B. and H.C. Govindu 1975. Studies on the pathogenicity of three isolates of *Helminthosporium sativum* Pam., King and Bakke on wheat at its different stages of growth in Karnataka State. *Mysore J. Agric. Sci.* **9**: 87-94.

Biswas, S.K., K.D. Srivastava, R. Aggarwal and D.V. Singh 2002. Evaluation of *Chaetomium globosum* as an inducer of resistance against spot blotch of wheat. *Indian Phytopath.* **55**: 510-512.

Biswas, S.K. 2000. *Biocontrol potential of Chaetomium globosum against Drechslera sorokiniana.* M.Sc. Thesis, IARI, New Delhi.

Biswas, S.K. 2002. *Biochemical changes in wheat induced by Chaetomium globosum against spot blotch pathogen.* Ph.D. Thesis, IARI, New Delhi.

Biswas, S.K., K.D. Srivastava, R. Aggarwal, S. Praveen and D.V. Singh. 2003. Biochemical changes in wheat induced by *Chaetomium globosum* against spot blotch pathogen. *Indian Phytopath.* **56**: 324-329.

Chang, N and W. Yousan. 1998. Incidence and current management of spot blotch of wheat in China. **In**: *Helminthosporium Blight of Wheat: Spot Blotch and Tan Spot* (Eds. E. Duveiller and associates), pp. 119-125, CIMMYT, Mexico, D.F.

Chattopadhyay, S.B. and K.B. Chakrabarti, 1968. Role of different pathogens in inciting the Helminthosporium disease in West Bengal. *Indian J. Agric. Sci.* **38**: 937-940.

Christesen, J.J. (1922). Studies on parasitism of *Helminthosporium sativum. Min. Agric. Expt. Sta. Tech. Bull.* **11**: 39.

Das Soma, R. Aggarwal and D.V. Singh 2003. Differential induction of defense related enzymes involved in lignin biosynthesis in wheat in response to spot blotch infection. *Indian Phytopath.* **56**: 129-133.

Das Soma, R. Aggarwal, P. Dureja and D.V. Singh. 1999. Leaf surface waxes in relation to resistance against spot blotch (*Drechslera sorokiniana*) of wheat. *Acta Phytopathologica et Entomologica Hungarica* **34**: 75-84.

Das, A.M. (1962). *Further studies on foot rot of wheat incited by Helminthosporium sativum P.K. & B. and possible measures for its control.* Ph.D. Thesis. IARI New Delhi.

Das, A.M. and D.N. Srivastava 1971. Evaluation of some seed dressing fungicides against foot rot of wheat incited by *Helminthosporium sativum. Indian J. Agric. Sci.* **41**: 387-389.

Dastur, J.F. 1942. Notes on some fungi isolated from black point affected kernels in the Central Provinces. *Indian J. Agric. Sci.* **12**: 731-742.

de Mayo P., E.Y. Spencer and R.W. White. 1961. Helminthosporal, the toxin from *Helminthosporium sativum. I.* Isolation and characterization *Canad. J. Chem.* **39**: 1608-1612.

Directorate of Wheat Research. 1995. Report of the Coordinated Experiments, 1994-95. Vol. 1. Crop Protection (Pathology), AICWIP, DWR, Karnal, 206 p.

Drechsler, C. 1923. Some graminicolous species of *Helminthosporium. J. Agric. Res.* **34**: 641-739.

Dubin, H.J. and S. Rajaram 1996. Breeding disease resistant wheats for tropical highlands and lowlands. *Ann. Rev. Phytopathology.* **34**: 503-526.

Duveiller, E., I. Gracia, J. Toledo, J. Fanco, J. Crossa and F. Lopez. 1998. Evaluating spot blotch resistance of wheat: Improving disease assessment under controlled conditions and in the field. **In**: *Helminthosporium Blight of Wheat: Spot Blotch and Tan Spot* (Eds. E. Duveiller and associates), CIMMYT, Mexico, 171-181 pp.

Ginkel, M.V. and S. Rajaram. 1998. Breeding for resistance to Spot Blotch in Wheat: Global perspective. **In**: *Helminthosporium Blight of Wheat: Spot blotch* *and Tan Spot* (Eds. E. Duveiller and associates), pp. 162-170. CIMMYT, Mexico.

Grover, R.K. 1957. Folliculous fungi of Saugar. *Bull. Bot. Univ.* (Saugar). **1&2**: 9-13.

Hetzler, J., Z. Eyal, Y.R. Mehta, L.A. Campos, H. Febramann, V. Kushnir, O.J. Zakaria and L. Cohen. 1991. Interactions between *Cochliobolus sativus* and wheat cultivars. **In**: *Wheat for the Non-traditional Warms Area* (Eds. D.A. Saunders), pp. 146-164. CIMMYT, Mexico, D.F., 549 p.

Hosford, R.M. 1972. Propagules of *Pyrenophora trichostoma*. *Phytopathology* **62**: 627-629.

Jagdeesh, K.M. and K.G. Nema.1978. Cell wall degrading enzymes in pathogenesis of *Helminthosporium sativum* on wheat. All India Symposium on physiology of parasitism, JNKVV, Jabalpur, 37 p.

Jahani, M., R. Aggarwal, P. Dureja and K.D. Srivastava. 2006. Toxin production by *Bipolaris sorokiniana* and its role in determining resistance in wheat genotypes. *Indian Phytopath.* **59**: 340-344.

Joshi, A.K., R. Chand, S. Kumar and R.P. Singh. 2004. Leaf tip necrosis: a phenotypic marker associated with resistance to spot blotch disease in wheat. *Crop Sci.* **44**: 792-796.

Joshi, A.K., S. Kumar, R. Chand and G.O. Ferrara. 2004. Inheritance to spot blotch caused by *Bipolaris sorokiniana* in spring wheat. *Plant Breeding* **123**: 213-219.

Joshi, L.M., K.D. Srivastava, D.V. Singh, L.B. Goel and S. Nagarajan. 1978. *Annotated Compendium of Wheat Diseases in India*. ICAR, New Delhi, India, 332 p.

Kalekar, A.R. 1973. Varietal resistance in wheat to *Bipolaris tetramera* and *Helminthosporium rostratum* in Maharashtra State. *Indian J. Mycol. Pl. Pathol.* **3**: 71-75.

Kalekar, A.R. and P.L. Patil. 1973. A seed borne disease of wheat (*Triticum aestivum* L.) *PKV Res. J.* **2**: 16-20.

Kalekar, A.R. and U.B. Kulkarni, 1972. Leaf blight of wheat caused by *Bipolaris tetramera* in Maharashtra. *MPAU Res. J.* **3**: 144-145.

Karwasra, S.S., M.S. Beniwal and R. Singh (1998). Occurrence, cultivar reaction and yield losses due to leaf blight of wheat. *Indian Phytopath.* **51**: 363-364.

Lapis, D.B. 1985. Insect pests and diseases of wheat in the Philippines. **In**: *Wheats for More Tropical Environmentas*. (Eds. Villareal, R.L. and Klatt, A.R.). pp 152-153. Proceedings of the International Symposium, CIMMYT, Mexico, 354 pp.

Ludwig, R.A. 1957. Toxin production by *Helminthosporium sativum* and its significance in disease development. *Canad. J. Bot.* **35**: 291-304.

Mahto, B.N. 1999. Assessment of yield loss done to foliar blight of wheat in Nepal. *Ann. Agric. Res.* **20**: 212-215.

Mahto, B.N. 2001. Effect of Helminthosporium leaf blight on yield components and determination of resistance in selected wheat varieties. *Ann. Agric. Res.* **22**: 177-181.

Mahto, B.N. and H.P. Bimb (1996). Foliar blight of wheat in Nepal. Plant Pathological Research Review Meeting held at NARC, Khumltar, Kathmandu, Nepal, May 29-30, 1946.

Mahto, B.N., D.V. Singh, K.D. Srivastava and R. Aggarwal. 2002. Mycoflora associated with leaf blight of wheat and pathogenic behaviour of spot blotch pathogen. *Indian Phytopath.* **55**: 319-322.

Maity, S.S. , R.P. Sanyal and S. Das 2002. Effect of inorganic nutrients on leaf blight severity in wheat caused by *Helminthosporium sativum. Ann. Pl. Protec. Sci.* **10**: 106-110.

Mujeeb Kazi, R.L. Villareal, L.A. Gilchrist and S. Rajaram. 1996. Registration of five wheat germplasm lines resistant to Helminthosporium leaf blight. *Crop Sci.* **36**: 216-217.

Mandal, S. 1995. *Effect of some antagonists on Drechslera sorokiniana, the causal agent of spot blotch of wheat.* M.Sc. Thesis, IARI, New Delhi. 74 p.

McRae, W. 1922. Report of the Imperial Mycologist. *Sci. Rep. Agric. Res. Inst., Pusa,* 1930-31, 73-86 pp.

McRae, W. 1924. Economic Botany, Part III, Mycology. *Rep. of Board of Scientific Advice,* India, 1922-23, 31-35 pp.

Meenakumari, R.V.S. 1985. *Comparative studies on some leaf spots and blights of wheat.* M.Sc. Thesis, IARI, New Delhi, 94 p.

Mehta, Y.R. 1981. Indentification of races of *Helminthosporium sativum* of wheat in Brazil. *Pesquisa Agropecuaria Brasiliera* (Brasilla) **16**: 311-336.

Mehta, Y.R. 1993. Spot blotch (*Bipolaris sorokiniana*) **In**: *Seed Borne Diseases and Seed Health Testing of Wheat* (Eds. S.B. Mathur and B.M. Cunfer), Institute of Seed Pathology for Developing Countries, Copenhagen, Denmark. 168 p.

Mehta, Y.R. and C.A. Gaudenico (1991). The effect of tillage practice and crop rotation on the epidemiology of some wheat diseases. **In**: *Wheat for the Nontraditional Warm Area* (Ed. D.A. Saunders), pp. 266-283, CIMMYT, Mexico, 549 p.

Misra, A.P. and R.A. Singh 1971. Helminithosporium diseases of wheat in India. *Second Int. Symp. Pl. Path.* IARI, New Delhi, Group Discussion at Kanpur, p. 5 (Abstr.).

Misra, A.P. 1973. *Helminthosporium* species occurring on cereal and other Grainineae. Final report on PL 480 Project, Tirhut College of Agriculture, Dholi, Muzaffarpur (Bihar), India.

Misra, A.P. and R.A. Singh 1972. Pathogenic differences amongst three isolates of *Helminthosporium tritici repentis* and performance of wheat varieties against them. *Indian Phytopath,* **25**: 350-353.

Misra, A.P., S.P. Pandey, A.K. Misra and J. Jha 1981. Pathogenic difference in isolates of *Bipolaris sorokiniana* Shoemaker (*H. sativum* P.K. & B.) from wheat, barley and triticales from different geographical regions. *3rd Inter. Symp. Plant Path.*, IARI, New Delhi, pp. 131-132. (Abstr.).

Mitra, M. 1930. A comparative study of the species and strains of *Helminthosporium* on certain Indian cultivated crops. *Trans. Brit. Mycol. Soc.* **15**: 254-293.

Mitra, M. and R.D. Bose 1935. *Helminthosporium* diseases of barley and their control. *Indian J. Agric. Sci.* **5**: 449-484.

Nagarajan, S. and J. Kumar 1998. Foliar blight of wheat in India: Germplasm improvement and future challenges for sustainable high yielding wheat production. **In**: *Helminthosporium Blights of Wheat: Spot Blotch and Tan Spot.* Proceedings of an international Workshop (Eds. E. Duveiller, H.J. Dubin, J. Reeves and A. Mc Nab), pp. 52-58. CIMMYT, Mexico, D.F.

Nair, N.G. (1962). Behaviour of *Bipolaris sorokiniana* on the rhizosphere of wheat *Proc. Indian Acad. Sci.*, B **55**: 290-295.

Nakajima, H., K. Isomi, T. Hamasaki and M. Ichinoque. 1994. Sorokinianin: A novel phytotoxin produced by the pathogenic fungus *Bipolaris sorokiniana Tetrahedron Lett.* 35: 9597-9600.

Nakajima, H., Y. Toratsu, Y. Fujii, M. Ichinoque and T. Hamasaki. 1998. Biosynthesis of sorokinianin, a phytotoxin of *Bipolaris sorokiniana*: Evidence of mixed origin from the sesquiterpene and TCA pathways. *Tetrahedron Lett* **39**: 1013-1016.

Narain, A., S.K. Sinha, B.K. Mahapatra, D.T. Sechler and J.M. Poehlman. 1973. Incidence of Helminthosporium blight of wheat in Orissa State, India. *Plant Dis. Reptr.* **57**: 278-280.

Nema, K.G. and L.M. Joshi. 1971a. The spot blotch disease of wheat caused by *Helminthosporium sativum*. *Proc. 2nd Int. Symp. Plant Path.*, IARI, New Delhi, 42 pp.

Nema, K.G. and L.M. Joshi. 1971b. Symptoms and diagnosis of the spot blotch and leaf blight diseases of wheat. *Indian Phytopath.* **24**: 418-419.

Nene, Y.L., S.C. Saxena, P.D. Tyagi, A.B. Misra, D.V. Singh, A.P. Misra, S.V. Pandit and L.M. Joshi 1970. Fungicide and nematicide test results of 1969. *News American Phytopath. Soc.* No. 25.

Nisikado, Y. 1928. Studies on the *Helminthosporium* diseases of Gramineae in Japan. *Special Report*, Ohara Inst., Landw, Forsch., 4, iv+384+11 pp.

Olbe, M., M. Sommarin, M. Gustaffon and T. Lundborg. 1995. Effect of fungal pathogen *Bipolaris sorokiniana* toxin pre-helminthosporal on barley root plasma membrane vesicles. *Plant Pathol.* **44**: 625-635.

Pandian, C. 2004. *Epidemiological studies on spot blotch of wheat caused by Drechslera sorokiniana,* Ph.D. Thesis, IARI, New Delhi.

Patel, M.K., M.N. Kamat and Y.A. Padhye. 1953. A new species of *Helminthosporium* on wheat. *Indian Phytopath.* **6**: 15-26.

Rahman, M.L., M.Z. Rahman, I.H. Main and K.B. Alam. 1996. Effect of different level of temperature and pH on growth of *Bipolaris sorokiniana* and method of its inoculation into wheat. *Bangladesh J. Pl. Pathol.* **12**: 51-53.

Raemaekers, R.H. 1988. *Helminthosporium:* Disease complex on wheat and sources of resistance in Zambia, **In**: *Wheat Production Constraints in Tropical Environments* (Ed. A.R. Klatt), pp. 175-186, CIMMYT, Mexico, 410 p.

Reddy, D.B. 1960. Outbreak and new records. *FAO Plant Prot. Bull.* **8**: 113-114.

Saari, E.E. 1986. Wheat disease problems in South-East Asia, **In**: *Problems and Progress of Wheat Pathology in South Asia* (Eds. Joshi, L.M., D.V. Singh and K.D. Srivastava), pp. 31-40. Malhotra Publishing House, New Delhi, India, 401 p.

Selvakumar, R., K.D. Srivastava, R. Aggarwal and D.V. Singh. 2001. Biocontrol of spot blotch of wheat using *Chaetomium globosum*. *Ann. Pl. Protec. Sci.* **9**: 286-291.

Shaner, G. 1981. Effect of environment of fungal leaf blight of small grains. *Ann. Rev. Phytopathol.* **19**: 273-296.

Sharma, R.C., H.J. Dubin, M.R. Bhatta asnd R.N. Devkota. 1997. Selection for spot blotch resistance in four spring wheat populations. *Crop Science* **34**: 432-435.

Shoemaker, R.A. 1959. Nomenclature of 6 *Drechslera* and *Bipolaris* grass parasites segregated from *Helminthosporium*. *Can. J. Bot.* **37**: 879-887.

Singh, A. and R.C. Rai. 1986. Chemical control of wheat diseases in India. **In**: *Problems and Progress of Wheat Pathology in South Asia* (Eds. L.M. Joshi, D.V. Singh and K.D. Srivastava), pp. 344-374. Malhotra Publishing House, New Delhi, 401 p.

Singh, D.P. 2008. Disease problems of wheat and management approaches **In**: *A Compendium of Lectures on Integrated Pest Management in Wheat based*

Cropping system, DWR Compendium No. 2 Directorate of Wheat Research, Karnal, India, 285 p.

Singh, D.V. and K.D. Srivastava. 1997. Foliar blights and Fusarium scab of wheat: Present status and strategies for management. **In**: *Management of Threatening Plant Diseases of National Importance* (Eds. V.P. Agnihotri, A.K. Sabhoy and D.V. Singh), pp 1-16. Malhotra Publishing House, New Delhi, 314 p.

Singh, R.P. 1992. Association between gene *Lr34* for leaf rust resistance and leaf tip necrosis in wheat. *Crop Sci.* **32**: 874-878.

Singh, R.P. and S. Rajaram 1992. Genetics of adult plant resistance to leaf rust in 'Frontana' and three CIMMYT wheats. *Genome.* **35**: 24-31.

Singh, R.V., A.K. Singh and S.P. Singh. 1998. Distribution of pathogens causing foliar blight of wheat in India and neighbouring countries. **In**: *Helminthosporium Blight of Wheat: Spot Blotch and Tan Spot.* (Eds. E. Duveiller and associates), pp. 59-62. CIMMYT, Mexico, D.F.

Singh, R.V., A.K. Singh and S.P. Singh. 1995. Progress of work under National Centre on Foliar Diseases during 1994-95. Paper presented in the 34th All India Wheat Research Workers' Workshop, AICWIP, ICAR, at USA, Dharwad, August 1995.

Singh, R.V., A.K. Singh, R. Ahmad and S.P. Singh. 1998. Influence of agronomic practices on foliar blight and identification of alternate hosts in the rice – wheat cropping system **In**: *Helmithosporium Blight of Wheat: Spot Blotch and Tan Spot* (Eds. E. Duveiller and associates), 346-348 pp. CIMMYT, Mexico, D.F.

Singh, S.D. and Amar Singh. 1971. Wheat blight incited by *Helminthosporium speciferum* and its control. *Second Int. Symp.* IARI. New Delhi, pp. 76, (Abstr.).

Singh, S.D. and Amar Singh 1968. Helminthosporium blight of wheat and its control. *Proc. Ist Summer Institute on Plant Dis. Control*. IARI, Delhi.

Singh, S.K. 2003. *Studies on foliar blight complex of wheat and its management.* Ph.D. Thesis, IARI, New Delhi, 135 p.

Singh, S.K., A.K. Dikshit, K.D. Srivastava, R.S. Tanwar and P. Dureja. 2003. Evaluation of tebuconazole: bioefficacy against spot blotch of wheat and residue studies. *Pestology* **27**: 13-17.

Singh, S.K., K.D. Srivastava and D.V. Singh. 2004. Pathogenic behaviour of leaf blight organisms on wheat. *Indian Phytopath* **57**: 319-322.

Srivastava, K.D., S.K. Singh, D.V. Singh and R. Aggarwal. 2004. Studies on foliar blight of wheat and its management. **In**: *Wheat for tropical Areas* (Eds. K.A. Nayeem, M. Sivasamy and S. Nagarajan), pp. 131-134. IARI. Regional Station, Wellington (Nilgiris), T.N. (India).

Srivastava, K.D. and A.K. Tiwari. 2002. Fungal diseases of wheat and barley: Foliar diseases. **In**: *Diseases of Field crops* (Eds. V.K. Gupta and Y.S. Paul), pp. 58-78. Indus Publishign Co., New Delhi, 464 p.

Srivastava, O.P., J.K. Luthra and P.N. Narula. 1971. Inheritance of seedling resistance of leaf blight wheat. *Indian J. Genet.,* **31**: 209-211.

Subramanian, C.V. 1962. Foot rot disease of wheat. *Curr. Sci.* **31**: 46-49.

Subramanian, C.V. and B.L. Jain. 1966. A revision of some graminicolous *Helminthosporia. Curr. Sci.* **35**: 352-355.

Tandon, J.P. 1985. Wheat improvement programs for the hotter parts of India, **In**: *Wheats for More Tropical Environments* (Eds. Villareal, R.L. and A.R. Klatt), pp. 63-67. Proceedings of the International Symposium. CIMMYT, Mexico, 354 p.

Velazquez, C. 1994. *Genetics de la Resistencia a Bipolaris sorokiniana on Trigoes Harineros*. Ph.D. Thesis, Monterillo, Mexico, 84 pp.

Xiao Zhimin, SunLianfa and Xin Wenli. 1998. Breeding for foliar blight resistance in Heilongjiang Province, China. **In**: *Helminthosporium blight of Wheat: Spot blotch and Tansport* (Eds. E Duveiner and associates) pp. 114-119. CIMMYT, Mexico, D.F.

Zillinsky, F.J. 1983. *Common Diseases of Small Grain Cereals: A guide to identification*. CIMMYT Mexico, 141 p.

2.2.3 Alternaria Leaf Blight

The existence of Alternaria leaf blight in India was reported about 84 years ago in 1924 from Bihar (McRae, 1924) and Maharashtra (Kulkarni, 1924). Subsequent to McRae's observations, no further report of the occurrence of this disease was made by any investigator till 1950. Mehta (1951) in 1950 recorded severe infection of *Alternaria* sp. in wheat varieties Bansi and Bansipalli 808 at Kanpur. Later, Mathur (1956) felt that *A. tenuis*, as seed borne pathogen was associated with the disease. Similar observations of Alternaria blight were also made from Cuttack and Andhra Pradesh (Bose, 1956; Bener Raj, 1960; Govinda Rao, 1961). Alternaria blight was never considered to be destructive until failure of Kenphad series of wheat bred for resistance to stem rust in Maharashtra in 1960–61. Prasada and Prabhu (1962) established that the leaf blight in Kenphad wheats was caused by a new species of *Alternaria* which they named as *Alternaria triticina*.

Monitoring of the disease in different wheat growing regions of the country has revealed the occurrence of Alternaria blight in Delhi, Haryana, Jammu & Kashmir, Uttar Pradesh, Bihar, Rajasthan, Maharashtra and Gujarat, though its frequency of distribution is comparatively less than spot blotch caused by *D. sorokiniana* (Anon., 1995). According to Singh and Srivastava (1997) present day wheat cultivars possess fairly good degree of resistance to the pathogen because susceptible germplasm like Kenphad is not involved in breeding programmes. Outside India, the disease is prevalent in Middle East, North Africa, North America, Western Asia and South Italy (Frisullo, 1982; Singh *et al.*, 1986; Aggarwal *et al.*, 1993).

The disease can cause substantial yield losses ranging from 35-60% (Prabhu and Singh, 1974). Sokhi and Joshi (1974) experimentally proved that maximum damage of 35% is caused when the second and third top leaves are severely blighted. In 1983, an epidemic of the disease caused 75% reduction in grain yield in Vidarbha region of Maharashtra (Raut *et al.*, 1983). The loss in yield in wheat varieties is due to reduction in ear-head length, number of grains per ear-head and 1000 grain weight (Ram and Joshi, 1978).

Symptoms: The first symptoms can be seen when the plants are 7–8 weeks old. Seedlings have never been found to be infected. The disease initially makes its appearance as small, oval, chlorotic lesions, irregularly scattered on the leaf blade. The lesions as they enlarge become irregular in shape and take up dark brown to grey colour. A bright yellow marginal zone is sometimes seen around the spots (Plate 2B). As the disease progresses, several lesions coalesce and cover large areas resulting in the death of entire leaf. The leaf starts drying up from the tip and drops off pre-maturely. The lower most leaves are always the first to show the signs of infection which gradually appears to the upper leaves. In severe infection, leaf sheath, ears, awns and glumes are also affected. Heavily infected fields present a burnt appearance and since this is the conspicuous phase of the disease, it is called as leaf blight rather than leaf spot. Infected seeds are shrivelled and have a brown discolouration (Prabhu and Prasada, 1966).

Causal Organism: *Alternaria triticina* Prasada & Prabhu produces velvety and olive-buff colony with a smooth to rough surface with characteristic zonation on standard nutrient agar medium. The mycelium is hyaline which turn dark grey to black in colour at maturity. The conidiophores are olivaceous brown, septate, usually unbranched and erect 30 µm long and arise singly or in small groups through stomata. The conidia are borne on conidiophores solitary or in short chain of 2–4 spores. They are obclavate, rostrate, golden brown, smooth, 20–30 × 8–30 µm in size and light brown to dark olive-buff that becomes darker with age. Conidia have 1–10 transverse septa and 0-5 longitudinal septa (Fig. 30). Their size is variable, 15–89 × 7–30 µm including the beak (Prasada and Prabhu, 1962).

The pathogen *A. triticina* differs from *A. tenuis* (*A. alternata*) widely as the conidia of the former are bigger in size and form short chain where as in the later they are short and stubby in long chains. Thus, *A. tenuis* is grouped under section Longicatinatae but *A. triticina* falls in Brevicatinatae. Besides, *A. triticina* and *A. tenuis, A. triticola* and *A. tenuissima* are also described on wheat (Rao, 1964; Bhadkamkar, 1965). Kumar and Arya (1975) studied the taxonomic characters of all the four species and thought that all the four species of *Alternaria* may be ecological types of *A. tenuis*.

The presence of toxic principles in culture filtrate of *A. triticina* has been demonstrated. This filtrate could induce symptoms on wheat leaves similar to those induced by the fungus (Janardhan and Hussain, 1977; Vijay Kumar and Rao, 1979). Sokhi and Joshi (1972) tested eight monoconidial isolates of the fungus and reported two races on the basis of host-pathogen interactions.

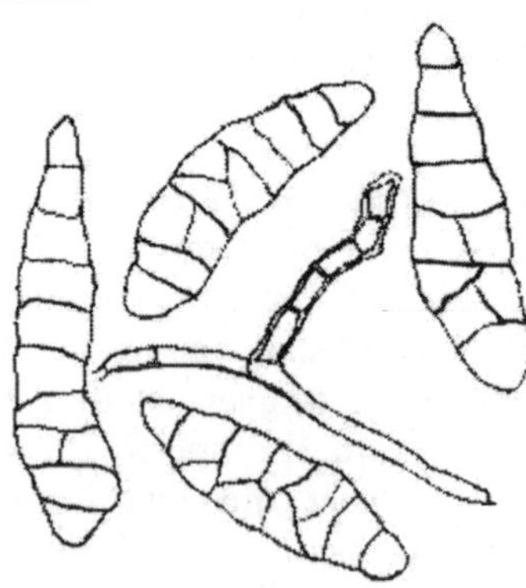

Figure 30: Conidiophore and conidia of *Alternaria triticina*

Disease Cycle and Epidemiology: The pathogen survives in soil and seed (Prabhu, 1962; Raut *et al.*, 1983; Sharma *et al.*, 1983). Prabhu and Prasada (1970) noted that the fungus perpetuating on plant debris is carried in the soil and remains viable for more than 20 months. According to Kumar and Arya (1973) *A. triticina* survives in the form of conidia on seed surface as well as dormant mycelium inside the seed coat. Under favourable conditions, the conidia from different sources spread by means of air or rain drops and on reaching the lower most leaves they germinate and produce germtubes. The germtube penetrates the leaves directly through epidermis or stomata. The optimum temperature for the germination of conidia is between 15–27°C with more than 90% relative humidity. The lesions are produced 4–5 days after establishment of infection. A large number of conidia are produced on lesions which are air dispersed and act as source of secondary inoculum for infection on upper foliage (Fig. 31). Temperature plays an important role during infection and incubation period. Maximum disease development takes place at 20–25°C, the optimum temperature being 25°C and minimum 19°C (Prabhu and Prasada, 1966). Moisture saturated atmosphere prevailing at least for 48 hours helps in quick disease development.

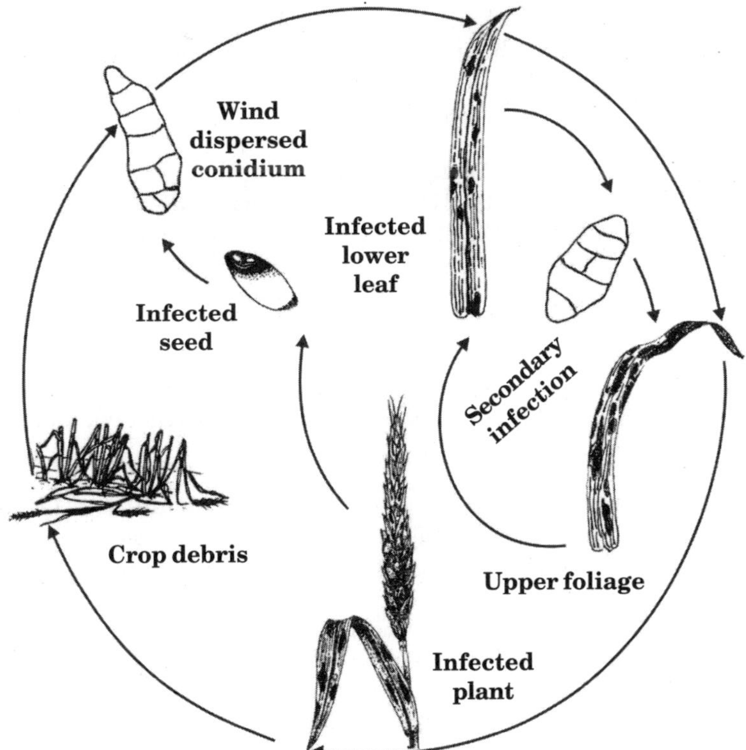

Figure 31: Disease cycle of Alternaria leaf blight (*Alternaria triticina*) of wheat

Management

Resistance: Inheritance of resistance to Alternaria leaf blight has been studied. Narula and Srivastava (1971) found that resistance in wheat was governed by two pairs of recessive genes. Sokhi and Joshi (1974) reported that resistance to *A. triticina* was conditioned by a pair of recessive genes in Agra local, NP 54 and two pairs in NP 824 and NP 809. In variety Timenstein x Mida (E 1912) the resistance was controlled by three pairs of dominant genes and in Kenya 184 P A.I.F (E 581) by one pair. Kulshrestha and Rao (1976) found that susceptibility to NP 830 was controlled by a dominant gene and that of NP 891 by two complementary genes. Srivastava *et al.* (1981) studied genetics of resistance of two parents *viz.* E 5477 and Gabo against *A. triticina* and noticed that the single gene (s) carried by both the resistant parents were non-allelic. These results, however, differ from those of Kulshrestha and Rao (1976).

Host resistance to foliar blights in India is not at high level so breeding for disease resistance commands high priority. Genes for several sources including alien species of wheat have to be combined to achieve sufficient resistance. Resistant sources to *A. triticina* such as Kalyansona, Sonora 63, Chhoti Lerma, Sonalika (S 308) and S 227 were identified by Katiyar and Singh (1968) and Sokhi and Joshi (1973). Some other wheat cultivars namely Lal Bahadur, Hira, UP 301, Ridley and HD 4502 also exhibited high degree of resistance (Sokhi, 1972).

Recently, Singh (2008) suggested that adoption of tolerant varieties like DWR 14, DDK 1009, HD 2189, HD 2687, HD 2733, HUW 234, HUW 533, HS 277, VL 738 and VL 804 can be useful for management of foliar blights.

Cultural Control: Use of healthy disease free seed and sanitation practices including burning of plant debris are recommended as these steps are helpful in reducing the initial inoculum. Restricted application of nitrogenous fertilizers helps in reducing infection (Srivastava and Tiwari, 2002).

Since pathogen is carried inside the seed, Prabhu and Prasada (1966) advocated hot water treatment of seed. Pre-soaking of seed in water for 4 hours followed by hot water treatment at 52–54°C for 10 minutes is effective in eradication of fungal infection.

Chemical Control: Seed treatment with thiram or anclor (carboxin + thiram) @ 2g/kg seed effectively controls seed-borne inoculum and increases seed germination (Raut *et al.*, 1983). Foliar application of mancozeb and zineb @ 0.2% at 10–15 days intervals can reduce disease severity significantly (Nene *et al.*, 1971; Singh *et al.*, 1979). The disease can also be controlled effectively through spray of propiconazole (Tilt 25 EC) at 0.1 percent (Misra *et al.*, 2005).

Microbial Control: Influence of phyllosphere mycoflora of wheat on the growth of *Alternaria triticina* was investigated by Kumar and Sinha, (1975). The preliminary report showed that *Curvularia lunata, Trichothecium roseum, Stachybotrys atra, Nigrospora spherica, Fusarium moniliforme, F. conclor, Aspergillus nidulans* and *A. terreus* inhibited the germination of conidia of *A. triticina* up to 60 percent.

REFERENCES

Agarwal, V.K., S.S. Chahal and S.B. Mathur. 1993. Alternaria leaf blight (*Alternaria triticina*) **In**: *Seed-borne Diseases and Seed Health Testing of Wheat* (Eds. S.B. Mathur and B. M. Cunfer), pp. 9-13. Danish Govt. Institute of Seed Pathology for Developing Countries, Copenhagen, Denmark.

Anonymous. 1995. Report of the Coordinated Experiments, 1994–95. Vol. 11. Crop Protection (Pathology), AICWIP, DWR, Karnal, 206 p.

Benerraj, R. 1960. Alternaria blight of wheat in Andhra Pradesh. *Andhra Agric. J.* **7**: 114-115.

Bhadkamkar, V.B. M.K. Desai and N.B. Kulkarni. 1965. Alternaria blight of wheat in Maharashtra State (India). *Sydowia* **19**: 247-249.

Bose, S.R. 1956. Alternaria within the pericarp of wheat seed. *Nature* (Lond.) **178**: 640-641.

Frisullo, S. 1982. Fungal parasites of plants in Southern Italy. I. *Alternaria triticina* Prabhu and Prasada on durum wheat. *Phytopathologia Mediterranea* **21**: 113-115.

Govinda Rao, P. 1961. Foliar blight of wheat. *Proceedings of All India Wheat Research Workers' Workshops,* IARI, New Delhi.

Janardhan, K.K. and A. Husain. 1977. Isolation and partial purification of a phytotoxin from *Alternaria triticina. Indian J. Exp. Biol.* **12**: 195-198.

Katiyar, R.L. and D.V. Singh. 1968. Resistance of wheat varieties to Alternaria blight. *Indian J. Microbiol.* **8**: 55-56.

Kulkarni, G.S. 1924. Report of the work done in Plant Pathology Section during the year 1922-23. *Rep. Dept. Agric. Bombay-Presidency for the year 1922-23,* 167-171 pp.

Kulshreshtha, V.P. and M.V. Rao. 1976. Genetics of resistance to an isolate of *Alternaria triticina* causing leaf blight of wheat. *Euphytica.* **25**: 769-775.

Kumar, S. and S. Sinha. 1975. Phyllosphere mycoflora of wheat and its influence on the development of *Alternaria triticina* Prasada and Prabhu, causing leaf blight. *Proc. 62nd Indian Sci. Cong.* Part III, 49-50 pp.

Kumar, V.R. and H.C. Arya 1973. Certain aspects of perpetuation and recurrence of leaf blight of wheat in Rajasthan. *Indian J. Mycol. Plant Pathol.* **3**: 93-94.

Kumar, V.R. and H.C. Arya. 1973. Taxonomy of the fungus causing leaf blight of wheat. *Proc. 62nd Sci. Cong.* Part III. 40-41 pp.

Mathur, R.S. 1956. Alternaria leaf spot of wheat in Uttar Pradesh. *Agric. Anim. Husb.* **6**: 12-14.

Mc Rae, W. 1924. Economic Botany, Part III, Mycology. *Rep. of Borad of Scientific Advice,* India 1922-23, 31-35 pp.

Mehta, P.R. 1951. Some new diseases of plants of economic importance of Uttar Pradesh. *Plant. Prot. Bull.* (New Delhi) **2**: 50-51.

Narula, P.N. and O.P. Srivastava. 1971. Genetics of *Alternaria* resistance in wheat. *Indian J. Genet.* **31**: 105-107.

Nene, Y.L., S.C. Saxena, P.D. Tyagi, A.B. Misra, D.V. Singh, A. P. Misra S.V. Pandit and L.M. Joshi. 1970. Fungicides and nematicide tests. Results of 1969. *News American Phytopathological Society.* No. 25.

Prabhu, A.S. 1962. *Leaf blight of wheat caused by a new species of Alternaria.* Ph. D. thesis, IARI, New Delhi.

Prabhu, A.S. and Amar Singh, 1974. Appraisal of yield loss in wheat due to foliage diseases caused by *Alternaria triticina* and *Helminthosporium sativum. Indian Phytopath.* **27**: 632-634.

Prabhu, A.S. and Prasada R. 1970. Investigations on the leaf blight diseases of wheat caused by *Alternaria triticina* **In**: *Plant Diseases Problems.* Indian Phytopathological Soc., IARI, New Delhi, 17-27 pp.

Prabhu, A.S. and R. Prasada. 1966. Plant pathological and epidemiological studies on leaf blight of wheat caused by *Alternaria triticina. Indian Phytopath.,* **19**: 95-112.

Prasada, R. and A.S. Prabhu. 1962. Leaf blight of wheat caused by a new species of *Alternaria. Indian Phytopath.* **15**: 292-293.

Ram, B. and L.M. Joshi. 1978. Spray schedule of Fytolan for leaf blight of wheat and its effect on yield components. *Indian Phytopath.* **31**: 348-351.

Rao, V.G. 1964. An undescribed sp. of *Alternaria* on wheat from India. *Mycopath. Mycol. Appl.* **23**: 311-313.

Raut, J.G., S.M. Gurdhe and P.D. Wangikar. 1983. Seed brone infection of *Alternaria triticina* in wheat and its control. *Indian Phytopath.* **36**: 274-277.

Sharma, S.C., H.S. Randhawa and H.L. Sharma. 1983. Seed infection in relation to the susceptibility of wheat to *Alternaria triticina* and *Cochliobulus sativus. Indian Phytopath.* **36**: 372-374.

Singh, D.P. 2008. Disease problems of wheat and management approaches. **In**: *A Compendium of Lectures. on Integrated Pest Management in Wheat based Cropping system.* DWR Compendium No. 2, Directorate of Wheat Research, Karnal, India. 258 p.

Singh, D.V. and K.D. Srivastava. 1997. Foliar blights and Fusarium scab of wheat: Present status and strategies for management. **In**: *Management of Threatening Plant Diseases of National Importance* (Eds. V.P. Agnihotri, A.K. Sarbhoy and D.V. Singh), pp. 116. Malhotra Publishing House, New Delhi, 314 p.

Singh, D.V., L.M. Joshi and K.D. Srivastava 1986. Foliar blight and spot of wheat in India. *Indian J. Genet.* **46**: 217-245.

Singh, V.K., P.C. Verma and B.M. Khanna. 1979. Fungicidal spray schedule for the control of Alternaria blight of wheat. *Indian J. Mycol. Plant Pathol.* **9**: 200-204.

Sokhi, S.S. 1972. *Studies on leaf blight of wheat caused by Alternaria triticina with special reference to the disease resistance.* Ph.D. Thesis, IARI, New Delhi.

Sokhi, S.S. and L.M. Joshi. 1972. Physiological specialization in *Alternaria triticina. Indian J. Microbiol.* **12**: 209-210.

Sokhi, S.S. and L.M. Joshi. 1973. Reactions of wheat varieties to leaf blight. *Indian J. Genet.* **33**: 423-425.

Sokhi, S.S. and L.M. Joshi. 1973. Sugars and phenols in relation of Alternaria leaf blight of wheat. *Phytoparasiticia* **1**: 117-118.

Sokhi, S.S. and L.M. Joshi. 1974. Estimation of losses in yield due to leaf blight disease of wheat caused by *Alternaria triticina*. *Indian J. Mycol. Plant. Pathol.* **4**: 29-33.

Srivastava, K.D. and A.K. Tiwari. 2002. Fungal diseases of wheat and barley: Foliar diseases. **In**: *Diseases of Field Crops* (Eds. V.K. gupta and Y.S. Paul), pp. 58-78. Indus Publishing Co., New Delhi, 464 p.

Srivastava, O.P., R.A. Singh and M.V. Rao. 1981. Note on genetics of seedling resistance to Alternaria leaf blight of wheat. *Indian J. Agric. Sci.* **51**: 810-811.

Vijaya Kumar, C.S.K. and A.S. Rao. 1979. Production of phytotoxic substances by *Alternaria triticina*. *Canad. J. Bot.* **57**: 1255-1258.

2.2.4 Septoria Diseases

Speckled leaf blotch and glume blotch are two major Septoria diseases of wheat that cause problems in many parts of the world. These diseases have the greatest impact on global wheat production. There have been epidemic outbreaks of these diseases in several countries (Saari and Wilcoxson, 1974). Annual yield losses world-wide due to both diseases are estimated at about 9 million metric tons (Eyal *et al.*, 1987). The increase in the economic importance of Septoria leaf blotch is attributed to replacement of local wheat cultivars with early maturing, semi-dwarf cultivars that are susceptible to the pathogen (Stewart *et al.*, 1972). The situation of Septoria diseases in India is not alarming (Srivastava, 1986).

Speckled Leaf Blotch

Speckled leaf blotch also known as Septoria leaf blotch, is considered as major leaf disease in moist regions of the world. It generally appears in epidemic form in South America, Mediterranean basin and in certain regions of Africa, Asia and Australia. In India, the occurrence of the disease was recorded by Butler (1918) and subsequently by Luthra *et al.* (1937) from Punjab. This disease is mainly restricted in some humid-cooler regions of Punjab and is ranked as minor disease. In late sixties, speckled leaf blotch in epidemic proportions, was reported from North-western Punjab during 1967-68 crop season (Tyagi *et al.*, 1969).

Symptoms: The symptoms appear first on the lower leaves as small, chlorotic spots which elongate with time and change into reddish brown, irregular in shape that, being restricted by the veins of the leaf tends to develop longitudinally. As the disease progresses the lesions enlarge, they loose their dark borders and eventually coalesce and cover most of the leaf area, often resulting in complete necrosis. The disease becomes less aggressive as the crop begins to mature. A large number of tiny black dots like pycnidia arranged in rows between the veins appear on both the leaf surfaces exhibiting blotches as speckled (Plate 2C). Hence, the disease is commonly called as speckled leaf blotch. By holding the leaf against the light, pycnidia embedded in the leaf tissue, can often be seen. In case of heavy infection, blotches may also appear on culms, floral bracts and glumes but the spots are smaller and less conspicuous, with scattered pycnidial development with in the lesions. In extreme case of infection, affected plants are stunted and ear-head either do not develop or may remain sterile.

Causal Organism: The causal agent, *Septoria tritici* was found on wheat in 1842 by Desmazieres, which he described and attributed to Roberge as *Septoria tritici* Rob. ex Desm. (Shearer and Wicoxson, 1978). The fungus is characterized by the production of conidia, termed pycnidiospores, which are produced in semi closed fruiting bodies known as pycnidia. Pycnidia bearing pycnidiospores represent asexual state of the fungus. The pycnidia are embedded in the epidermal cells and mesophyll tissue. They are brown unilocular, subglobose to elliptical having a papillate ostiole and measure 60–200 µm (generally 100–75 µm). The pycnidiospores of *S. tritici* are present in two forms within the pycnidium: macroconidia (macropycnidiospores) and microconidia (micropycnidiospores). The macroconidia are hyaline, slender with round end, curved, 3–5 septate and measure 35–98 × 1–3 µm (Fig. 32). The microconidia may be produced in association with macroconidia at low temperature. They are curved, hyaline, aseptate and measure 8–10.5 × 0.8–1 µm. Both spore forms are equally able to infect wheat (Shipton *et al.*, 1971).

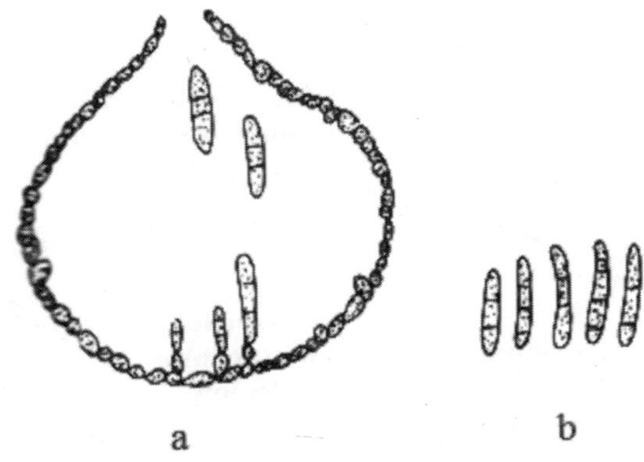

Figure 32: *Septoria tritici* (a) Pycnidium (b) Pycnidiospores

The asexual form of the fungus *Mycosphaerella grminicola* (Fiicel) Schroeter was described by Sanderson in 1972 in New Zealand (Sanderson and Hampton, 1978). The asexual state has also been identified in Australia, Brazil, Netherlands, United Kingdom and U.S.A. Butler (1918) speculated *Leptosphaeria tritici* as sexual state of the fungus but could not define its role in the perpetuation of the disease in India. Psuedothecia produced on leaf surface are sub-epidermal up to 115 µm broad, asci 30–40 × 11–14 µm having 8 bi-celled ascospores which are elliptical in shape and measure 9–16 × 2.5–4.0 µm. The ascospores are reported to provide primary inoculum for speckled leaf blotch in New Zealand, Australia (Eyal, 1981). Physiologic races of *S. tritici* have not been identified as yet, although there are conflicting reports on the issue of physiologic specialization in U.S.A., Australia, Uruguay and Israel (Eyal, 1981).

Disease Cycle: The sexual state of the fungus wherever it occurs, is one of the sources of inoculum. In countries where sexual stage has not been found, it is asexual state, *S. tritici*, which is chiefly associated with disease development. Pycnidiospores which remain viable in pycnidia for several months under adverse climatic conditions on infested wheat straw and stubble, serve as primary source of inoculum in several countries (Hilu and Bever, 1957). Fournet (1969) observed that pycnidiospores are produced in a thick, sticky matrix containing a high concentration of preserving sugars and proteins. This preserving medium or ooze permits the spores to remain viable during periods of dry weather.

Luthra *et al.* (1938) emphasised that in India, wheat straw is stored in stacks during off season of the crop and the pycnidia remain viable on infested debris from June to October. Although the viability of the pycnidiospores decreases considerably, the remaining viable spores are capable of causing infection to the next crop. They ruled out the possibility of survival of the fungus on infected debris buried in the soil during summer months. Thus infection on wheat straw has been shown to carry over the disease. Infection processes occurs best on rainy cloudy days with low temperatures. The straw gets drenched with rainy water during winter so the embedded pycnidia swell and pycnidiospores ooze out through ostiole.

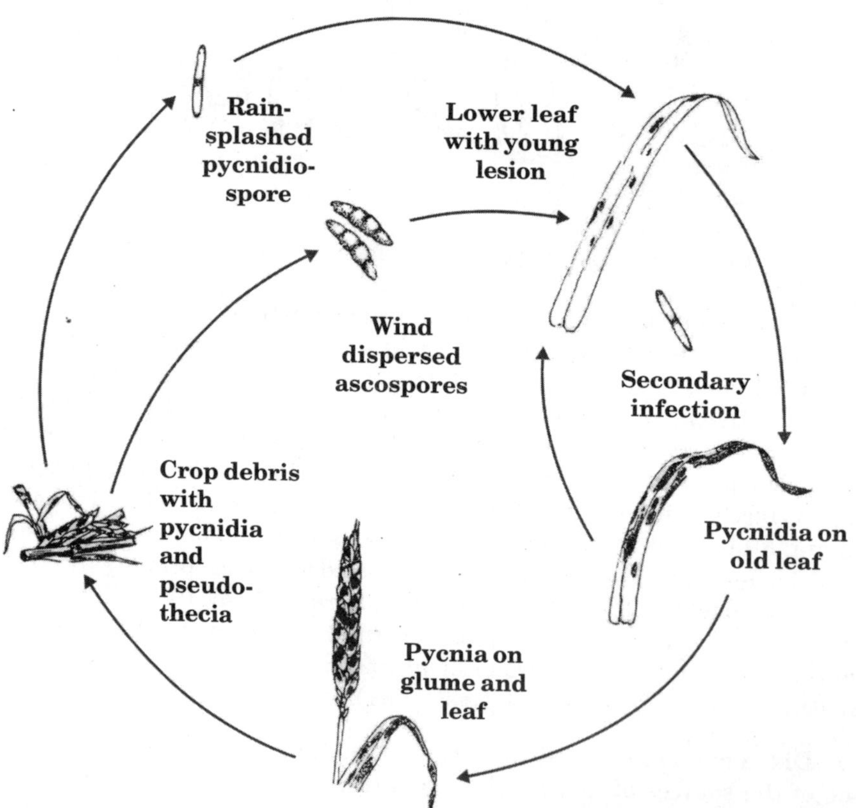

Figure 33: Disease cycle of speckled leaf blotch (*Septoria tritici*) of wheat.

These spores are disseminated by wind, water splashing and insects and the crop in vicinity gets infected (Fig. 33). Wind also helps in dissemination of inoculum indirectly by spreading dried diseased plant debris which lodge on healthy wheat plants. Luthra *et al.* (1938) noted that during rainy days the spores are carried to some distance in the form of fine spray.

Pycnidiospores germinate following release from the pycnidia, when the plants are wet. Spores begin to germinate with in 12 hours and leaf penetration occurs after 24 hours. The fungus may penetrate the leaf through stomata or directly through the cell walls of the epidermis. Cardinal temperatures reported for germination of conidia are a minimum of 2–3°C and maximum of 33–37°C with an optimum of 20–25°C. Infection can be delayed in the field if the temperature falls below 7°C during two consecutive nights. Moisture is required for all stages of infection (Shaner and Finney, 1976; Eyal *et al.*, 1987). Normally first symptom of infection is visible on lower leaves after 14-21 days under sufficient moist conditions. A moist period of only 24 hours is insufficient to produce disease symptoms. After infection on lower leaves, the vertical progress of Septoria from lower to upper leaves is affected by the distance between consecutive levels, called ladder effect. The distances between the first emerging three to four leaves are similar in case of dwarf and tall cultivars. But the distance between each leaf is greater towards the flag leaf in tall varieties. The movement of splashed pycnidiospores from lower leaves is there by made simpler in dwarf cultivars than they do on leaves of taller cultivars. As a result pycnidia often appear earlier on upper plant parts of dwarf wheats.

Graminaceous hosts like *Poa pretensis, P. annua, P. scunda* and *Agropyron repens* get infection of *S. tritici* (Brokenshire, 1975) but all these grasses may not act as alternate host. However, Prestes and Hendix (1978) felt that few grasses like *Stellaria media* which are naturally infected with *S. tritici,* may play some role in epidemiology of speckled leaf blotch.

Glume Blotch

Glume blotch is another Septoria disease of wheat, closely allied to speckled leaf blotch. It is especially important in warm, moist growing areas in Europe, U.S.A., Australia, South America and Africa. In India, the disease was first recorded by Chona and Munjal (1952) from Nilgiri hills in 1951 on *Triticum dicoccum* wheat variety Samba. The disease in the country is localised and prevalent mainly in South Indian hills. Unlike *S. tritici* the glume blotch pathogen (*S. nodorum*) does not make appearance in North Indian plains. However, Joshi *et al.* (1971) recorded blotch incidence in Kumaon hills (Uttranchal Pradesh) in some experimental plots.

The disease is capable of implicating heavy yield reductions, affecting significantly the world wheat production (Saari and Wilcoxson, 1974). Sometimes the yield losses may be up to 50 percent. In Germany, head infection was stated to be the main cause of yield reduction (King *et al.*, 1983). Since glume blotch in India is not a threatening disease, the losses have not been estimated.

Symptoms: The blotches occur not only on the glumes but also on leaf blade, leaf sheaths, nodes and internodes. Early leaf symptoms are yellowish green oval or lens-shaped spots. These spots later turn purplish brown with a yellow-green border surrounding the necrotic area, varying from 4 to 9 mm in length. At maturity, the lesions become greyish or straw coloured studded with black pycnidial dots. Pycnidia may or may not appear with in the centre of the lesions on the leaves, but are more common on nodes, leaf sheaths and glumes (Plate 2D). The pycnidia develop on both sides of the leaves but on glumes they are formed on the external surface only. Whenever nodes are infected, the pathogen may cause distortion and bending of the tiller with a possibility of lodging and breakage at the node. Dwarfing of culms and ears and sometimes infertility of spikes may result due to severe early infection of glume blotch.

Infected seeds produce seedlings which may show light to dark brown streaks on the coleoptile. Coleoptiles are shortened, turn brown and may be greatly distorted with knob-like swellings depending on wheat cultivars.

Tan spot (*Drechslera tritici repentis*/*Pyrenophora tritici repentis*) produces symptoms similar to leaf and glume blotch. In regions where both diseases occur, it is often necessary to observe fruiting structures to differentiate the two pathogens. The same is true where *S. nodorum* and *S. tritici* occur together (Wiese, 1987).

Causal Organism: Berkeley (1845) attributed the cause of glume blotch of wheat to *Depazea nodorum* but he regarded *Depazea* and *Septoria* synonyms. Later, Brekeley and Broome (1850) described the pathogen as *Septoria nodorum* (Berk.) Berk. in Berk & Broome. Castellani and Germano considered that *Stagonospora nodorum* and *S. nodorum*, both are synonyms and represent asexual state. Its ascomycetous sexual state *Leptosphaeria nodorum* Miller (syn. *Phaesosphaeria nodorum* (Miiler) Hedjaroude) was described by Miller in 1952 (Dickson, 1956). According to Chona and Munjal (1952) probably the perfect stage of the fungus resembling those of *Leptosphaeria* occur under Indian conditions on dried up leaves and leaf sheaths.

S. nodorum pycnidia are globose or sub-epidermal, yellowish brown to dark, 70–230 µm in diameter with a central sub-cuticular ostiole. Mature pycnidia exude milky white to buff cirrhi with pycnidiospores. Pycnidiospores released through the ostiole are hyaline, cylindrical, with 0–3 septa, upper end usually round and lower slightly tapering and 15-32 × 2–4 µm in size (Fig. 34). Each spore cell contains single nucleus.

Pseudothesia of *L. nodorum* are formed on host tissue, containing numerous club-shaped asci holding eight ascospores. They are light buff in colour and measure 160–250 µm. Asci are binucleate and cylindrical, 60–70 × 8–10 µm and contain 8 ascospores. Ascospores are hyaline to pale brown, spindle shaped and measure 19–26 × 3.5–4.0 µm. They have four cells with the two end cells tapered and one of the interior cells slightly swollen. The septum between the second and third cells is constricted (Scharen and Sanderson, 1985).

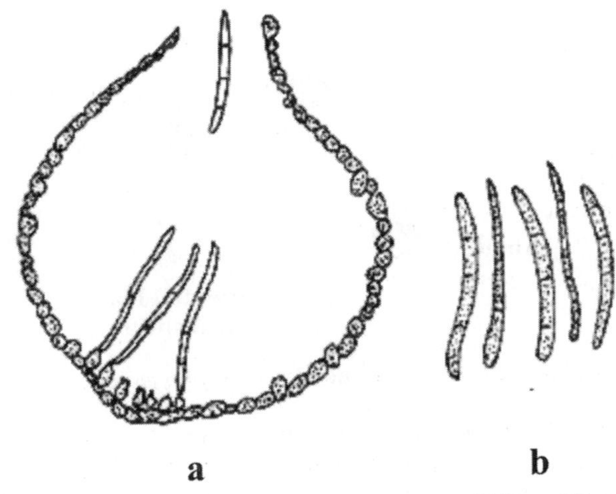

Figure 34: *Septoria nodorum* (a) Pycnidium (b) Pycnidiospores

The presence of classical races of *S. nodorum* remains unclear. It appears that term such as race, cultivar and isolate might not be meaningful (Griffiths and Ao, 1980; Eyal *et al.*, 1987). *S. nodorum* produces phytotoxic compounds such as septorin and ochracin which play a role in symptom development (Eyal *et al.*, 1987).

Disease Cycle: Pycnidia of *S. nodorum* remain viable on crop debris and are able to produce infective pycnidiospores after 1 year (Scharen, 1964). It appears therefore, that infected crop debris function as a source of primary infection to the next crop. The pycnidia liberate pycnidiospores in a moist environment on rainy and/or dewy days. The pycnidiospores after exudation are spread to the upper leaves and ear-head by splashing or wind blown rain (Fig. 35). Dispersal of spores in a droplet is found to occur when atmospheric temperature is more than 10°C with at least 5 mm rainfall followed by 10 mm or more rainfall during next 48 hours (Jordan and Overthrow, 1980). Wind greatly increases the dispersal of smaller droplets and spores in the downwind direction. The pycnidiospores

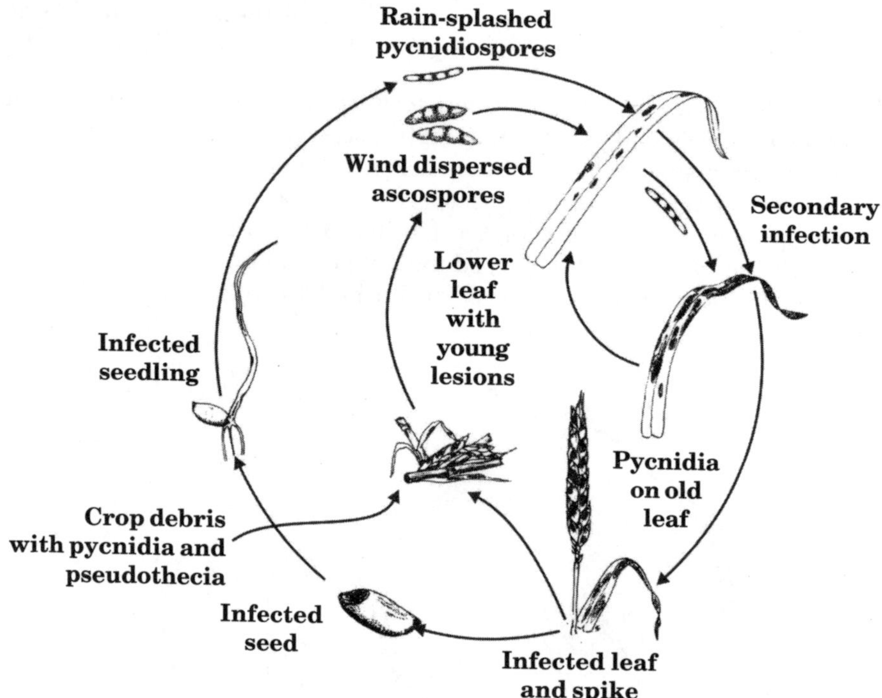

Figure 35: Disease cycle of Glume blotch (*Septoria nodorum*) of wheat

germinate with in 2 hours after emerging from the pycnidium. Hyphae growing from the germinating spores penetrate the young leaves to produce lesions on affected plant parts. Histological studies have shown that during mycelial invasion, the hyphal colonization in the leaf is both between and within the cells and the host cell walls are disorganised due to release of digestive enzymes (Baker and Smith, 1974; Margo, 1984). Spore germination and penetration are greatest between 15–25°C with a minimum of 6 hours wetness. The most favourable temperature for infection and disease development are 22–24°C with symptoms appearing after 7–14 days.

Seed-borne inoculum is another potential source of infection of glume blotch. *S. nodorum* remains viable in seeds up to 10 years or more depending upon storage conditions (Siddiqui and Mathur, 1989; Cunfer, 1991). Fungal hyphae grow from the germinating seed to coleoptile and then invade the young leaves (Aggarwal *et al.*, 1986). Conidia are disseminated by splashing rain to the upper leaves and ear. The fungus enters the glumes, and then penetrates the developing seed (Cunfer and Johnson, 1981). Coleoptile infection is increased at low temperature of about 10°C but decreased at temperatures around 20°C and high soil moisture (Holmes and Colhoun, 1971; Shearer, 1967).

Management

Resistance: Host resistance is the best approach of defense against Septoria diseases. But most of the high yielding wheat cultivars are susceptible to *Septoria tritici* and *Septoria nodorum*. Disease resistance has been difficult to incorporate in cultivars because not enough is known about the types of resistance, their mode of action, inheritance, manipulation and accumulation. Gough and Tuleen (1979) and Yechilvich-Auster, (1983) have detected resistance to *Septoria* among population and accessions of *Triticum monococcum boeoticum, T. turgidum dicoccoides, T. longissimum, T. speltoides* and *T. tauschii*. Efforts are being made to transfer resistance from such sources to bread wheat. Resistance to Septoria blotch from winter wheats Aurora, Bezostayal, Kavkaz and Trakia is available in agronomically suitable semi-dwarf cultivars developed by CIMMYT in Mexico (Eyal *et al.*, 1987). In India, some indigenous and exotic wheat varieties were found tolerant to *S. tritici* (Tyagi *et al.*, 1969), and Timgalin, Frondosa and S-331 to *S. nodorum* (Joshi *et al.*, 1971).

The dwarfing gene *Rht 2* has a slight effect on resistance to Septoria blotch (Scott and Benedikz, 1985). Resistance to *S. tritici* in some winter wheat cultivars is expressed by low pycnidial density which has been transferred to early-maturing dwarf wheats (Danon *et al.*, 1982). Combination of Septoria tolerance with slow Septoring or slow blotching may bring success (Bronniman, 1982).

Cultural Control: Cultural practices that reduce diseased plant debris through ploughing, burning, crop rotation etc., help in reduction of the major source of primary inoculum of the pathogen. Luthra *et al.* (1938) advocated successful control of the disease by destroying the wheat residue through deep ploughing of harvested field after first shower of rain. Burning of stubble and destruction of weed and volunteer wheat also help in checking the disease. Two years rotation followed with seed treatment with triadimefon @ 0.2% provides good control of Septoria blotch (Luke *et al.*, 1983).

Chemical Control: Foliar spray of dithiocarbamates such as maneb, manzate, mancozeb and zineb have proved effective in controlling Septoria blotch (Eyal and Wahl, 1975), but these protectant fungicides are not economical because repeated application of the chemical at 10–14 day intervals is required.

The systemic fungicides benomyl, prochloraz, triadimefon and propiconazole have provided effective control of Septoria blotch in several countries. Other new chemicals such as HWG 160, fenpropimorph and myclobuta have also been found to be effective. It has been observed that protectant fungicides reduce the selection pressure on the pathogen exerted by the systemic fungicides and expand the control spectrum. Application of triadimefon + manzate 200 provides good control of *Septoria nodorum* blotch, leaf rust and powdery mildew of wheat in U.S.A. (Anazole, 1986). Combinations of Bayleton + Dithane M-45, Tilt + Dithane M-45, Bayleton + Difolatan, Prochloraz + Dithane M-45 cause significant reductions in disease severity and increase yield (Eyal *et al.,* 1987).

The most effective fungicides for seed treatment are: thiabendazole (1.5 g/kg seed), triadimol (0.3 g/kg seed) and nuarimol (0.2 g/kg seed).

Epidermal coating with antitranspirants polymers such as Wiltpruf and Vapor Gard are most effective in controlling the disease (Ziv, 1983).

REFERENCES

Agarwal, K., T. Singh, D. Singh and S.B. Mathur 1986. Studies on glume blotch disease of wheat II. Transference of seed-borne inoculum of *Septoria nodorum* from seed to seedling. *Phytomorphology* **36**: 291-297.

Anzalone, L. 1986. Evaluation of fungicides for control of foliar diseases of soft red winter wheat. 1985. *Fungicide and Nematicide Test* **41**: 87-88

Baker, E.A. and I.M. Smith. 1974. Antifungal compounds in winter wheat resistant and susceptible to *Septoria nodorum*. *Ann. Appl. Biol.* **87**: 67-73.

Berkeley, M.J. 1845. Disease in the wheat crop. *Gard. Chron.* **35**: 601

Berkeley, M.J. and C.E. Broome. 1850. XXXIII. Notices of British fungi. *Ann. Mag. Nat. Hist.* 5: (Ser. 2): 365-380.

Brokenshire, T. 1975. The role of graminaceous species in the epidemiology of *Septoria tritici* on wheat. *Plant Pathol.* **24**: 33-38.

Bronnimann, A. 1982. Entwicklung der kenntisse under *Septoria nodorum Berk in* Hinblick auf die Toleranze- order resistenzzuchtung bei Eizen. *Neth. J. Agric. Sci.* **30**: 47-69.

Butler, E.J. 1918. *Fungi and Diseases in Plants.* Thacker Spinck and Co. Calcutta, 547 p.

Chona, B.L. and R.L. Munjal 1952. Glume blotch of wheat in India. *Indian Phytopath.* **5**: 17-20.

Cunfer, B.M. 1991. Long term viability of *Septoria nodorum* in stored wheat seed. *Cereal Res. Commun.* **19**: 347-349.

Cunfer, B.M. and J.W. Johnson 1981. Relationship of glume blotch symptoms on wheat heads to seed infection by *Septoria nodorum*. *Trans. Brit. Mycol. Soc.* **76**: 205-211.

Danon, T., J.M. Sacks and Z. Eyal. 1982. The relationships among plant stature, maturity class, and susceptibility to Septoria leaf blotch of wheat. *Phytopathology* **72**: 1037-1042.

Dickson, J.G. 1956. *Diseases of Field Crops*. McGraw Hill Book Co., Inc. New York, 517 p.

Eyal, Z. 1981. Research on Septoria leaf blotch: Research advances. *EPPO Bull.* **11**: 53-67.

Eyal, Z. and I. Wahl. 1975. Chemical control of Septroia leaf blotch disease of wheat in Israel. *Phytoparasitica* **3**: 76-77.

Eyal, Z., Scharen, A.L., J.M. Prescott. and M. Van Ginkel. 1987. *The Septoria disease of Wheat: Concepts and Methods of Disease Management,* Mexico D.F.: CIMMYT 52 p.

Gough, F.J. and N. Tuleen. 1979. Septoria leaf blotch resistance among *Agropyron elongatum* chromosomes in *Triticum aestivum* Chinese Spring. *Cereal Res. Commun.* **7**: 275-280.

Griffithe, E. and H.C. Ao. 1976. Dispersal of *Septoria nodorum* spores and spread of glume blotch of wheat in the field. *Trans. Brit. Mycol. Soc.* **67**: 413-418.

Hilu, H.M. and W.M. Bever. 1957. Inoculation, oversummering and suspect pathogen relationship of *Septoria tritici* on *Triticum species. Phytopathology* **47**: 474-480.

Holmes, S.J.I. and J. Colhoun. 1971. Infection of wheat seedlings by *Septoria nodorum* in relation to environmental factors. *Trans. Brit. Mycol. Soc.* **57**: 493-500.

Jordan, V.W.L. and R.B. Overthrow. 1980. Epidemiology and control of splash-dispersed and other cereal diseases, pp. 132-133. Report of Long Ashton Research Station for 1979.

Joshi, L.M., E.E. Saari and R.D. Wilcoxson 1971. Epidemics of glume blotch of wheat and reaction of wheat varieties to *Septoria nodorum. Indian Phytopath.* **24**: 413-415.

King, J.E., R.J. Cook and S.C. Melville. 1983. A review of Septoria diseases of wheat and barley. *Ann. Appl. Biol.* 103: 345-373.

Luke, H.H., P.L. Pfahler and R.D. Barnett. 1983. Control of *Septoria nodorum* on wheat with crop rotation and seed treatment. *Plant Dis.* **67**: 949-951.

Luthra, J.C., A. Sattar and M.A. Ghani. 1937. A comparative study of species of *Septoria* occurring on wheat. *Indian J. Agric. Sci.* 7: 271-289.

Luthra, J.C., A. Sattar and M.A. Ghani. 1938. Perpetuation and control of *Septoria* diseases of wheat in the Punjab. *Agric. Livest.* **8**: 17-25.

Margo, P. 1984. Production of polysaccharide-degrading enzymes by *Septoria nodorum* in culture and during pathogenesis. *Plant Sci. Lett.* **37**: 63-68.

Prestes, A.M. and W.J. Hendrix. 1978. The role of *Stellaria media* in the epidemiology of *Septoria tritici* on wheat. *Proc. Third Int. Pl. Congr.* Munchem, 336 p.

Saari, E.E. and R.D. Wilcoxson 1974. Plant disease situation of high yielding dwarf wheats in Asia and Africa. *Ann. Rev. Phythopathol.* **12**: 49-68.

Sanderson, F.R. and J.C. Hampton 1978. Role of the prefect stages in the epidemiology of the common *Septoria* diseases of wheat. *N.Z.J. Agric. Res.* **21**: 277-281.

Scharen, A.L. 1964. Environmental influences on the development of glume blotch in wheat. *Phytopathology* **56**: 580-581.

Scharen, A.L. and F.R. Sanderson. 1985. Identification, distribution and nomenclature of the *Septoria* species that attack cereals. **In**: *Proc. Workshop on Septoria of Cereals* (Ed. A.L. Scharen), August 2-4, 1983, Bozeman, M.T. USDA-ARS Publ. No. 12, 116 p.

Scott, P.R. and P.W. Benedikz. 1985. The effect of *Rhtz* and other height genes on resistance to *Sepotoria nodorum and Septoria tritici* in wheat. pp. 18-21. **In**: *Septoria of Cereals* (Ed. A.L. Scharen), August 2-4, 1983, Bozeman, M.T. USDA-ARS Publ. No. 12, 116 p.

Shaner, G. and R.E. Finney. 1976. Weather and epidemics of Septoria leaf blotch of wheat. *Phytopathology* **66**: 781-785.

Shearer, B.L. 1967. *Epidemiology of Septoria nodorum* Berk., B.Sc. Hons. Thesis, Univ. of Western Australia, 147 p.

Shearer, B.L. and R.D. Wilcoxson. 1978. Variations in the size of macropycnidiospores and pycnidia of *Septoria tritici* on wheat. *Can. J. Bot.* **56**: 742-746.

Siddiqui, M.R. and S.B. Mathur. 1989. Survival of *Septoria nodorum* Berk. in wheat seed stored at 5°C. *FAO / IBPGR Plant Genetic Resources New letter* 75/76: 7-8.

Stewart, C.M., A. Hafiz and T. Abdel Hak. 1972. Disease epiphytotic threats to high yielding and local wheats in the Near East. *F.A.O. Plant Prot. Bull.* **20**: 50-70.

Srivastava, K.D. 1986. Septoria diseases of wheat. **In**: *Problems and Progress of Wheat Pathology in South Asia* (Eds. L.M. Joshi, D.V. Singh and K.D. Srivastava), pp. 191-215. Malhotra Publishing House, New Delhi, 401 p.

Tyagi, P.D., L.M. Joshi and B.L. Renfro. 1969. Reaction of wheat varieties to *Septoria tritici* and report of an epidemic in North western Punjab. *Indian Phytopath.* **22**: 175-178.

Wiese, M.V. 1987. *Compendium of Wheat Diseases.* American Phytopathological Society, APS Press, St. Paul, Minnesota, 106 p.

Yechilevich-Auster, M.E. Levi and Z. Eyal. 1983. Assessment of interactions between cultivated and wild wheats and *Septoria tritici. Phytopthology* **73**: 1077-1083.

Ziv, O. 1983. Control of Septoria leaf blotch of wheat and powdery mildew of barley with antitranspirant epidermal coating materials. *Phytoparasitica* **11**: 33-38.

Ziv, O. and Z. Eyal 1976. Evaluation of tolerance to Septoria leaf blotch in spring wheat. *Phytopathology* **66**: 485-488.

2.2.5 Fusarium Leaf Blotch

Fusarium nivale causes leaf blotches on winter wheat in the cooler regions of sub-tropics. This pathogen is most destructive at temperatures just a few degree above freezing and the wheat crop is most likely to be damaged under prolonged snow cover. These conditions never exist in India during wheat crop season; therefore, the chances of existence of *F. nivale* in the country are very remote. Although, some *Fusarium* species like *F. dimerum, F. semitectum, F. avenaceum, F.graminearum* and *F.culmorum* have been reported from different areas in India but the prevalence of *F. nivale* is not in record from any region of the country (Joshi *et al.,* 1978).

Fusarium leaf blotch is an important disease prevalent on the high plateau of Mexico, in the high valleys of East Africa and Andean region of South America (Zillinsky, 1983). Larger areas of USSR, Japan, Canada, central Europe, Scandinavia and North-West USA suffer most damage from snow mold caused by *F. nivale* (Wiese, 1977). The major crop losses result from poor grain development following the leaf blotch disease.

Symptoms: Generally, the symptoms of leaf blotch appear at heading stage of the plant. Early symptom occurs as greyish green, mottled, oval or irregular spots on lamina. The spots enlarge rapidly, turn greyish brown and develop into oval shaped blotches with grey centres. The leaf tissue in the older lesion becomes distorted and tends to split or fold. Plants affected by snow mold become chlorotic with pinkish mycelium and visible spore accumulations on the necrotic leaves.

Causal Organism: *Fusarium nivale* is the most readily identified of the *Fusarium* species that infect cereal crops. Conidia of the fungus are short, curved, taper towards the ends, and foot cells are not well marked. Mature conidia measure 20–28 µm × 2.5–5.0 µm and usually have three septa. At maturity of the plants, perfect stage of the fungus namely *Calonectria nivalis* develops in the lesions in the form of perithecia. The perithecia look like pycnidia of the *Septoria,* but they can be differentiated by the asci and ascospores they contain. The ascospores are hyaline, elliptical, irregularly curved with 1–3 septa, and measure 10–18 µm × 3.5 µm.

Disease Cycle: The fungus persists as spore, mycelium or perithecia in association with host debris beneath deep snow for a year or more. These fungal structures under favourable conditions act as primary source of inoculum to re-start the disease cycle.

Control: Foliar spray of benomyl at two weeks interval, starting at the heading stage, keeps the disease in control.

2.2.6 Dilophospora Leaf Spot

This minor disease was recorded for the first time in India on wheat variety NP 4 in Kashmir valley (Reddy, 1960). It is believed that the fungus *Dilophospora alopecuri* was introduced in Germany either from France or Switzerland, transmitted through a nematode vector *Anguina tritici* (Atansoff, 1925). The disease is prevalent in USA, Canada and Europe.

Symptoms: The disease is characterised by the yellow, elliptic to elongate, sometime spindle shaped flecks which later turn tan coloured and develop crusts in the centre. These spots appear on both the surfaces of the leaf, being more common on the upper surface (Munjal and Kaul, 1961). The fungus is usually transmitted to the whorl through larvae of *A. tritici*. Leaf spotting can occur without nematode but the leaf twist symptom of the disease is extremely rare in its absence (Wiese, 1977).

Causal Organism: The fungus, *Dilophospora alopecuri* (Fr.) Fr. produces black, globose, pycnidia within host lesions. Mature pycnidia are 60–300 μm in diameter, have broad torn ostiole and simple, hyaline short conidiospores. Pycnidiospores are hyaline, cylindrical to ellipsoidal, 1.5–2.5 × 8–15 μm with claw shaped appendages at each end which measure 0.5 × 5.7 μm.

Disease Cycle: Mycelium of the fungus in plant debris and pycnidiospores on seed act as primary source of inoculum. Pycnidia are wind dispersed or splashed by rain to cause infection on upper plant parts. Claw-appendages help in attachment of pycnidiospores to seed, host plants and nematode larvae.

Control: Sanitation, crop rotation and use of clean seed can lower the level of disease incidence.

2.2.7 Leptosphaerulina Leaf Spot

A new leaf spot caused by *Leptosphaerulina trifolli* (Rostr.) Petrak was recorded in 1967 at Sabour (Bihar) on wheat variety NP 884 (Misra *et al.*, 1969). This is a minor disease.

The spots measured 1–2.5 × 0.5–1 mm and are brown in colour. *L. trifolli* on wheat is considered to be the first record from India or elsewhere.

The fungus produces perithecia in the lesions. Perithecia are dark, elongated, irregular in shape, papillate and measure 104–374 × 96–288 μm. Asci are clavate to cylindrical, slightly curved, somewhat round at the apex, paraphyses or paraphysoids lacking, typically 8-spored and 70–76 × 26–34 μm in size. Ascospores are muriform, light brown, both transversely and both longitudinaly septate and measure 24–34 × 8–12 μm.

2.2.8 Chaetomium Leaf Spot

A new leaf spot of wheat caused by *Chaetomium dolichotrichum* Ames was recorded in 1974 at Keylong (Lahul and Spiti valley of Himachal Pradesh, 8000 f.a.s.l). The disease appeared as irregular, elliptical, light brown or straw coloured spots on lower leaves. Lesions were more pronounced on matured leaves (Goel *et al.*, 1976).

2.2.9 Phoma Glume Blotch

The Phoma glume blotch was recorded from Madhya Pradesh in 1969–70 crop season on wheat varieties S 227, S 308, Kalyansona and RR 21 (Nema *et al.*, 1971). The disease was found to occur primarily on the glumes and awns, but in rare cases leaves were also affected by *Phoma insidiosa* Tassi.

The disease is not widely prevalent in the country and is economically insignificant, though Phoma glume blotch can reduce the grain weight from 3.7 to 53.8% depending on the severity of infection.

The fungus produces purplish brown lesion on glumes under humid conditions. Lesions are oval in shape and measure 5–7 × 2–2.5 mm in size. Pycnidia are present in the lesions. Later the lesions enlarge and the centre becomes light grey. Symptoms on leaves appear as faded green streaks on the upper surface and the pycnidia develop below the epidermis in rows. The grains that develop in the infected glumes are shrivelled and discoloured.

The fungus *P. insidiosa* grows well on PDA. The pycnidia formed in culture are globose and measure 112.2–165.0 µm in diameter. The pycnidiospores are elliptical, hyaline, single celled and measure 4.9 × 3.57 µm.

REFERENCES

Goel, L.B., L.M. Joshi, Ram Nath, D.V. Singh and K.D. Srivastava. 1976. A new leaf spot of wheat caused by *Chaetomium* sp. *Curr. Sci.* **45**: 277.

Joshi, L.M., K.D. Srivastava, D.V. Singh, L.B. Goel and S. Nagarajan. 1978. *Annotated Compendium* of *Wheat Diseases in India,* ICAR, New Delhi, 332 p.

Mishra, A.P., R.A. Singh and B. Misra. 1969. A new leaf spot disease of wheat in India. *Curr. Sci.* **38**: 471-472.

Munjal, R.L. and T.N. Kaul. 1961. Dilophospora lef spot of wheat in India. *Indian Phytopath.* **14**: 13-15.

Nema, K.G., G.S. Dave and H.K. Khosla 1971. A new glume btotch of Wheat. *Plant Disease Reptr.* **55**: 95

Wiese, M.V. 1977. *Compendium of Wheat Diseases*. American Phytopathological Society, APS Press, St. Paul Minnosota, 106 p.

Zillinsky, F.J. 1983. *Common Diseases of small Grain cereals.* CIMMYT, Mexico, 141 p.

2.2.10 Powdery Mildew

Powdery mildew is widely distributed disease in humid and sub-humid regions of the world. In India, the disease is mainly confined to northern and southern hills (Misra *et al.*, 2005). Sporadic appearance of the disease has also been reported in the plains and foot hills of the country. Butler and Bisby (1930) noticed heavy incidence of the disease at Dehradun as early as 1903. Powdery mildew occurred in Allahabad, Jodhpur, Bombay and parts of Karnataka (Mehta, 1930; Arya and Ghemawat, 1953; Patil *et al.*, 1969), and during 1968–69 the disease was also observed in severe form at Pantnagar, Ludhiana, Delhi and northern Rajasthan (Joshi *et al.*, 1986). Monitoring of the disease has revealed that powdery mildew has gained importance in the country and present day wheat cultivars grown in North-western plains, *viz.*, PBW 343, PBW 373, PBW 502 and HD 2687 are severely affected (Singh, 2008).

The disease is economically significant as it inflicts considerable losses in many parts of the world. Yield losses occur in relation to the intensity of powdery mildew attack and are measurable as reduced number of heads and low grain weight. When the crop is attacked during seedling stage, the losses may be up to 40% (Wiese, 1987). Bahadur and Aggarwal (1997) estimated that 1000 grain weight in wheat cultivar WL 711 reduced from 32–35 grams with varying disease severity.

Symptoms: The fungus usually develops on the upper surface of the leaf blade. Early symptoms appear as white to greyish-white colonies of fluffy superficial powdery mass. These white powdery patches later turn brownish and are seen with small, dot like black perithecia (Plate 3A). The underside of the infected leaves have yellowish necrotic spots at the infection sites. In case of severe infection, the affected leaves are crinkled, twisted or otherwise deformed. Under favourable conditions, the disease spread is very fast and the pathogen attacks leaf sheaths and spikes also. In such cases premature dying of the leaves is noticed and the ear-heads may totally be suppressed or emerge only partially.

Causal Organism: The pathogen responsible for the powdery mildew of wheat is the member of the family Erysiphaceae (Rilley, 1886). Yarwood (1957) and Junell (1965) have reviewed the taxonomy of the related members of this family. The fungus *Erysiphe graminis* DC f.sp.*tritici* E. Marchal (syn. *Blumeria graminis* f.sp. *tritici*) is heterothallic obligate parasite. The fungus produces epiphytous colonies of interlaced, septate hyphae having uninucleate cells, 5–10 µm wide. Conidiospores arising from the mycelium are short, simple, hyaline, 8–10 × 25–30 µm with a terminal generative cell and a swollen base. Conidia arise from the generative cell. They are vacuolated, granular, ovate, 8–10 × 20–35 µm and are borne basipetally and diuranally in long chains. Conidia fall on the leaf surface, germinate and infect the host by haustoria. The forms of haustorium are unique of this genus. They are formed in the epidermal cells and are elliptical with long finger shaped appendages radiating from both ends (Fig. 36). The haustorium forms a close physical association but does not penetrate the host cell plasmalemma.

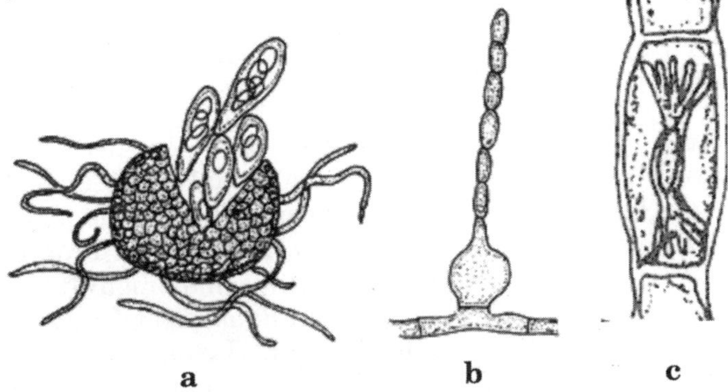

Figure 36: *Erysiphe graminis* (a) cleistothecium with asci and ascospores (b) conidia and condiophore (c) haustorium with finger shaped appendages

Aggarwal and Bahadur (1995) studied the conidial germination, appressorium and conidiospore formation in *E.graminis tritici* under scanning electron microscopy. They noticed that hyphae formed after germination of conidia cross the contours of host epidermis and form appresoria. The conidiophore development starts with the appearance of terminal swelling which produces a papilla. This papilla grows in length and divide to form a chain of conidia (Fig. 37).

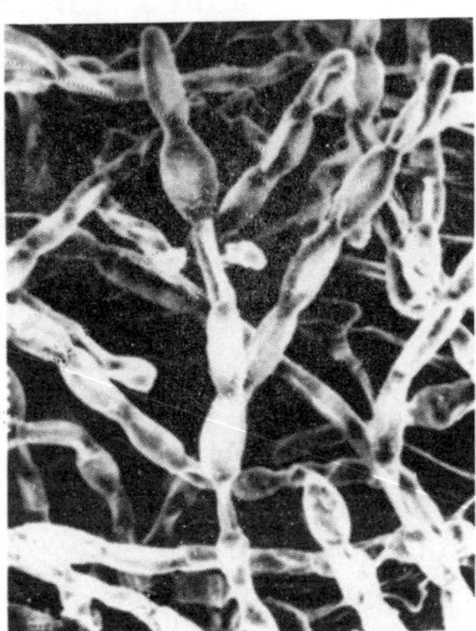

Figure 37: Scanning electron micrographs showing growth and sporulation of *Erysiphe graminis tritici*

The sexual fruiting structures *i.e.* cleistothecia are black, spherical, large, depressed, immersed in woolly mycelium, measure about 150–300 µm in diameter and have a very tough outer covering. The cleistothecia have simple or sparingly branched appendages on their surface. Appendages are short, few to many, sub-hyaline to pale brown. Asci are 8–25 in number, ovate to cylindrical, more or less pedicillate and 70–100 × 25–40 µm. Each ascus contains eight, elliptic, hyaline, thin walled ascospores measuring 12–27 × 9–15 µm.

Occurrence of physiologic races within the specialized forms of *E. graminis tritici* has been reported. In 1902 Marchal distinguished 7 *forma specialis* on the basis of principal host genera. Existence of 3 races of the fungus was demonstrated by Arya (1962). Prabhu and Prasada (1963) distinguished only 2 races. According to Sharma and Singh (1990) 17 races of *E. graminis* exist in Himachal Pradesh but Bahadur and Aggarwal (1997) recorded 41 pathotypes from Himachal Pradesh, Haryana and western Uttar Pradesh and 22 pathotypes from Nilgiri hills in South India. For the first time, they also reported 11 pathotypes from Mahabaleshwar.

Disease Cycle: In India, the mode of survival and source of primary inoculum has not been defined clearly. Mehta (1930) considered that the annual recurrence of powdery mildew at the foot hills and the plains is through wind blown conidia from the hills where the conidia seem to survive during summer season on self sown plants. Arya and Ghemawat (1953) were also of the view that cleistothecia are non-functional and wheat crop is re-infected annually by air-borne conidia possibly originating from Himalayas. However, in his subsequent work Arya (1964) has shown that the cleistothecia produced on wheat crop are not permanently sterile. He experimentally proved that when immature asci are subjected to alternating dry and wet conditions, ascospore formation is induced. Such conditions are available in low lying fields so Arya presumed that cleistothecia material found in nature may be responsible for the initiation of infection every year in the fields. Bahadur and Aggarwal (1994) have found that at higher altitude (600 a.s.l.) mature cleistothecia are produced with asci and ascospores. On the basis of these observations, it is believed that cleistothecial stage occurring in nature may be the principal means of annual recurrence of the disease.

During crop season the ascospores released from cleistothecium bring about primary infection on plant leaves. These lesions produce conidia in abundance within a short period. Secondary infection in the foot hills and adjoining plains occurs through wind borne conidia. The conidia are produced in large quantity under relatively cool and moist climate to repeat production cycles. The conidium on germination produces a germ tube with appressorium. A hyphal peg arises from the appressorium that penetrates the cuticular and sub-cuticular wall of the plant and forms a haustorium in the epidermal cell resulting in infection. At maturity of the crop, the fungus produces cleistothecia which remain on plant debris and act as a source of primary inoculum for the next crop season (Fig. 38).

Conidial germination is optimum at 100% relative humidity and a temperature of 15–20°C (Prabhu *et al.,* 1962). The conidia loose viability within 24 hours of their liberation if temperatures are higher than 20°C. Powdery mildew development is optimal between 20–22°C with a relative humidity between 24–75% and is markedly retarded above 30°C (Arya, 1962). Arya (1964) holds that ascospore formation takes place most rapidly at 22–27°C if the cleistothecial material is exposed to alternate drying and wetting in the soil. High plant density, early sowing, high nitrogen fertilization and prolonged high relative humidity with a number of rainy days favour disease development. Rani *et al.* (2008) observed that susceptibility of wheat to powdery mildew is higher in late sown crop (2nd week of December) because during this period, availability of inoculum is more and crop growth stage is vulnerable to infection and environmental conditions remain favourable for disease development in Punjab.

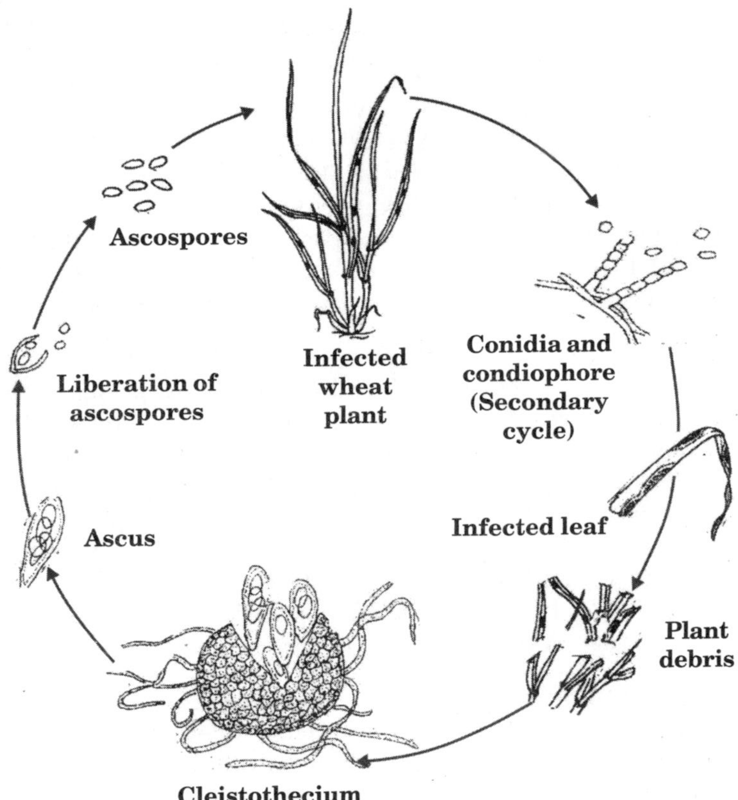

Figure 38: Disease cycle of powdery mildew (*Erysiphe graminis tritici*) of wheat

Management

Resistence: Resistence to powdery mildew is reported to be conditioned by single dominant gene. This report of Waterhouse (1930) on genetics of resistance paved the way for identification of resistance genes in different wheat varieties. Six genes namely *Mlt, Mlu, Mls, Mlc, Mlb* and *Mla* were identified in Australia (Carter, 1954; Pugsley, 1961). Briggle (1966) developed isogenic lines *Pm 1, Pm 2, Pm 3a, Pm 3b,* and *Pm 4*. Upadhyay and Kumar (1974) used combinations of single gene lines *Pm 3a, Pm 3b, Pm 3c, Pm 8* and *PmMa* and increased latent period of the pathogen. The isogenic lines *Pm 1, Pm 2* and *Pm 4* and cultivars Thew, TD 16 and PBW 91 showed resistance through out the growing period while single gene lines *Pm 3b, Pm 3c* and *PmMa* and cultivar HD 2189 exhibited resistance in adult plants (Sharma *et al.,* 1921). Bahadur and Aggarwal (1994) reported that effectiveness of gene *Pm 8* to powdery mildew flora of India was about 80-97%. Resistance genes *Pm 8, Pm 3a, Pm 3c, Pm 5, Pm 1* with *Pm 2, Pm 3c, Pm 5, Pm 6, Pm 7* or *Pm 8* or *Pm 2* with *Pm 3c, Pm 5, Pm 7* or *Pm 8* etc. have been identified which incorporate resistance to powdery mildew in wheat. Sharma *et al.* (1998) have suggested the use of somaclones for the development of resistant varieties.

Wheat varieties – MACS 1967, DWR 162, DWR 195, DWR 698, DW 776, DW 777, HD 2667 etc. have been reported to possess high level of resistance to the disease (Hasabnis *et al.,* 1997; Bahadur *et al.,* 1998). Slow mildewing has been recorded in cultivars HS 208 and CPAN 1922 (Sharma and Singh, 1990). Disease tolerant varieties, *viz.* HPW 251, HS 240, HS 277, HS 420, VL 616, VL 738, VL 829 and VL 829 are suitable for cultivation to keep the disease under control in northern hill zone (Singh, 2008).

Chemical Control: Sulphur was used to control powdery mildew in the 19th Century. Use of yellow sulphur dust or sulphur preparations like Karathane and Miltox effectively reduced the disease (Yarwood, 1950; Pathak and Joshi, 1972; Singh and Saxena, 1974). Foliar spray of systemic fungicides like triadimenol, triadimefon, , tebuconazole + triademenol or triademorph + triademenol @ 0.15-0.20 percent first at appearance of the disease and second after 2-3 weeks has been found to be quite effective (Aggarwal *et al.,* 1997). The disease could be controlled through spray of propiconazole (Tilt 25 EC @ 0.1%) on disease appearance of the symptoms till grain filling stage at 15 days intervals (Singh, 2008).

Cultural Control: Burning of the diseased plant debris and ploughing during summer is an effective cultural practice to eliminate the primary source of inoculum. Since heavy dose of nitrogen affects the disease, reduced use of nitrogenous fertilizers and increased use of phosphorous have been found useful. Avoidance of late sowing, reduced plant density and well drained soil also help in the reducing the disease severity (Singh and Srivastava, 1992).

REFERENCES

Aggarwal, R. and P. Bahadur. 1995. Host pathogen interaction in powdery mildew of wheat through SEM. *Cereal Rusts and Powdery Mildew Bull.* **23**: 1-6.

Aggarwal, R., P. Bahadur and S.K. Jain. 1997. Efficacy of some systemic fungicides against powdery mildew (*Erysiphe graminis* f. sp. *tritici*) of wheat. *Indian J. Plant Prot.* **25**: 84-87.

Arya, H.C. 1962. Studies on physiologic specialization and varietal reaction of wheat to powdery mildew in India. *Indian Phytopath.* **15**: 127-132.

Arya, H.C. 1964. Studies on the physiology of ascospore formation in powdery mildew of wheat. *Indian Phytopath* **17**: 27-34.

Arya, H.C. and M.S. Ghemawat. 1953. Occurrence of powdery mildew of wheat in neighborhood of Jodhpur. *Indian Phytopath* **6**: 123-130.

Bahadur P. and R. Aggarwal. 1994. Quantifying the level of variability in *Erysiphe graminis* f. sp. *tritici* and effectiveness of *Pm* genes. Final Reports: Cess Fund Project, Div. of Plant Pathology, IARI, New Delhi 1-35 p.

Bahadur, P. and Aggarwal, P. 1997. Powdery mildew of wheat- A potential disease in North western India. **In**: *Management of Threatening Plant Diseases of National Importance*, pp. 17-26. Malhotra Publishing House, New Delhi, 314 p.

Bahadur, P., D.V. Singh, K.D. Srivastava, R. Aggarwal and S.K. Jain. 1998. Multiple disease resistance in wheat and triticale. *Indian Phytopath.* **51**: 58-71.

Briggle, L.W. 1966. Transfer of resistance to *Erysiphe graminis* f. sp. *tritici. Crop Sci.* **6**: 401-469.

Butler, E.J. and G.R. Bisby. 1930. *The Fungi of India.* Imp. Council of Agric. Res. Scientific Monograph I, XVIII, 237 p.

Carter, M.V. 1954. Additional gene in *Triticum vulgare* for resistance to *Erysiphe graminis tritici. Australian J. Bio. Sci.* **7**: 411-414.

Hasabnis, S.N., S. Kulkarni, R.V. Wuike and R.R. Hanchinal 1997. Reactions of wheat varieties to powdery mildew caused by *Erysiphe graminis* DC f. sp. *tritici.Karnataka J. Agric. Sci.* **10**: 1235-1237.

Joshi, L.M. , D.V. Singh and K.D. Srivastava 1986. Wheat and wheat diseases in India. **In**: *Problems and Progress of Wheat Pathology in South Asia* (Eds. L.M. Joshi, D.V. Singh and K.D. Srivastava), pp 1-19. Malhotra Publishing House, New Delhi, 401 p.

Junell, L. 1967. Erysiphaceae of Sweeden. *Symbolac Botanica Upsallensis* **19**: 1-117.

Mehta, K.C. 1930. Studies on annual recurrence of powdery mildew of wheat and barley in India. *Agric Jour.,* India **25**: 283-286.

Mishra, B. and associates. 2005. Cost effective and sustainable wheat production technologies. *Tech. Bull.* No. 8, Directorate of Wheat Research, Karnal, 36 p.

Sharma, T.R. and B.M. Singh 1990. Physiologic races of *Erysiphe graminis tritici* in Himachal Pradesh. *Indian Phytopath.* **43**: 33-37.

Sharma, K.D., B.M. Singh and R.S. Chauhan 1998. Variation in Karnal bunt and powdery mildew resistance among somaclones and doubled haploids of bread wheat cv. Sonalika. *Indian Phytopath.* **51**: 324-328.

Pathak, K.D. and L.M. Joshi 1972. Chemical control of powdery mildew of wheat. *Indian Phytopath.* **25**: 139-141.

Patil, D.K., R.K. Hegde and H.C. Govindu. 1969. New record of powdery mildew on wheat in Mysore state. *Mysore J. Agric. Sci.* **3**: 238-239.

Prabhu, A.S. and R. Prasada. 1963. Physiological races of wheat powdery in Simla and Nilgiri hills. *Indian Phytopath.* **16**: 201-204.

Prabhu, A.S., V. Rajendran and R. Prasada. 1962. Moisture requirement for the germination of conidia of *Erysiphe graminis tritici* E. Marchal. *Indian Phytopath.* **15**: 280-286.

Pugsley, A.T. 1961. Additional resistance in *Triticum vulgare to Erysiphe graminis tritici. Aus. J. Bio. Sci.* **14**: 70-75.

Rani U., G.D. Munnshi, I. Sharma and K.Chand. 2008. Opportune period for screening wheat germplasm against powdery mildew in Punjab. *Indian Phytopath.* **61** : 75-78.

Rilley, C.V. 1886. The mildew of grape-wine and an effectual remedy. *Pro. Amer. Pom. Soc.* **1886**: 49-54.

Singh, A. and S.C. Saxena. 1974. Chemial control of powdery mildew of wheat. *Indian J. Mycol. Plant Pathol.* **3**: 202-203

Singh, D.V. and K.D. Srivastava. 1992. Wheat disease control through IPM strategies. **In**: *Farming System and Integrated Pest Management* (Eds. J.P. Verma and A. Varma), pp. 159-176. Malhotra Publishing House, New Delhi, 332 p.

Singh, D.P. 2008. Disease problems of wheat and management approaches. **In**: *A Compendium of Lectures on Integrated Pest Management in wheat based cropping system,* DWR Compendium No. 2, Directorate of Wheat Research, Karnal, India, 258 p.

Yarwood, C.E. 1950. The effect of temperature on the fungicidal action of sulfur. *Phytopathology* **40**: 173-180.

Yarwood, C.E. 1957. Powdery mildews. *Bot. Rev.* **23**: 235-301.

Upadhyay, M.K. and R. Kumar. 1974. Field reaction to powdery mildew strains of wheat lines possessing known genes for resistance. *Indian J. Genet.* **34**: 150-150.

Watershouse, W.L. 1930. Australian rust studies III. Initial studies of breeding for rust resistance. *Proc. Linn. Soc. N.S. Wales* **55**: 596-636.

Wiese, M.V. 1987. *Compendium of Wheat Diseases*. Amerixan Phytopathological Society, APS Press, St. Paul, Minnesota, 106 p.

2.3 SEED-BORNE DISEASES

Stinking smut or common bunt (*Tilletia spp.*) and loose smut (*Ustilago segetum tritici*) are centuries old seed borne diseases of wheat. It is evident from historic references that the bunt of wheat was one of the first seed borne diseases, investigated during second half of 18th century. The contagious nature of bunt was shown by Tillet in 1755 and then Prevost provided the first proof and interpretation of the cause of this disease in the classic studies in 1807. In 1847, Tulasne brothers and later Anton de Bary (1884) confirmed the observations of Prevost with regard to causal organism of wheat bunt. Loose smut of wheat is another classic disease which was studied with great interest more than 100 years ago. Jensen (1888) evolved hot water treatment for eradication of seed borne inoculum of loose smut pathogen, which was first used by Swingle in 1892 and subsequently by Freeman and Johnson in 1909. Apart from smut and bunts, many other seed-borne diseases still continue to be important worldwide (Mathur and Cunfer, 1993).

In India, herbarium records show that loose smut was first reported from Punjab in 1897 (HCIO NO 7675). It is ranked next only to rusts in lowering the crop productivity and has shown harmful trends over years (Singh *et al.*, 1993). A new bunt of wheat, popular as Karnal bunt (*T. indica*) was recorded as a minor disease in 1930 but now it has assumed significance internationally. Karnal bunt reduces not only the quality of seed but also influences exchange of germplasm and global trading of wheat due to quarantine restrictions. Although hill bunt (*T. caries*/*T. foetida*), downy mildew (*Sclerophthora macrospora*) and black point are localized in distribution but they are equally important diseases in the country.

2.3.1 Loose Smut

Loose smut of wheat has a long history of its existence. There is mention of the disease which shows that smut was present in pre-Christian era. The Greek philosopher, Theophratis has described smut during 384-332 BC. Long ago there was great awareness and concern about the disease among Romans. They termed smut as *Ustilago*, a Latin word that means burn. This term was later used in many languages as a common name of smut fungi. An illustration of loose smut is given in Hieronymns Bock's Herbal, published in1556 and an accurate symptomatology is given in Fabricius text of 1717 (Nielsen and Thomas, 1996). By 1890, it was learned that wheat is infected by loose smut pathogen via ovary (Maddox, 1896).

The disease occurs throughout the world wherever wheat is cultivated. It is prevalent in all wheat growing states of India; however, the incidence of loose smut is relatively more in cool and moist areas of northern plains and hills than dry southern-peninsular region of the country (Goel *et al.*, 1977). Prior to introduction of Kalyansona in 1967, the incidence of the disease was quite high in traditional tall varieties like C 306, C 591 etc. Because of extensive cultivation of morphologically resistant variety Kalyansona, the incidence of loose smut in the country gradually declined to less than 0.5 percent (Joshi *et al.*, 1973). In recent years, the disease has once again aggravated due to cultivation of susceptible genotypes. Survey data indicate 3–4 percent incidence of loose smut in northern region of the country (Srivastava *et al.,* 1992). The incidence of the disease is reported to vary from 1–2% in North-western India with 5–7% in isolated cases (Karwasra, 2008).

Loose smut is very destructive as almost every ear-head of the affected plant is converted into black powdery mass of smut spores and there is no grain formation. The yield loss is, therefore proportional to percentage of infected ear-heads (Weibel, 1958). It has the potential to cause 100% loss (Persons, 1954). In India, substantial yield losses due to loose smut have been estimated. Patel *et al.* (1950) noted that annual loss in Bombay State was four million rupees considering the average loss to the crop at 5 percent. In Madhya Pradesh, Mishra and Singh (1969) estimated 3–6% yield losses. At the rate of 3–4% Jhooty (1985) recorded the loss ranging from 203 to 207 crores of rupees during 1984 crop season. The national average loss to the crop from 1985 to 1989 crop season ranged from 1.5 to 3.5% and it amounted for Rs 105 crores at 2.5 percent infection level in North-western India during 1988–89 (Srivastava *et al.,* 1992). Currently, the overall losses due to loose smut in North India is about 0.6 million tonnes in wheat yield (Singh, 2008).

Symptoms: The infection of loose smut is internally seed-borne. The infected seeds appear normal in shape and colour, and look like healthy ones. The systemic infection proceeds soon after germination of infected seed and the symptoms are visible only at the time of heading. The blackish spore mass can be seen through the covering of the boot several days before ear emergence. The spikes emerged from the boot contain dark black powdery mass of smut spores. The spore mass is initially covered by a smooth, delicate greyish membrane which soon ruptures and releases the chlamydospores (Plate 3B). Subsequently, the spores are dislodged by wind leaving behind the nacked rachis.

In diseased plants, a few or all the spikes are affected but in most of the cases entire spike gets converted into smut sori, though partially infected spikes can often be seen. Narrow linear sori may be formed rarely on the flag leaf and leaf sheath (Batts and Jeater, 1958). In some cultivars, flag leaf of diseased plants show characteristic yellowing and chlorotic streak which ultimately turn necrotic prior to emergence of infected spikes (Aggarwal *et al.,* 1982). The disease reduces number and height of tillers and length of the rachis and peduncle. The lower internodes of the tiller are usually longer and the upper ones shorter than in healthy plants (Gothwal, 1972).

If a spike is partially infected, the upper florets can be infected by ergot, head blight and Karnal bunt. Both flag smut and loose smut have been observed on the same plant or culm (Aujla and Sharma, 1977; Nielsen and Thomas, 1996). Association of nematode *Anguina tritici* and downy mildew with loose smut affected ears is also observed (Bedi *et al.*, 1959; Singh and Bedi, 1983).

Causal Organism: *Ustilago tritici* (Pers.) Rostr. was described as causal agent of both wheat and barley loose smuts on account of morphological similarity of chlamydospores and biology of the fungus. However, it was noted that one form of the fungus affects wheat not barley and another form infects barley not wheat, therefore Fisher (1943) considered these forms as specialized variety and proposed separate designation for causal organism of barley loose smut as *Ustilago nuda* (Jens.) Rostr. Considering host speciality as a criterian, Schaffnit (1926) named *U.nuda* f. sp. *tritici* for the pathogen responsible for loose smut of wheat. But Anisworth and Sampson (1950) thought that taxonomically the description of loose smut pathogen based on varietal status should be *U. nuda* var. *tritici* (Pers.) Rostr. Later, Fisher and Shaw (1953) named pathogen as *U. nuda* var. *tritici* for wheat and *U. nuda* var. *hordei* for barley pathogens. Vanky (1985) has proposed new nomenclature and placed loose smut pathogen of wheat as *Ustilago segetum* (Pers.) Roussel var. *tritici* Jens. Neilsen (1987) expressed different opinion regarding designation of two pathogens. He observed certain differences in the two fungi such as different morphological features of promycelium and monokaryotic and dikaryotic hyphae after germination of chlamydospores; covering of sori of *U. nuda* by a thin membrane, where as those of *U. tritici* are normally naked and pronounced variations in the polypeptide and isozymes between the two species, suggesting genetic variability. Hence Neilsen and Thomas (1996) justified *U. nuda* and *U. tritici* as valid nomenclature of loose smut of barley and wheat respectively. *U. tritici, U. nuda* var. *tritici* and *U. segetum* var. *tritici* are accepted as synonymous of causal organism of loose smut of wheat.

The chlamydospores of the fungus are sub-spherical to spherical, echinulate, pale yellow brown, lighter in colour on one side and 5-9 µ in diameter. Electron microscopy has shown that smut spores consist of two layers, the exosporium and endosporium. The exosporium is partially thickened and echinulate. The echinulate is dense on thinner side and sparce on the thicker side of the exosporium. The endosporium is also thickened (Khanna *et al.*, 1971). The chlamydospores germinate to form four celled promycelium or basidium. There is no sporidia formation in *Ustilgo tritici*. The fusion of compatible cells by means of short or long conjugation tube yields infection dikaryotic hyphae.

The fungus grows well on malt agar and PDA medium (Sen and Munjal, 1964). Mondal (1994) used brown standard synthetic medium for culturing of the fungus. The pathogen is slow growing fungus. Sen and Munjal (1968) added glycine, pectine and L-sucrose to the culture media to support maximum vegetative growth. Selvakumar (1997) used modified brown standard synthetic medium by addition of 5% orange juice for profuse growth of the fungus. The temperature ranging from 20–25°C and 3.9 pH of the medium favours the growth of *U. tritici.*

Pathogenic variability exists in *U. tritici.* New races of the fungus arise by recombination of pre-existing virulence genes or by mutation at loci responsible for virulence. Neilsen (1987) selected 19 differential hosts in such a way that each of these hosts identifies either different combinations of known gene or a previously undetected gene for virulence. Neilsen and Dyck (1988) gave a number to each differential line and designated these differentials as TD series (TD 1 to TD 19) which can detect all the 44 prevalent races of loose smut pathogen (Table 9).

Table 9. Differential hosts to differentiate races of loose smut (*Ustilago tritici*) of wheat

Differential	Cultivar, line or pedigree	Canada number
TD-1	Mindum	1795
TD-2	Renfrew	1796
TD-3	Florence/Aurore	1797
TD-4	Kota	1798
TD-5	Little Club/Reward	18129
TD-6	PI 69282	
TD-7	Reward	1801
TD-8	Carmal/Reward	18130
TD-9	Kearney	1803
TD-10	Red Bobs	1804
TD-11	Pentad	1811
TD-12	Thatcher/Regent/Reward	18131
TD-13	PI 298554/CI 7795	
TD-14	Sonop	1814
TD-15	H44/Marquis	1815
TD-16	Marroqui 588	1816
TD-17	Marquillo/Waratah	
TD-18	Manitou*2/Giza 144	1818
TD-19	Wakooma	1819

In India, first indication of the existence of pathogenic races of loose smut of wheat was given by Padwick (1941). Pal and Mundkur (1945) identified two races namely, L1 and L2. Dastur (1946) also suggested the presence of two physiologic races. Gothwal and Pathak (1977) detected 8 loose smut races LSR 1 to LSR 8 from the North-western region of India. Rewal and Jhooty (1986) used Neilsen's set of 19 differentials and identified three distinct races of the pathogen *i.e.* T1, T10 and T11 from Punjab. The existence of race T1 from Delhi, Ludhiana and Jallandhar, T3 from Alwar and T11 from Ghaziabad was reported by Srivastava *et al.* (2001). Recently, Padmja *et al.* (2006) reported the occurrence of race T1 of loose smut in Himachal Pradesh, T7 and T11 in Haryana, T7 and T4 in Rajasthan and T1 and T11 in Punjab.

These observations reveal that races of loose smut do not follow any definite pattern of distribution because the virulence pattern does not depend on its geographical origin, but is determined by the cultivar on which it occurs (Neilsen, 1987). Karwasra *et al.* (2002) made an attempt to characterised loose smut fungus using RAPD, ISSR and AFLP markers but failed to reveal polymorphism. However, Padmaja *et al.* (2006) proved by RAPD assay that genetic variability exists among 23 isolates of *U. segetum tritici.*

Disease Cycle: The infection of loose smut is internally seed borne. The fungus perpetuates inside the embryo of infected wheat seeds as dormant mycelium which resumes its activity, and invades all parts of the plant, without causing injury to it. The active hypha just behind the growing point keeps pace intercellularly through the young seedlings until it reaches apex of the shoot. When the ear formation is in progress the mycelium infects all the young spiklets, where it grows intercellularly and destroys most of the tissues of the ear, except rachis. It is reproductive phase of the life cycle of the fungus, forming numerous spores and disorganising the spikelets and embryo. After hyphae proliferation in the ear, the binucleate hyphae segments swell, round off and become thick walled as they are converted into chlamydospores.

The chlamydospore mass (smut sori) in the diseased spikes is held together within a delicate membrane of the host tissue. Eventually, the black smut spores are released from the spikes after the thin papery membrane breaks and blown by wind to fall on the flowers where they cause fresh infection. The germinating spores produce a promycelium or basidium. At this stage, meiotic division takes place, with the result four nuclei are formed and promycelium becomes four celled, each cell having a haploid nucleus. Dikaryotisation takes place at this time between any of the two cells of promycelium through anastomosis, the infective hypha is produced from dikaryotic cells which enters in and between the cells and pass down the lower part of the style, and then penetrates through the ovary wall, usually at the brush end (Batts, 1955; Shinohara, 1976). Rewal (1983) found that spores which germinate on the ovary wall produce foot like appressoria and cause direct penetration rather than through style. The process requires 5 to 7 days. Once in the testa, the hyphae grow intracellularly; but in the integument and nucleus, the fungus grows intercellularly, mainly on the dorsal side of the developing caryopsis. The mycelium enters the upper and side parts of the scutellum 10-15 days after penetration and grows through the hypocotyl into the plumular bud, or growing point of the embryo. Thus in about 3 weeks the fungus establishes itself as dormant mycelium in different parts of developing seed including embryo (Neilsen and Thomas, 1996). In the embryo, the mycelium remains in dormant form till sowing of the seed in the next season. The infected seed looks like healthy, resting period of the fungus begins at this stage (Fig. 39).

Transmission of loose smut pathogen from infected seed is dependent on prevailing weather conditions. Excessive heat or dry air reduces spore germination and retards growth of infection hyphae, and delays penetration

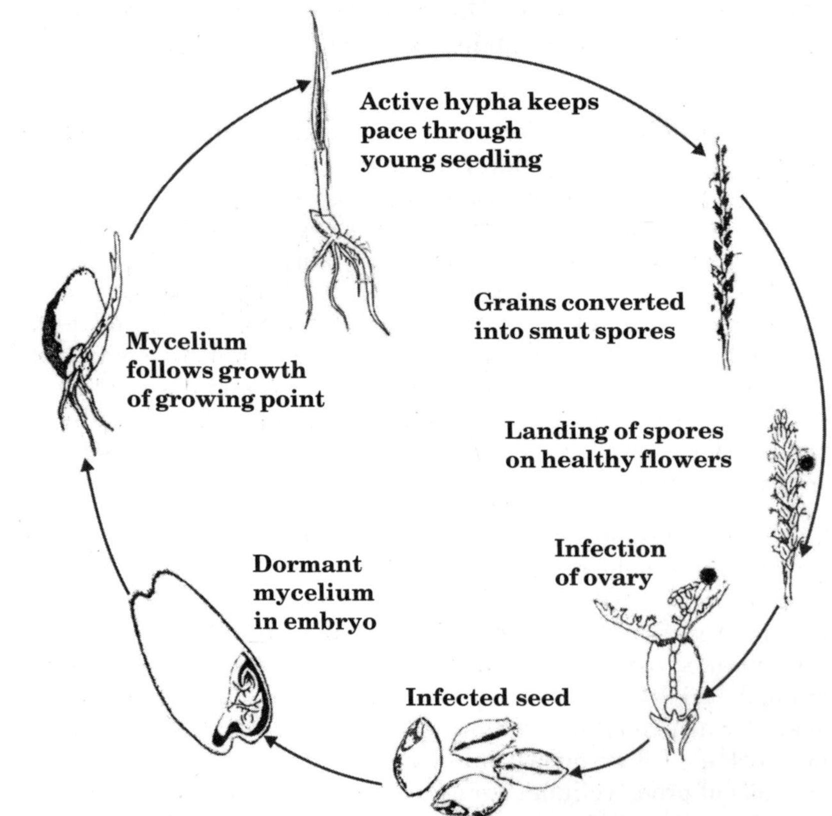

Figure 39: Disease cycle of Loose smut (*Ustilago tritici*) of wheat

of the ovary. It is likely that under warm dry conditions the embryo is dehydrated so the infection of the plumular bud is prevented. Poor disease incidence has been recorded on wheat cultivars in warm and dry climate than in cooler and more humid one (Atkin *et al.,* 1963). According to Dean (1969) maximum disease development takes place at 23°C, however, temperature lower than that and higher than 29.5°C leads to decrease in disease incidence. It is noticed that disease is less in areas where humidity at flowering stage is low. High humid condition with 60–80% RH is conducive for good infection (Tapke, 1929). It is because of this reason the loose smut is considered a serious disease of humid and semi-humid regions in India (Joshi *et al.,* 1978).

Several workers have established relationship between disease incidence in the field with embryo or seedling infection. Maddox (1896) and Ruttle (1934) indicated that embryos of the resistant cultivars did not contain mycelium. On the other hand Ohms and Bever (1955) found that embryo infection of resistant cultivars was as frequent as those of susceptible cultivars. Munjal and Chatrath (1964) showed the presence of interseminal mycelium and reported direct relationship between embryo infection and disease expression in the field. Rewal and Jhooty (1971) also observed that the embryos of the field resistant cultivars and field susceptible cultivars were equally infected with the fungus. They correlated the disease expression in the field based on percent embryonic infection. They estimated high effective infection and thought that the detection of mere presence of internal mycelium in the embryo may not always give the realistic picture of disease forecasting. Similarily, Popp (1951) suggested that the presence of mycelium in the sub-crown internodes of heavily infected seedlings of a susceptible cultivar, was indicative of the development of the smut in adult plants and the seedlings with light infection were indicative of smut free plants. The findings of Rewal and Jhooty (1971) revealed that if the crown nodes of three week old seedlings had extensive branched mycelium, it was indicative of emergence of smutted ears in the field, whereas presence of aggluminated or restricted mycelium in the crown node indicated smut free plants. Ram *et al.* (1983) also demonstrated the expression of loose smut in the field by clipping the seedlings at different growth stages.

A major factor for establishment of infection is the growth stage. It is essential because long time is required by the mycelium of the fungus to reach the growing point in the maturing embryo, together with the thickened cuticle and epidermis. According to Gothwal (1972) maximum infection occurs when the ears just emerge out of the leaf sheath and anthers are preferably green. But Rewal (1983) reported that the ears remained equally susceptible to loose smut from pre-anthesis to post-anthesis stage. Loria *et al.* (1982) found that for natural infection, susceptibility is determined by opening of flower. These differences pertaining to susceptible growth stage are attributed to environmental factors.

Wind velocity plays positive role in disseminating the chlamydospores from smutted ears to cause infection on healthy flowers. Oort (1947) showed that the smut spores may be disseminated and cause infection up to 100 meter from the point of origin if wind velocity is high. Singh (1983) noted that the chalmydospores can travel a distance up to 60 meters at normal wind speed. Agarwal (1983) recommended that seed production area should be at least at 150 meter distance from commercial feilds.

Management

Resistance: Genetical studies for resistance to loose smut have been in progress since 1934 (Tingey and Tolman, 1934; Heyne and Hansing, 1955; McIntosh, 1983; Dhitaphichit *et al.,* 1989). The details of these studies summarised by Neilsen and Thomas (1996) suggest that resistance to smut can be dominant or recessive; a gene can impart complete resistance or partial resistance; gene effects may be additive; different genes may condition resistance to a single race; resistance to one race can be conditioned by different single genes; genes may stop pathogen development at one or several specific sites in the ovary, embryo or seedling and the genes of the embryo usually determine whether the fungus is arrested in the ovary or embryo, resulting in resistance.

Studies on the inheritance of resistance in India also indicate that resistance is governed by either single dominant gene or by fewer genes. Pal and Mundkur (1945) revealed that resistance was controlled by one dominant gene in variety NP 824 and NP 790 whereas two dominant factors conditioned resistance in variety NP 798. Mathur (1985) reported resistance to be dominant in four varieties, namely E 220, E 957, E 1847 and E 2102. All varieties except E 220 carried one resistant gene. E 220 had dominant duplicate gene. Saini and Sharma (1985) indicated that Florance × Aurore, NP 824 and Col. 222 each had dominant gene. Sanop had two dominant and independently inherited genes, one of which was allelic and closely linked to the dominant gene in Col. 222 and NP 824. However, Indian wheat breeding efforts have not been able to keep a track of loose smut resistance genes. Therefore, one of the most urgent areas of study in relation to loose smut resistance is the search for effective genes that condition genetical resistance. It is now clear that gene for gene relationship exists in *Ustilago tritici / Triticum* system (Oort, 1963). In this genetic background breeding procedures can be effectively utilised for incorporating resistance. According to Gill (1985) pedigree as well as backcross methods offer great scope in breeding varieties resistant to loose smut.

In some cultivars morphological resistance to loose smut is conferred by differing floral habits like limited floret opening, smaller angle of opening for palea and lemma during flowering, difference in the extursion of anthers etc. (Tyler, 1965; Paril, 1973; Pandey and Gautam, 1988). These traits for the resistance may be selected and exploited in breeding programme.

Durum wheat in general possesses greater resistance than bread wheat. Resistant sources to loose smut have been identified from time to time (Chatrath and Mohan, 1973; Grewal *et al.,* 1973; Joshi *et al,.* 1985; Srivastava *et al.,* 1992). Multilocation tests have shown high degree of resistance in A-9-30-1, HD 4502, HI 8381, K 8027, MACS 2046, N 59, PBW 226, PBW 34, PBW 65, PDW 233, RAJ 1555, and WH 896 genotypes (Mishra *et al.,* 2002). Resistance may be transferred from these sources into an alredy established cultivar or other agronomically superior genotype. Resistant bread wheat varieties like HS 227, VL 829, PBW 34, Halna and durum wheat varieties like PDW 233, WH 896, HI 8498 and RAJ 1555 have been found suitable for cultivation (Singh, 2008; Karwasra, 2008).

Seed Certification: Certification of seed specifies the incidence of loose smut infected plants at inspection of a field producing foundation and certified seeds. In United Kingdom, central seed regulations are strictly followed to meet the standards for seed quality. The standard for loose smut for basic seed is 0.2% and 0.5% for first and second generation certified seed (Aggarwal *et al.,* 1993; Rennie *et al.,* 1983). The tolerance limit for foundation and certified seed in U.S.A. is zero to near zero. The permissible maximum limit of loose smut infection at 0.1% and 0.5% as a certification standard for foundation and certified seeds has been fixed in India by the Central Seed Certification Board (Tunwar and Singh, 1988). Prescribed limits for seed certification are achieved by seed dressing with carboxin (Mathre *et al.,* 1982). An isolation distance up to 150 meters between seed production plots in the seed certification programmes is desirable.

Heat Therapy: The hot water treatment was developed by Jensen (1888) for eradication of internally seed-borne infection of loose smut. The method involves soaking of seed in normal water for 4–6 hours, then dipping in hot water at 49°C for 2 min followed by drying before sowing. The underlying principle is that during pre-soaking, the dormant mycelium becomes active and when the seeds are further dipped in hot water the activated mycelium is killed making the seeds free of mycelium. Treatment in hot water may adversely affect seed viability if there is increase in either temperature or duration of soaking of seed. Because of this problem in the method, a significant change from hot water treatment was demonstrated by Luthra and Sattar (1934) where solar energy was used for controlling loose smut of wheat in North-western India. In this method seeds are first soaked in water for hours in the morning and at noon soaked seeds are spread on a cemented floor in thin layers for drying. The solar heat treatment is effective during bright summer day in the months of May-June. Subsequently, certain modifications were suggested either in duration of exposure time or drying of seed (Mitra and Taslim, 1936; Patel *et al.,* 1950;

Bedi, 1952). Recently Duhan and Beniwal (2006) proposed soaking of seed in water (1:1 w/v) in galvanized tub tightly covered with polythene sheet and keeping the seed in sun.

Chemical Control: Seed dressing with systemic fungicide is one of the most promising practice for control of seed borne diseases. This is quite amenable to integration with other 1DM strategies. Excellent control of loose smut has been achieved by seed treatment with carboxin @ 2.5 g/kg seed (Chatrath *et al.*, 1969; Nene and Saxena, 1971). Seed dressing with carboxin at lower dose (2 g/kg seed) is also reported to be effective (Thomas and Chatrath, 1975; Tyagi, 1972). Ram *et al.* (1984) modified the seed dressing method with reduced dose of carboxin @ 0.5–1.0 g/kg seed. The chemical treatment is based on the seed activation by soaking in ordinary water for 12 hrs and then dressing with the fungicide prior to sowing . Rewal *et al.* (1990) have shown that uptake of carboxin by the seed requires at least 72 hrs for effective control of the disease. Carboxin in combination with Thiram (Vitavax 200) @ 4 g/kg seed also provides effective control of loose smut of wheat (Tyagi *et al.*, *et al.*, 1976). A new formulation Carboxin 40 SC having more sticking capability, being in slurry form, can check the infection of the disease effectively (Aggarwal *et al.*, 1993). Chemotherapy with other systemic fungicides like benomyl (@ 2.5 g/kg seed), carbendazim , fenfuram, triademefon, dichlopentazole, Baytan, Raxil @ 2g/kg seed etc. also provide good control of the disease (Aggarwal *et al.*, 1993; Joshi *et al.*, 1975; Khnzada and Mathur, 1984; Srivastava *et al.*, 1982, 1991, 1998, Srivastava and Yadav, 2006)

Biological Control: The biological control being eco-friendly, helps in reducing the dependency on chemical fungicides. Loose smut infection was reduced through antagonistic organisms such as *Trichoderma viride, Gliocladium deliquescens, G. virens* and *Bacillus subtilis* (Aggarwal and Nagarajan, 1992, Srivastava and Yadav, 2006). Biocontrol potentiality of *T. viride* strain TV-5 was proved against the disease by Mondal *et al.*, 1995. However, this strain did not show compatible response to widely used seed dressing fungicide carboxin. Thereafter, carboxin tolerant mutants of *T. viride* (TV-5) developed by Selvakumar *et al* (2000) showed better bio-efficiency and more competitive saprophytic ability. Multilocation field test using *T. viride* were conducted which showed that application of the biocontrol agent along with half dose of carboxin (Singh *et al.*, 2000) is effective. Aggarwal *et al.* (2001) have also found that TV5-2 strain of *T. viride* could be used as seed dresser using *Aloes* gum as the sticking agent. Seed treatment with *T. viride* @ 4 g/kg seed + Vitavax 75 WP (carboxin @ 1.25 g/kg) seed is recommended to take care of loose smut disease (Sharma, 2008).

Cultural Control: The regular inspection of the crop at the early heading stage and careful rouging of smutted ears helps to prevent spread of the spores of the pathogen. Generally, the smutted ears are covered with a paper or plastic bag and ears are plucked with scissors. Then collected diseased ears are destroyed by either burning or burying in a soil pit.

Agarwal (1983) has proposed an integrated approach for production of pathogen-free seed: (1) adoption of relatively resistant cultivars, (2) production of seed in areas isolated about 150 m from commercial fields, (3) field inspection to meet the requirements for certification, (4) roughing of smutted plants, (5) testing of seeds by the embryo count method (Agarwal *et al.,* 1978) to identify heavily infected seed lots, and (6) seed treatment with systemic fungicides.

REFERENCES

Agarwal, V.K. 1983. An integrated approach for the control of loose smut of wheat. *Fourth Intern. Cong. Plant Pathol.*, Melbourne, Australia, 101 p.

Agarwal, V.K. 1983. Quality seed production at Pantnagar, India. *Seed Science & Technol.* **11**: 1071-1078.

Agarwal, V.K., H.S. Verma and S.B. Singh. 1978. Technique for the detection of loose smut infection in wheat seeds. *Seed Technology News* **8**: 1.

Agarwal, V.K., H.S. Verma, M. Agarwal and R.K. Gupta. 1982. Studies on loose smut of wheat III. Effect on plant morphology and control through seed treatment with carboxin. *Seed Res.* **10**: 79-86.

Agarwal, V.K., S.S. Chahal and S.B. Mathur. 1993. Loose smut. **In**: *Seed-borne Diseases and Seed Health Testing of Wheat* (Eds. S.B. Mathur and B.M. Cufer),pp. 59-68. Danish Govt. Institute of Seed Pathology for Developing Countries, Copenhagen, Denmark, 168 p.

Aggarwal, R. and S. Nagarajan. 1992. Possible biocontrol of loose smut of wheat (*Ustilago segetum* var. *tritici*) *J. Biol. Control.* **6**: 114-115.

Aggarwal, R., K.D. Srivastava and D.V. Singh. 1993. Raxil-a potent fungicide to control loose smut of wheat. *Indian Phytopath.* 46: 172-173.

Aggarwal, R., K.D. Srivastava and D.V. Singh. 2001. Biological control of loose smut of wheat: Seed treatment with *Trichoderma viride* and its influence on plant growth. *Ann. Pl. Protec. Sci.* **9**: 63-67.

Aggarwal, R., K.D. Srivastava, P. Bahadur and D.V. Singh. 1998. Resistance to loose smut and rust in wheat. *Rachis* 17: 63-64.

Aggarwal, R., K.D. Srivastava and D.V. Singh. 1992. Note on the efficacy of carboxin 40 SC against loose smut of wheat. *Seed Res.* **20**: 57-58.

Ainsworth, G.C. and K. Sampson. 1950. *The British Smut Fungi (Ustilaginales).* CMI, Kew, UK.

Atkins, I.N., O.G. Merkle, K.B. Porter, K.A. Lahar and D.E. Weibel. 1963. The influence of environment on loose smut percentages, reinfection and grain yields of winter wheats at four locations in Texas. *Plant Dis. Rep.* **47**:192-196.

Aujla, S.S. and Y.R Sharma. 1977. Simultaneous occurrence of *Ustilago nuda tritici* and *Urocystis agropyri*. *Indian Phytopath.* **30**: 262.

Aujla, S.S., A.S. Grewal, G.S. Nanda and I. Sharma. 1990. Identification of stable resistance to loose smut. *Indian Phytopath.* **43**: 90-91.

Batts, C.C.V. 1955. Infection of wheat by loose smut (*Ustilago tritici*) *Nature* (London) **175**: 467-468.

Batts, C.C.V. and A. Jeater. 1958. The development of loose smut (*Ustilago tritici*) in susceptible varieties of wheat and some observations on field infection. *Trans. Brit. Mycol. Soc.* **41**: 115-125.

Bedi, K.S. 1952. Loose smut of wheat and its control in Punjab on a large scale by solar heat. *Punjab Frm.* 4: 302.

Bedi, K.S., J.S. Chohan and D.S. Chahal. 1959. Simultaneous occurrence of *Ustilago tritici* (Pers.) Rostr., and *Anguina tritici* (S) G. Ben. in a single ear of wheat. *Indian Phytopath.* **12**: 187.

Chatrath, M.S. and M. Mohan. 1973. Studies in Indian cereal smuts. XI. Varietal resistance of wheat to loose smut. *Indian Phytopath.* **26**: 257-259.

Chatrath, M.S., B.L. Renfro, Y.L. Nene, R.K. Grover, M.K. Roy, D.V. Singh and S.M. Gandhi. 1969. Control of loose smut of wheat with systemic fungicides. *Indian Phytopath.* **22**: 183-187.

Dastur, J.F. 1946. Report of the Imperial Mycologist. Sci. Rept. Agric.Res. Inst., New Delhi. 1944-45, 66-72 pp.

De Bary, A. 1884. Vergleichende Morphologie and Bacterien, W. Engelmann, Leipzig, 558 p.

Dean W.M. 1969. The effect of tempeature on loose smut of wheat (*Ustilago nuda*). *Ann. App. Biol.* **64**: 75-83.

Dhitaphichit, P., P. Jones and E.M. Keane. 1989. Nuclear and cytoplasmic gene control of resistance to loose smut (*Ustilago tritici* (Pers.) Rostr). in wheat (*Triticum aestivum* L.). *Theor. Appl. Genet.* **78**: 897-903.

Duhan, J.C. and M.S. Beniwal. 2006. Improved solar energy treatment for the control of loose smut, flag smut and Karnal bunt diseases of wheat. *Seed Res.* **32**: 184-188.

Fischer, G.W. 1943. Some evident synonymous relationships in certain graminicolous smut fungi. *Mycologia* **35**: 610.

Fischer, G.W. and Shaw, C.G. 1953. A proposed species concept in the smut fungi with application to North American species. *Phytopathology* **43**: 181-188.

Freeman, E.M. and E.C. Johnson. 1909. The loose smuts of wheat and barley. *USDA Bureau Plant Intro. Bull.* 152: 48 p.

Gill, K.S. 1985. Sources of resistance to Karnal bunt and loose smut and their use in breeding resistant wheat varieties. *Proc. 3rd Nat. Seminar on Genetics & Wheat Imp.* held at IARI Regional Station. Flowerdale, Shimla from May 8-10, pp. 10-12 (Abstr.)

Goel, L.B., Singh, D.V., Srivastava, K.D. Joshi, L.M. and Nagarajan, S. 1977. *Smuts and Bunts of Wheat in India.* Indian Agril. Res. Inst. New Delhi. 38 p.

Gothwal, B.D. 1972. Effect of loose smut on the growth and morphology of spring wheat. *India J. Mycol Plant Pathol.* **2:** 171.

Gothwal, B.D. and V.N. Pathak. 1997. Pathogenic races of *Ustilago tritici* in India. *Indian Phytopath.* **30**: 311-314.

Grewal, A.S., S.S., Aujla, A.S., Minhas, and L.M., Joshi. 1973. Screening of different varieties of wheat to loose smut in the Punjab State. *J. Res.* (Ludhiana) **10**: 398-400.

Heyne, E.G. and E.D. Hansing. 1955. Inheritance of resistance to loose smut of wheat in the crosses of Kawvale × Clarkan. *Phytopathology* **45**: 8-10.

Jhooty, J.S. 1985. Smuts of wheat. *Indian J. Mycol. Pl. Pathol.* **15**: 1-30.

Joshi, L.M., D.V. Singh and K.D. Srivastava 1985. Status of rusts and smuts during dwarf wheat area in India. *Rachis* **4**: 10-16.

Joshi, L.M., K.D. Srivastava, D.V. Singh, L.B. Goel and S. Nagarajan. 1978. *Annotated Compendium on wheat Diseases in India*, ICAR, New Delhi, 331 p.

Joshi, L.M., P.D. Tyagi, D.V. Singh, M.R. Siddiqui, L.K. Joshi, R.P. Gupta. 1975. Control of loose smut of wheat with benomyl. *Indian Phytopath.* **28**: 417-419.

Karwasra, S.S. 2008. Smut and bunts of rice-wheat system and management under current scenario. **In**: *A Compendium of Lectures on Integrated Pest Management in Wheat based Cropping System*. DWR Compendium No. 2, Directorate of wheat Research, Karnal, India. 258 p.

Khanna, A., M.M. Payak and N. Prakash. 1971. Teliospore morphology of some smut fungi. III. *Ustilago nuda. Indian Phytopath.* **24**: 481-486.

Khanzada, A.K. and S.B. Mathur. 1984. Control of loose smut of wheat by carboxin, fenfuram and triadimenol. *Seed Science & Tehnol.* **11**: 947-949.

Loria, R., M. Wiese and A.L. Jones. 1982. Effect of free moisture, head development and embryo accessibility on infection of wheat by *Ustilago tritici. Phytopathology* **72**: 1270-1272.

Luthra, J.C. and Sattar, A. 1934. Some experiments on the control of loose smut of wheat. *India J. Agric. Sci.* **4**: 177-199.

Maddox, F. 1896. Smut and bunt. *Agric. Gaz.*, Tasmania **4**: 92-95.

Mathre, D.E., S.G. Metz and R.H. Johnson. 1982. Small grain cereal seed treatment in the post-mercury era. *Plant Dis.***66**: 526-531.

Mathur, H.C. 1985. Genetics of resistance to loose smut in bread wheat. **In**: *Genetics and wheat improvement* (Eds. A.K. Gupta, S. Nagarajan and R.S. Rana), pp. 99-100. ICAR, New Delhi, 362 p.

Mathur, S.B. and B.M. Cunfer. 1993. *Seedborne Diseases and Seed Health Testing of Wheat*. Institute of Seed Pathology for Developing Countries, Danish Govt., Denmark. 168 p.

Mishra, R.P. and J.N. Chand. 1970. Efficacy of different fungicides against foot-and root-rots caused by *Sclerotium rolfsii* of wheat. *PANS* **16**: 327-330.

Mitra, M. and M. Taslim. 1936. The control of loose smut of wheat in North Bihar by the solar energy and sun heated water methods. *Agric. Live Stk. India* **6**: 43-47.

Mondal, G., K.D. Srivastava and R. Aggarwal. 1995. Antagonistic effect of *Trichoderma* species on *Ustilago segetum tritici* and their compatibility with fungicides and biocides. *Indian Phytopath.* **48**: 466-470.

Munjal, R.L. and M.S. Chatrath. 1964. Preliminary screening of wheat varieties against loose smut by the embryo test. *Indian Phytopath.* **17**: 238-240.

Nene, Y.L. and S.C., Saxena. 1971. Present status of research of fungicidal control of wheat rusts and loose smut in India. *Proc. Second Int. Symp.* New Delhi pp. 83 (Abstr.)

Nielsen, J. 1987. Races of *Ustilago tritici* and techniques for their study. *Can J. Plant. Path.* **9**: 91-105.

Nielsen, J. 1987. Reaction of *Hordeum* species to the smut fungi *Ustilago nuda* and *U. tritici. Can. J. Bot.* **65**: 2024-2027.

Nielsen, J. and P. Thomas. 1996. Loose smut. **In**: *Bunt and Smut Disease Management*. (Eds. R.D. Wilcoxson and E.E. Saari), CIMMYT, Mexico, 66 p.

Nielsen, J. and P.L. Dyck. 1988. Three improved differential hosts to identify races of *Ustilago tritici. Can. J. Plant. Path.* **10**: 327-331.

Ohms, R.E. and Bever, W.M. 1955. Types of seedling reaction of Kawvale and Wabash winter wheat to three physiologies races of *Ustilago tritici. Phytopathology* **45**: 513-516.

Oort, A. J.P. 1963. A gene- for gene relationship in the *Triticum-Ustilago* system, and some remarks on host-pathogen combinations in general. *Tijdschr. Plantenz.* **69**: 104-109.

Oort, A.J.P. 1947. Specialization of loose smut of wheat – a problem for the breeder. Tijdschr. *Plantenz.* **53**: 25-43.

Padmaja, N., K.D. Srivastava and R. Aggarwal. 2006. Genetic variability in *Ustilago segetum* f. sp. *tritici* isolates based on random amplified polymorphic DNA. *Ann. Plant Prot. Sci.* **14**: 126-130.

Padmaja, N., K.D. Srivastava, R. Aggarwal and D.V. Singh. 2006. Pathogenic variability in Indian isolates of *Ustilago segetum* f. sp. *tritici* causing loose smut of wheat. *Ann. Plant Prot. Sci.* **14**: 141-142.

Padwick, G.W. 1941. Report of the Imperial Mycologist. *Sci. Rep. Agric. Res. Inst.* New Delhi 1939-40, 49-101 pp.

Pal, B.P. and B.B. Mundkar. 1945. Further studies in varietal resistance of India and other wheats to loose smut. *India J. Agric. Sci.* **15**: 106-108.

Pandey, D.K. and P.L. Gautam. 1988. Morphology of spike in relation to resistance to loose smut (*Ustilago tritici*) in wheat (*Triticum aestivum*). *Indian J. Agric. Sci.* **58**: 713-714.

Paril, I.F. 1973. Infection of wheat and barley with loose smut depending on the type of inflorescence. *Set. Semenovodstvo* **38**: 32-33.

Patel, M.K., G.W. Dhande and Y.S. Kulkarni. 1950. A modified treatment against loose smut of wheat. *Curr. Sci.* **19**: 324-325.

Person, T.D. 1954. Destructive outbreak of loose smut in Georgia wheat field. *Plant Dis. Reptr.* **38**: 422.

Poppy, W. 1951. Infection in seeds and seedlings of wheat and barley in relation to development of loose smut. *Phytopathology* **41**: 261-275.

Ram, B., I. Hooda, S. Singh and R.K. Grover. 1983. A technique for increased expression of loose smut in wheat. *Indian Phytopath.* **36**: 553-555.

Ramarao, P. and U. Raju. 1980. Effect of soil moisture on development of foot rot and root rot of wheat and on other soil microflora. *Indian J. Mycol. Plant Pathol.* **10**: 17-22.

Rennie, W.J., M.J. Richardson and M. Noble. 1983. Seed-borne pathogens and the production of quality cereal seed in Scottland. *Seed Science & Technology* **11**: 115-1127.

Rewal, H.S. 1983. *Studies on loose smut of wheat.* Ph.D. Thesis, Punjab Agril. University, Ludhiana.

Rewal, H.S. and J.S. Jhooty. 1982. Correlation between embryo, seedling and field infection of loose smut of wheat. *Indian Phytopath.* **35**: 571-573.

Rewal, H.S. and J.S. Jhooty. 1986. Physiologic specialization of loose smut of wheat in the Punjab State of India. *Plant Dis.* **70**: 228-230.

Rewal, H.S., G.D. Munshi and S.S. Sokhi. 1990. Effect of washing of carboxin from treated wheat seed and irrigation on loose smut control. *Plant Dis. Res.* **5**: 90-92.

Ruttle, M.L. 1934. Studies in barley smut and loose smut of wheat. *Tech. Bull. N.Y. Agric. Expt. Sta.*, 221-239 pp.

Saini, R.G. and S.C. Sharma. 1985. Diversity for loose smut resistance genes in wheat and pathogen virulences. **In**: *Genetics and Wheal Improvement* (Eds. A.K. Gupta, S. Nagarajan and R.S. Rana), pp. 101-107. ICAR, New Delhi, 362 p.

Schaffnit, E. 1926. Zur Physiologie von *Ustilago hordei* Kell. and S.W. Ber. Deutschen *Bot. Gez.* **44**: 151-156.

Selvakumar, R. 1996. *Biocontrol potentialities of carboxin tolerant mutants of Trichoderma viride against Ustilago segetum tritici.* M.Sc. Thesis. IARI, New Delhi.

Selvakumar, R., K.D. Srivastava, R. Aggarwal, D.V. Singh and Prem Dureja. 2000. Studies on development of *Trichoderma viride* mutants and their effect on *Ustilago segetum tritici*. *Indian Phytopath.* **53**: 185-189.

Sen, B. and R.L. Munjal. 1964. Nutritional requirement of *Ustilago nuda tritici* Schaf- The causal agent of loose smut of wheat. *Sci. Cult.* **31**: 196.

Sen, B. and R.L. Munjal 1968. Studies on the physiology of *Ustilago nuda tritici* Schaf. *Indian Phytopath.* **21**: 416-422.

Singh, D.P. 2008. Disease problems of wheat and management approaches. **In**: *A Compendium of Lectures on Integrated Pest Management in Wheat based Cropping System*. DWR Compendium No. 2, Directorate of Wheat Research, Karnal, India. 258 p.

Singh, D.P. and Associates. 2000. Efficacy of *Trichoderma viride* in controlling the loose smut of wheat caused by *Ustilago segetum* var. *tritici* at multilocation. *J. Biol. Control,* **14**: 35-38.

Singh, D.V., K.D. Srivastava, R. Aggarwal and P. Bahadur. 1993. Wheat disease problems: The changing scenario. **In**: *Pest and Pest Management in India – The Changing Scenario.* (Eds. H.C. Sharma and M.V. Rao), pp. 116-120Plant Protection Association of India, Hyderabad.

Singh, M. 1983. Studies on the isolation distance for loose smut in wheat. *Proc. 3rd All India Seed Technology Workshop,* Bangalore, May 1983.

Srivastava, J.P. and R.D.S. Yadav. 2006. Transmission and management of major seed-borne diseases of wheat (*Triticum aestivum* L.). *Crop Prot.* **31**: 175-178.

Singh, P.J. and P.S. Bedi. 1983. Synchronous occurrence of downy mildew and loose smut in wheat plant. *Indian Phytopath*. **36**: 734.

Srivastava, K.D., D.V. Singh and L.M. Joshi 1982. Efficacy of new fungicides for the control of loose smut of wheat. *Seed Res.* **10**: 66-68.

Srivastava, K.D., D.V. Singh and R. Aggarwal. 1998. Loose smut disease of wheat **In**: *IPM System in Agriculture* (Eds. R.K. Upadhyay, K.G. Mukerji and R.L. Rajak), pp. 291-303. Aditya Book Pvt. Ltd., New Delhi, Vol. 3 (Cereals).

Srivastava, K.D., D.V. Singh, L.M. Joshi and S. Nagarajan. 1979. Field evaluation of some new fungicides for the control of loose smut of wheat in India. *Pesticides* **13**: 41-42.

Srivastava, K.D., D.V. Singh, R. Aggarwal, A.K. Dixit and P. Bahadur. 1997. Bioefficacy and persistence of tebuconazole against loose smut of wheat. *Indian Phytopath.* **50**: 434-436.

Srivastava, K.D., D.V. Singh, R. Aggarwal, P. Bahadur and S. Nagarajan. 1991. Control of loose smut of wheat with dichlopantazole. *Rachis* **10**: 32.

Srivastava, K.D., D.V. Singh, R. Aggarwal, P. Bahadur and S. Nagarajan. 1992. Occurrence of loose smut and its sources of resistance in wheat. *Indian Phytopath*. **45**: 111-112.

Swingle, W.T. 1892. Treatment of smut of oats and wheat. *USDA Farmers Bull*. No. 5.

Tapke, V.F. 1929. The role of humidity in the life cycle, distribution and control of loose smut fungus in wheat. *Phytopathology* **19**: 1173-1178.

Tingey, D.C. and B. Tolman. 1934. Inheritance of resistance to loose smut in certain wheat crosses. *J. Agric. Res.* **48**: 631-655.

Tunwar, N.S. and S.V. Singh. 1988. Indian minimum seed certification standards. The Central Seed Certificate Board, Department of Agriculture and Cooperation, Ministry of Agriculture, Govt. of India, New Delhi, 388 p.

Tyagi, P.D., M. Singh and M.S. Chauhan. 1976. Comparative efficiency of some systemic fungicides for controlling loose smut of wheat. *Pesticides* **10**: 26-27.

Tyler, L.J. 1965. Failure of loose smut to build up in winter wheats exposed to abundant inoculum naturally disseminated. *Plant Dis. Reptr.* **49**: 239-241.

Vanky, K. 1985. Carpathian Ustilaginales Symbolae Botanicae Upsalienses 24(2), Tegel bruk svagen. Galgnef, Sweden, 309 pp.

Wiley, H.B. and T. Kommedahl. 1981. Biological seed treatment in sweet corn and wheat as a component of crop management. *Phytopathology* **71**: 265.

2.3.2 Karnal Bunt

The Karnal bunt is native to South Asia and was detected in experimental wheats grown at the Botanical Station (Regional Station, IARI), Karnal, Haryana in 1930 (Mitra, 1931). However, it is felt that the pathogen was prevalent much earlier. Mitra (1931) suspected that it might be the same bunt described by Howard (1909) from Layallpur (now Faizlabad, Pakistan) in 1909. Since disease specimens of Layallpur are not available, Mitra's report is taken as the first one. The disease is known by various names such as new bunt (Mitra, 1931), Karnal bunt (Mundkur, 1943), partial bunt (Bedi *et al.,* 1949). Some foreign workers wrongly refer to it as Kernel bunt (Munjal, 1966).

The disease which was confined to isolated pockets in the Indian sub-continent till early 1970's, has now occupied more areas in North-western region of the country. It occurs in an endemic form in the states of Punjab, Haryana, Jammu region of J&K, parts of Himachal Pradesh, Uttar Pradesh, Delhi, northern Rajasthan, Bihar and West Bengal. So far the disease has not been recorded in south of Madhya Pradesh, Maharashtra, Orissa, Assam, Karnataka, Andhra Pradesh, Tamil Nadu and Kerala (Joshi *et al.,* 1983). Besides India and Pakistan, Karnal bunt is reported from Syria (William, 1983), Afghanistan (Loke and Watson, 1975), Nepal (Singh *et al.,* 1989), Iran (Torabi *et al.,* 1996), Mexico (Duran, 1972), USA (Ykema *et al.,* 1996), South Africa (Crous *et al.,* 2000) and Iraq (Mathur, 1968; CMI, 1974). It has also been intercepted in wheat samples imported into India from Lebanon, Sweden and Turkey (Nath *et al.,* 1981).

The occurrence of Karnal bunt is sporadic in nature but becomes serious in epidemic years and causes substantial losses to wheat crop. McRae (1933) estimated loss up to 20% in a number of wheat cultivars. Loss estimates recorded in India reveal that the disease reduced annual yield of wheat by 0.2% in Punjab and Jammu (Munjal, 1976). These estimates of losses have been confirmed by data collected through Wheat Disease Surveys conducted by IARI, New Delhi since 1975. Data have shown that even during worst years of epidemic, the total damage to the crop was not more than 0.2–0.5% of the total production (Joshi *et al.,* 1983). According to Singh (1994) an estimated 1.0% can be lost due to Karnal bunt in an epidemic year on account of grain quality and yield losses alone. Brennan *et al.* (1990) estimated the economic losses at 0.12% in North-western Mexico. Hassan (1973) recorded that this disease causes 2–3% loss of grain in Pakistan.

The yield loss is related to the size of bunt sori produced by different varieties. Wheat varieties having small sorus suffer a loss of 5.2% in yield by weight, whereas the loss is nearly 19.9% and 51.5% in medium and large sori respectively (Aggarawal *et al.*, 1989). Moreover, seeds with small sorus produce normal seedlings, while those with large sori have poor germination (Rai and Singh, 1978; Bansal *et al.*, 1984). Increase in disease severity results in proportional decrease in seed weight (Singh, 1980; Bedi *et al.*, 1981).

The quality of wheat is adversely affected due to bunt infection. Flour milled from seed with 10% infection has dark colour. Since the Karnal bunt produces a volatile substance-trimethylamine, the palatability of 'chapaties' is reduced due to fishy odour and perceptible discolouration. Even at 1% infection of fresh seed lot, the palatability of chapaties is affected though very little, but the chapaties prepared from milled flour with 3% infected seed lot are unpalatable (Mehdi *et al.*, 1973). On the other hand, Sekhon *et al.* (1992) reported that bread, cookies and chapaties could be made from flour with 10% infection, if seeds are thoroughly washed and steeped. There is also a loss in flour recovery and chemical changes in composition of flour and gluten content (Gopal and Sekhon, 1988). Phenolic acid and phosphorous contents of the infected grains are slightly higher, and the total lysine content is nearly 20–25% lower than of healthy grains (Bhat, 1980; Shyama *et al.*, 1988).

The importance of Karnal bunt *vis-à-vis* export of wheat is well understood. There is quarantine restriction by several countries on movement of wheat grains from affected areas. Agriculture Department of Canada has maintained zero tolerance for the pathogen (Martin, 1986). European Union has also devised quarantine procedures to prevent introduction of the pathogen (Anon. 1991). With in North America, zero tolerance level to Karnal bunt is followed to prevent movement of contaminated seed (Babadoost, 2000). The movement of wheat grain from disease prone areas to other parts of Mexico is prohibited unless fumigated with methyl bromide (Davila, 1996). On account of such implications of disease, export in India suffers considerably (Singh and Srivastava, 2000).

Symptoms: Karnal bunt pathogen infects wheat at the flowering stage prior to seed formation; hence the symptoms are visible only when the grains have fully developed in ear heads. In a stool all the ear heads are not affected and also all the grains in a spike. A careful examination of individual ear reveals bunt infection in the field. In the standing wheat crop, the infected spike can be detected by the shiny silvery black spikelets, with glumes spread apart and swollen ovaries (Plate 3C). The spikes of infected plants generally are reduced in length and in number of spikelets (Mitra, 1937).

The pathogen converts the infected ovary into a sorus where a mass of dark brown coloured teliospores are produced. Infection of grains varies from small sori to completely bunted seeds. Small sori are generally developed in longitudinal furrow, leaving the dorsal side of the seed and endosperm unaffected. If the host is highly susceptible, the whole endosperm material may be converted into a large sorus, and seed looks hollow leaving only the pericarp and the aleurone layer (Plate 3D). In such cases the shrivelled embryo is dead. The grains are usually partially bunted and completely infected ones are rare (Mitra, 1935; Chona *et al.*, 1961). Therefore, the disease is also referred as 'Partial bunt' (Bedi *et al.,* 1949). The infected ears emit a fishy odour due to trimethylamine, a volatile compound produced due to pathogenesis.

Causal Organism: The pathogen of Karnal bunt was first described by Mitra (1931) as *Tilletia indica* Mitra. Later Mundkur (1940) argued that the fungus should have been assigned to genus *Neovossia,* as it produces numerous non-fusing sporidia, hence he renamed the fungus as *N. indica* (Mitra) Mundkur. Fischer (1953), limiting the species of *Neovossia,* again referred it to *T. indica* given by Mitra. Duran (1972) also proposed that *T. indica* is correct taxonomic name of the fungus.

Teliospores of *T. indica* are dark brown to black, globose to sub-globose in shape having hyaline sheath of 2–4 µm thickness and measure 22–49 µm in size, average being 35 µm in diameter. Mixed with the spores are globose to elongate, yellowish sterile cells which have smooth wall. They are smaller in size (15–28 µm) than normal teliospores. Scanning electron microscopy of teliospore shows three distinct layers: the perisporium (sheath), the episporium and the endosporium. The perisporium is a fragile, fractured structure and the episporium is reticulated having numerous curved projections with blunt margins, due to which the surface of the spore looks rough (Fig. 40). With the advancing maturity the perisporium ruptures at a few places and projections become visible. The surface projections are 3.1–4.0 µm thick in mature teliospore. An individual

Figure 40: Scanning electron micrograph of Karnal bunt spores

projection is composed of two double strands which are near the apex. The endosporium is thick and lamellate structure (Khanna *et al.,* 1968; Gardner *et al.,* 1983; Aggarwal *et al.,* 1998).

Fresh teliospores have a period of dormancy (McRae, 1932). A dormancy period of 1-6 months is needed prior to spore germination (Prescott, 1984). In comparison to freshly harvested spores old teliospores germinate favourably. The highest germination occurs with year-old teliospores (Mathur and Ram, 1963; Bansal *et al.,* 1983). Pre-soaking of teliospores for 4 days enhances the germination. Soaking of the spores in tap water, farm yard manure extract, soil extract, wheat straw extract and certain chemicals is reported to influence dormancy and the germinability of teliospores (Holton, 1949; Mathur and Ram, 1963; Rai and Singh, 1979; Dhiman and Bedi, 1984; Krishna and Singh, 1983; Gupta and Singh, 1983). Exposure of spores to a temperature of –196°C in liquid nitrogen for 15 minutes also provides higher germination (Bansal *et al.,* 1984).

The germination of teliospores is sensitive to temperature and light conditions. Munjal (1971) obtained maximum germination at 20°C with pH 6.0 while Krishna and Singh (1962) found that spores germinate significantly at 20–25°C under alternate light and darkness at pH 4.9. Zhang *et al.* (1984) found that optimal germination was at 15–22°C, extremely low at 2°C and no germination after prolonged exposure at 35°C in darkness. According to Similanick *et al.* (1985) highest germination of teliospores is obtained after 3 weeks incubation at 15–20°C under continous light at pH 6.0–9.5. Aujla *et al.* (1986) noted that fresh spores put in plain water and kept in darkness provide 50% germination.

On germination, teliospore produces a stout promycelium measuring 10–190 μm length and 6–13 μm breadth. The promycelium may branch but only one branch bears a whorl of 60-185 primary sporidia (Plate 4A) at the tip (Mitra, 1931; Krishna and Singh, 1981). The primary sporidia are sickled shaped and have mean length and width ranging from 64.4 to 78.8 μm and 1.6 to 1.8 μm, respectively (Peterson *et al.,* 1984). When teliospores germinate, the single diploid nucleus undergoes meiosis. Subsequent rapid mitosis gives rise to a large number of haploid nuclei which migrate from hypobasidium into the promycelium and primary sporidia, each of which receives one nucleus. The sensitive, short lived, mononucleate primary sporidia germinate terminally or laterally in free water, giving rise to thick mat of hyphae. In general, the colonies of the fungus are brittle, crustaceous, umbonate with wavy margins. Subsequently, from a cushion like structure, two types of secondary sporidia *i.e.* falcate (allantoid) and filiform sporidia are produced (Plate 4B). Most secondary sporidia are mononucleate. Mycelial cells that originate from either type of secondary sporidia are banana shaped, 11.9–13 μm long and 2–2.03 μm wide, and are forcibly discharged. These spores are the only infective entities. The filiform sporidia serve as the reproductive bodies to raise allantoid sporidia in successive generations (Dhaliwal and Singh, 1989).

T. indica is a heterothallic fungus with bipolar incompatibility, controlled by multiple alleles at one locus (Duran and Cromarty, 1977; Krishna and Singh, 1983). The heterothallism demands fusion between compatible secondary sporidia because monokaryotic sporidia originating from germinated teliospores are, individually non-infective. Therefore, dikaryotization between compatible types is a pre-requisite. Bonde *et al.* (1977) suggested that dikaryotization occurs prior to penetration on glume surface. However, Sharma *et al.* (2008) have observed that the dikaryotization takes place inside the host tissue.

Physiologic forms of *T. indica* are known to occur in India. Mitra (1935) reported two forms of the pathogen based on spore size but these variations were subsequently shown to be due to environmental effect (Mundkur, 1943). In agreement with Mundkur's observation, Bansal *et al.* (1984) also confirmed that size of the teliospores can not be taken as a criterion for differentiating physiological races of *T. indica*. Aujla *et al.* (1987) differentiated four pathotypes K1, K2, K3 and K4 from different regions of Punjab and Himachal Pradesh on the basis of host pathogen interactions on 17 differentials. However, Singh *et al.* (1991) felt that race or pathotype concept in *T. indica* is not valid. They argued that in each annual cycle of the pathogen, nuclear fusion takes place between heterothallic secondary sporidia so the pathogenic isolate is a population, not a race. It is appropriate to classify the variation in isolates as aggressiveness. Keeping this argument in view, five aggressive isolates designated as KBAg-1, KBAg-2, KBAg-3, KBAg-4 and KBAg-5 have been identified (Singh *et al.*, 1995). Still the issue of physiologic specialization or existence of races in *T. indica* is unresolved.

Examination of cross section of infected grains demonstrates that hyphae of the pathogen proliferate in the space formed by the disintegration of middle layer of parenchyma in the pericarp where they produce teliospore. In this process of proliferation, the nuceller projections get ruptured and hamper the flow of nutrients to pericarp. The consequence is atrophy of seed and due to disruption of normal flow of nutrients it starves first the endosperm and then the embryo. The endosperm is shrunken and the space thus created gets filled with teliospores of the pathogen. The embryo is free from infection except under very severe infection (Cashion and Luttrell, 1988; Aggarwal *et al.*, 1994).

Roberson and Luttrell (1987) have studied the ultrastructure of teliospore ontogeny and found that teliospore of *T. indica* arise directly from a thin hymenial layer of hyphae. Teliospores appear as small, smooth walled circular bodies with 9 μm size at 10 days after inoculation. The size of the developing spores starts at increasing at 15 DAI but the spore wall still remains smooth. The surface of teliospores shows small protuberances having round ends at 20 DAI and the size of the spore at this stage is around 25 µm. the rodlets become prominent at 25 DAI showing further increase in spore size from 34 µm to 40 µm. The surface ornamentation is fully developed at 28 DAI (Aggarwal *et al.,* 1999).

Toxicology: No presence of mycotoxins in Karnal bunt infected grains has been recorded and toxicological tests also indicate absence of acute toxicity to laboratory animals. Tests for ergot alkaloids were also negative. Toxicity study in weanling rats by feeding them up to 50% bunt infected grains over a period of 45 days did not show any adverse effect including their feed intake. Histopathological examinations of tissues from breast, lung, liver, spleen, pancreas, kidney, thymus, ovaries, testis, adrenals, stomach, duodenum, jejunum, ileum and colon of male and female rats did not show any abnormality. Chicks given oral doses of various extracts from Karnal bunt affected wheat also had no adverse acute effect (Bhat *et al.,* 1980). Monkey used as an experimental animal also showed that consumption of as high as 70% diseased wheat grains in diet was not toxic (Bhat *et al.,* 1981). Yet, the report by Shyama *et al.* (1988) indicated that animals fed on 10% Karnal bunt infected wheat grains resulted in reduced liver weight, white blood cells count and monocyte count. Rai *et al.* (1991) also observed that feeding of diseased grains to albino rats was not safe because in such animal liver and renal insufficiency was observed. Toxicological tests conducted by Singh *et al.* (1992) revealed adverse biological effect of the diet containing infected grains and trimethylamine. The test showed that gains in body weight of rats and protein efficiency ratio were lowest for the feeding diet containing 10 m.e.q. of trimethylamine. Packed cell volume and neutrophils and monocytes and eosinophils increased. However, the adverse effects of Karnal bunt on the haematological parameters were largely overcome by the peeling and debranning followed by washing (Sekhon *et al.,* 1992).

Disease Cycle: The pathogen survives in the form of teliospores which fall in the soil during harvesting, threshing and winnowing of the crop. The spores adhere to the surface of healthy seed and act as contaminant. The spores on the seed create problem only when they are transmitted from the seed and deposited on the soil surface where infection occurs. Obviously soil borne inoculum is the primary source of annual recurrence of the disease. Nagarajan *et al.* (1997) considered soil as the bank of teliospores of the pathogen. The resting spores buried in the soil germinate from the middle of February to middle of March when soil temperature and moisture are suitable. The germinated spores produce a germtube (promycelium) to come up to the soil level bearing tuft of 110–185 sickle-shaped primary sporidia. While still on germ-tube, the primary sporidia produce two types of secondary sporidia *i.e.* sickle shaped or filiform like primary sporidia and banana shaped or falcate type allantoid sporidia (Dhaliwal and Singh, 1988). The filiform sporidia serve as the reproductive bodies to raise allantoid sporidia in successive generations. These spores are the only form that infects the wheat ear head. Allantoid sporidia are released forcibly (Dhaliwal and Singh, 1988, 1989) and are carried by wind or rain splash to get deposited on the lower leaves of the host (Perscott, 1986). They germinate in the presence of leaf wetness and produce secondary crop of spores on wheat and gramineous hosts. As the leaf surface dries, these spores get dispersed to upper leaves of wheat plant. Having reached the higher leaves through monkey jumps, they settle on the flag leaf during ear head stage. Occasionally the air borne allantoid sporidia get lodged on spike at anthesis, germinate on the glume surface and the fungus becomes partially systemic in rachis and the rachilla. Subsequently, the hyphae spread to the adjacent florets and spikelets around the infection site. Hyphae of the fungus then grow through the base of glume into sub-ovarian tissue and enter the pericarp through the funiculus (Goates, 1988). Further growth of hyphae is entirely within the pericarp. These hyphae become sporogenous and produce teliospores in wheat seed to cause Karnal bunt infection in the next crop season (Fig. 41).

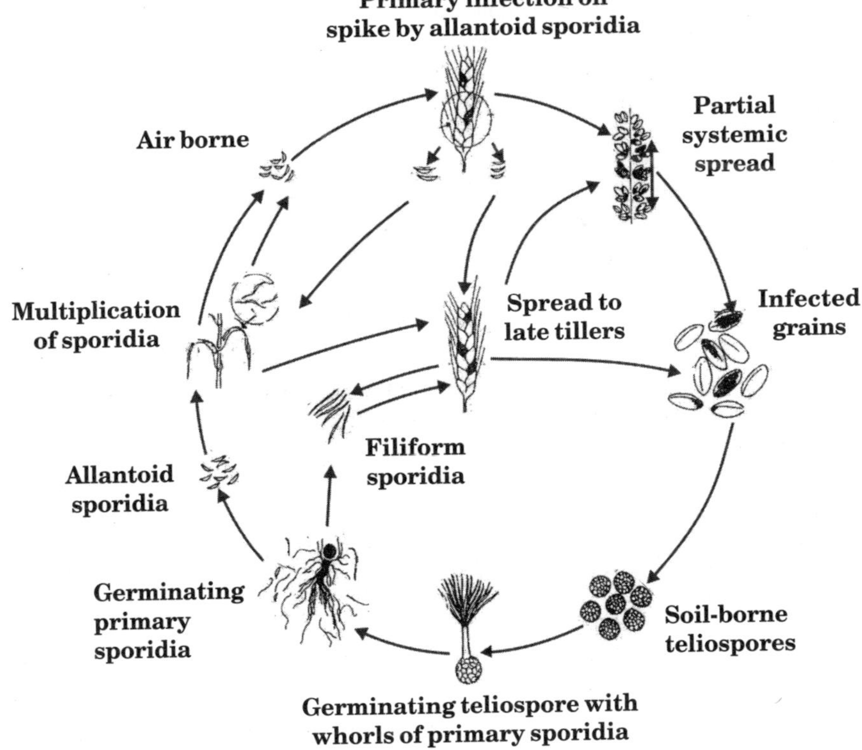

Figure 41: Disease cycle of Karnal bunt (*Tilletia indica*) of wheat

Epidemiology: Since Karnal bunt spreads through soil, seed and air, the disease development is influenced by a number of factors. The spores remain viable for a number of years in the soil lying 15 and 25 cm deep (Dhiman and Bedi, 1989). According to Nagarajan *et al.* (1997) the soil acts as a bank and soil borne teliospores have dormancy and retain their viability for more than eighteen months when buried in soil at a depth of 5 cm. Teliospores buried very deeply in soil loose their viability due to low availability of oxygen and moisture. Moreover, microbial disintegration also affects the germinability of the spores (Rattan and Aujla, 1990; Sidhartha, 1992). It has been recorded that high level of inoculum in the soil is positively correlated with severe disease incidence at disease hot spots like Gurdaspur in Punjab (Singh and Srivastava, 1992).

The spores on soil surface germinate from the middle of February to middle of March, if soil remains moist for a period of 7–10 days (Aujla *et al.*, 1990). Krishna and Singh (1982) recorded that for teliospore germination optimum temperature range is between 20–25°C. Sporidia formed after germination become air-borne. Hyphae that originate from primary or secondary sporidia produce large number of allantoid sporidia and comparative fewer filiform sporidia (Fuentes-Davila, 1984). Filiform sporidia are more or less ineffective and cause very less infection (Warham and Burnett, 1990; Sidhartha, 1992). Since there is rapid vegetative multiplication of allantoid sporidia, the probability of successful fusion of opposite mating types increases, resulting in severe infection (Nagarajan *et al.*, 1997).

The awn emergence stage corresponding to Z49-52 (Zadoks *et al.*, 1974) is the most vulnerable stage for establishment of Karnal bunt. Nagarajan (1991) hypothesized that at this stage, the flag leaf intercepts the allantoid sporidia and facilitate the run down of water droplets with sporidia. The aerobiology of the pathogen reveals that maximum sporidia in air are trapped at 30 cm and minimum at 90 cm, suggesting the possibility of spatial spread above the crop canopy by wind current. Hence, it appears that after release from the germinating teliospores, the sporidia first rest and multiply on lower leaves of the host plant and then move upward. Kumar and Nagarajan (1998) later confirmed that the flag leaf provides base for allantoid multiplication and inoculum build up. Diurnal periodicity in the sporidial release was shown by Bains and Dhaliwal (1989). Spore trapping indicates that maximum sporidia are produced in the early morning hours between 2–6 a.m. (Sidhartha *et al.*, 1995).

The establishment of the pathogen is highly dependent on favourable weather conditions. The prevalence of lower maximum (19–23°C) and higher minimum (8–10°C) followed by high relative humidity and intermittent rains lead to well established infection of Karnal bunt (Joshi *et al.*, 1981). These meteorological conditions prevail in the North-western region but absent in the central India, thereby restricted the development in of Karnal bunt in Madhya Pradesh and Rajasthan (Singh, 2005). On the other hand, the mid and higher altitudes of Himalayas, where snow fall occurs every year, Karnal bunt can not establish and survive. In this region, the snowing and thawing reduce the viability of spores in soil. Possibly snowing induces cold dormancy and therefore, the chilled spores require more energy to germinate (Sidhartha *et al.,* 1995). These spell of snow can delay the process of teliospore germination, sporidial proliferation and ultimately result in disease escape. According to Singh and Srivastava (1997) the chances of survival of teliospores are more under the irrigated conditions of rice-wheat cropping system in North-western region. They emphasised that the spores surviving in rice field are brought to the soil surface from different depths and serve as one of the sources of inoculum for fresh infection of Karnal bunt during wheat season.

Management

The control of Karnal bunt is difficult due to its mode of perpetuation through soil and seed. Hence, an integrated approach for tackling the disease is considered most ideal. Some of the approaches which influence the management of the disease are as follows-

Resistance: Genetic studies reveal that Karnal bunt resistance is polygenic and partially dominant (Gill *et al.,* 1993). Bag *et al.* (1999) have shown that resistance in wheat genotypes HD 29, HP 1531 and W 485 is conditioned by a single recessive gene and all the three accessions carry a non-allelic gene for resistance. Nanda *et al.* (1995) have also observed that dominant alleles are involved with resistance. Resistance to Karnal bunt is governed by two genes in genotypes Luan, Allila, Vee 7/Bow, Star, Weaver, Milan, Sasia and Taracio/Chil, and monogenic resistance in Cettia, Irena, Turacio, Opata, Picus and Yaco. Singh *et al.* (1995) found that digenic genotypes possess higher level of Karnal bunt resistance than those lines with a single gene. Sharma *et al.* (2005, 2007) found that HD 29, W 485 and ALDAN 'S'/ IAS 58 each carry two resistance genes whereas 3 genes are involved in H 567. 71/3 PAR. Genetic analysis shows that HD 2329 when used as the recurrent parent in backcross programme, gives more resistant progenies. Monosomic analysis using three resistant lines HD 29, WL 6975 and WL 2328 and the susceptible WL 711 indicates that the gene governing resistance to Karnal bunt is located on chromosomes 1D, 2D, 3B, 3D, 5B and 7A. The backcross involving some allien sources and cultivar Kalyansona yielded some Karnal bunt resistant derivatives (Sawhney and Sharma, 1996). Sirari *et al.* (2008) proposed use of mixture of most virulent isolates to identify Karnal bunt resistance genes.

Wild species of wheat are considered a valuable source of Karnal bunt resistance. Some synthesized amphiploids involving Karnal bunt resistant accessions of *Triticum monococcum, T. boeticum* and *Aegilops squarrosa* were free from Karnal bunt, indicating that the resistance is expressed in the presence of *durum* complement (Multani *et al.,* 1988). Some accessions like *Triticum tauschii, T. dicoccoides, T. spelta album, T. spelta grey, T. durum / Agropyron elongatum,* Chinese spring/*A. elongatum, Chinese junceum* and Chinese spring accessions have been found to be immune to Karnal bunt (Tomar *et al.,* 1991). *Triticum araraticum* (wild form of *Triticum timopheevi*, 2n = 4x = 28 AAGG) is reported to be a useful resistance source of *Tilletia* species and to a number of other wheat pathogens (Bijral and Sharma, 1995).

A large collection of wheat germplasm under field inoculated condition is being tested in India using standardized techniques and a numerical rating system (Aujla *et al.,* 1980). As a result of this rigorous multilocation testing of germplasm (Sharma *et al.,* 2002; Singh *et al.,* 2003), a number of genotypes possessing a coefficient of infection value of < 10 could be rated as resistant sources to Karnal bunt (Table10).

Currently popular wheat cultivars like PBW 502, HS 365, PBW 34, HP 1731, NW 1014, Raj 1555, HD 4672, WH 365 etc. (Singh, 2008) possess fairly good degree of disease resistance. Consequently, the extent of Karnal bunt damage has reduced in recent years.

Cultural Control: Crop rotation following non-cultivation of wheat crop for two consecutive years as an agronomic adjustment has been advocated by Mitra (1937) and Padwick (1939). Since the higher dose of nitrogenous fertilizers increases susceptibility of plants, adjustment in water and fertilizers in disease affected areas is recommended (Bedi, 1949; Singh and Singh, 1985). Biological mulching, such as inter-cropping wheat with chickpea and covering the inter row space with transparent polythene, reduces the disease severity (Singh *et al.,* 1992). Avoiding early sowing of wheat also results in poor incidence of Karnal bunt (Kalha *et al.,* 1987). Zero-tillage has shown a reduced incidence of Karnal bunt in comparison to furrow raised bed (FIRBS) and conventional tillage system. If zero tillage is followed for a few years, it may help in reducing the effective soil inoculum and thereby reducing the disease incidence over times (Sharma *et al.* 2007).

In North-west India, a variety of crop sequences are followed and field gets vacated at different times. The differing planting time delays wheat sowing and straggers the heading emergence stage and thus creates a temporal hurdle for *T. indica.* Coincidence of winter rainy days and ideal temperature conditions do not prevail over such a long heading period.

Table 10. Genotypes resistant to Karnal bunt (*Tilletia indica*) of wheat

Genotype	Pedigree
DL 896-2	TR 380–27* 4/3 Ag 3/4* KAL
HD 2281(D)	HD 2160/7/36896//CJ 54/P 4160 E/3/HUAR/6/KAL SIB/5/ SLSIB/NP 852/4/PJ SIB/P 14//KT 54 B/3/K 65
HD 2385	HI 687/HD2268
HD 2580	TTR 'S'/JCN 'S'
HD 29	HD 2160–HD 1977/HD 1949–HD 1944/HD 2136
HD 30	HD 2160–HD 1977/HD 1949–HD 1944/HD 2136
HD 4502	PI 'S' BY 2//TC/Z–B–W
HDR 132	JUPATECO 73/II 12300–TOB//CNO/4//HD 2315
HI 1077	GU/AUST 261-159//CNO/NO/3/KAL/BB
HI 1384	MRL 'S'/SVC 'C'/VA–1
HP 1531	HD 1257/C 306
HP 1633	RL 6010/6* SKA
HS 346	DL 20–9/HS 86/HB 501
HW 502	E 6039/NP 881/HW 131
HW 730	S 220/Bb/CNO "S"–PI 62
HW 737	S 308–TIMGALEN/CNO 'S'-S 220
HW 740	HW 202–HW 153/HD 2009
ISWRN 191	WGA 434
K 9006	CPAN 168/HD 2204
LEE	H/GB–AUS
PBN 51	BUC 'S'/FLK 'S' CM 00070–24 Y–IM–IY–0Y
PBW 225	WL 920/HD 2160
PBW 299	Bb/Kal//WL 711/PBW 65
PDW 215 (D)	DWL 5031/DWL 5002
PDW 227 (D)	PBW 70/PBW 71
RAJ 1707	COCORIT//RAJ 911
RAJ 1710	COCORIT//RAJ 911
VL 616	SKA/P 46
W 485	WL 923/HD 2160//UP 368
WH 805	GGO VZ 394–CIT'S'/21563–AA 'S'/FG 'S'
WL 1786	POLK/UP 301
WL 6975	WL 1535/UP 291//HD 2116/HD 2177
WI 1786	POLK/UP 301

Chemical Control: Seed dressing fungicides such as aureofungin, ethyl-mercury chloride, ethyl-mercury phosphate, fentin acetate, triphenyltin chloride, fentin hydroxide, indar, benomyl, carbendazim thiuram disulphide etc. have been found to restrict seed-borne inoculum of the pathogen but these chemicals have little effect on soil-borne teliospores (Singh and Rai, 1986). In this context, Cashion (1984) reported that application of maneb and mancozeb, can delay the sporidial germination at 5 and 20 g a.i/ ml but at 100 g a.i/ ml they are fungitoxic. PCNB, thiram and captan were also fungistatic at 5, 20 and 100 a.i/ ml.

Foliar sprays with systemic fungicides give effective control of the air-borne allantoids sporidia on wheat foliage. Among several fungicides evaluated for spray application, propiconazole (Tilt 250 EC) at the heading stage provides 71.4–100 percent disease control (Aujla *et al.,* 1989; Singh *et al.,* 1989). No residues of propiconazole were detected in wheat grain or straw when sprayed @ 250 ml/ha (Singh *et al.,* 1991; Sharma *et al.,* 1994). Some other fungicides such as bitertanol (baycor), tebuconazole (folicur) and cyproconazole (SAN 619F) have also potential for commercial exploitation at times of dire need (Singh *et al.,* 1985; Singh and Srivastava, 1992).

Nagarajan (1991) developed a linear model ($Y = 0.4381 + 2.97a - 2.77b - 0.09c + 0.13d$) for predicting Karnal bunt severity. A linear model ($Y = 128.6 - 0.58\,x_1 + 0.00057\,x^1 + 0.00619\,x^2 + 4.14\,x^3$) developed by Singh *et al.* (1990) can also predict the frequency and severity of the disease in the field. These prediction models are of fundamental importance for successful and economical use of chemicals against Karnal bunt.

Biological Control: Antagonistic potential of *Trichoderma viride, T. harzianum* and *Gliocladium deliquescens* on germination of teliospores of *N. indica* was observed by Aggarwal *et al.* (1996). Amer (1995) also evaluated some fungal and bacterial micro-organisms and recorded that bio-control agent *T. lignorum* causes significant decline in teliospore germination when used as soil inoculant under field and glasshouse conditions. Scanning electron microscopic investigation demonstrated over growth of *T. lignorum* on mycelium of *N. indica.* Ultra-structural observations showed that the antagonist mediated host mycelium by penetration pegs and resulted in disintegration of the host mycelium (Aggarwal, 1996).

Quarantine: Since Karnal bunt is prevalent only in a few countries around the world, there is general quarantine imposed against the disease. Most countries insist on a zero tolerance on shipment of wheat. In Mexico there is a domestic quarantine restricting the movement of wheat material from one area to another. European and Mediterranean Plant Protection Organisation (EPPO) has presented a list of A-1 quarantine pests including *T. indica.* In China, wheat consignment is accepted from other countries only after the wheat lots are declared as having zero infection. USA has declared *T. indica* as a quarantine pest after introduction of Karnal bunt in Arizona, Texas and New Mexico. Canada has maintained a rigid position on the disease and has imposed zero tolerance (Singh, 2005).

In India, there is no domestic quarantine. However, seed certification is desired for production of disease free seed, especially in seed multiplication programme. The National Seed Project lays down a maximum of 0.05 and 0.25% level of Karnal bunt infection for foundation and certified seeds respectively (Agarwal and Verma, 1983). Washing test is an efficient method for quarantifying the externally seed-borne spores in seed lots. If teliospores contamination is above 25 spores/gram, seed lot is rated as contaminated (Agarwal *et al.,* 1973).

Seed contamination with Karnal bunt spores is possible at the time of threshing of wheat crop. Smoke caused by burning of wheat stubble/straw in the field, combine and threshing machines, transportation by trucks and railway wagons are reported to promote spread of teliospores over long distance (Bonde *et al.,* 1987). Therefore, as part of the Karnal bunt risk management one must desist from setting fire to standing crop stubble and threshing machine/grain wagon should be thoroughly cleaned (Malick and Mathre, 1998).

The FAO and other global organisations are of the view that quarantine should be based on risk analysis using appropriate testing procedures (Singh, 2005). Nagarajan (1991) has developed a computer based pest risk analysis model which can be put for analysing the possibility of Karnal bunt occurrence in any geographic situation. The GEOKB is suitable for assigning probability value of disease occurrence at locations in different latitudes and agronomic conditions of wheat growing there and the second part of the software KBRISK takes care of climatic data to evaluate probability of Karnal bunt occurrence or establishment. Scientists and regulators are prospective clients for these PRA models.

REFERENCES

Aggarwal, R., D.V. Singh and K.D. Srivastava. 1994. Host pathogen interaction in Karnal bunt of Wheat. *Indian Phytopath.* **47**: 381-385.

Aggarwal, R., D.V. Singh and K.D. Srivastava. 1996. The potential of antagonists for biocontrol of *Neovossia indica* causing Karnal bunt of wheat. *Indian J. Biol. Control* **9**: 69-70.

Aggarwal, R., D.V. Singh and K.D. Srivastava. 1999. Studies on ontogeny of teliospore ornamentation of *Neovossia indica* observed through scanning electron microscopy. *Indian Phytopath.* **52**: 417-419.

Aggarwal, R., K.D. Srivastava and D.V. Singh. 1996. Mechanism of antagonism between plant pathogens and antagonists. Annual Meeting of IPS, PAU, Ludhiana, 14-16th Feb., 1996.

Aggarwal, R., K.D., Srivastava and D.V. Singh. 1998. Detection of bunt and smut pathogens of wheat through scanning electron microscopy. *Indian Phytopath.* **51**: 190-193.

Aggarwal, R., L.M. Joshi and D.V. Singh. 1989. Inoculum production in relation to varieties in Karnal bunt disease of wheat. *Indian Phytopath.* **42**: 569-571.

Agarwal, V.K. and H.S. Verma. 1983. A simple technique for the detection of Karnal bunt infection in wheat seed samples. *Seed Research* **11**: 100-102.

Agarwal, V.K., O.V. Singh and A. Singh. 1973. A note on certification standard for Karnal bunt disease of wheat. *Seed Research* **1**: 96-97.

Amer, G.A.M. 1995. *Biocontrol of Karnal bunt of wheat.* Ph.D. Thesis, Division of Mycology and Plant Pathology, IARI, New Delhi.

Anonymous. 1991. Quarantine procedure No. 37 *Tilletia indica. EPPO Bulletin* **21**: 265-266.

Aujla, S.S., I. Sharma and K.S. Gill. 1990. Effect of soil moisture and temperature on teliospore germination of *Neovossia indica. Indian Phytopath.* **43**: 223-225.

Aujla, S.S., I. Sharma and B.B. Singh. 1986. Method of teliospore germination and breaking of dormancy in *Neovossia indica. Indian Phytopath.***39**: 574-577.

Aujla, S.S., I. Sharma and B.B. Singh. 1987. Physiologic specialization of Karnal bunt of wheat. *Indian Phytopath.* **40**: 333-356.

Aujla, S.S., I. Sharma, P. Singh, G. Singh, H.S. Dhaliwal and K.S. Gill. 1989. Propiconazole-a promising fungicide against Karnal bunt of wheat. *Pesticides* **11**: 35-38.

Aujla, S.S., A.S. Grewal, K.S. Gill and I. Sharma. 1980. A screening technique for Karnal bunt disease of wheat. *Crop Improvement* **7**: 145-146.

Babadoost, M. 2000. Comments on the zero-tolerance quarantine of Karnal bunt of wheat. *Plant Dis.* **84**: 711-712.

Bag, T.K., D.V. Singh and S.M.S. Tomar. 1999. Inheritance of Karnal bunt (*Neovossia indica*) resistance in some Indian bread wheat (*T. aestivum*) lines and cultivars. *J. Genet. & Pl. Breed.* **53**: 67-72.

Bains, S.S. and H.S. Dhaliwal. 1989. Release of secondary sporidia of *Neovossia indica* from inoculated wheat spikes. *Plant and Soil* **115**: 83-87.

Bansal, R., D.V. Singh and L.M. Joshi. 1983. Germination of teliospores of Karnal bunt of wheat. *Seed Research* **11**: 258-261.

Bansal, R., D.V. Singh and L.M. Joshi. 1984. Comparative morphological studies in teliospores of *Neovossia indica*. *Indian Phytopath.* **37**: 355-357.

Bansal, R., D.V. Singh and L.M. Joshi. 1984a. Effect of Karnal bunt pathogen [*Neovossia indica* (Mitra) Mundkur] on weight and viability of wheat seed. *Indian J. Agric. Sci.* **54**: 663-666.

Bansal, R., D.V. Singh and L.M. Joshi. 1984b. Effect of liquid nitrogen treatment on germination of teliospores of Karnal bunt. *Indian Phytopath.* **37**: 368-369.

Bedi, K.S., M.R. Sikka and B.B. Mundkur. 1949. Transmission of wheat bunt due to *Neovossia indica* (Mitra) Mundkur. *Indian Phytopath.* **2**: 20-26.

Bedi, P.S., M. Meeta and J. S. Dhiman. 1981. Effect of Karnal bunt of wheat on weight and quality of the grains. *Indian Phytopath.* **34**: 330-333.

Bhat, R.V., Y.G. Deosthale, D.N. Roy, M. Vijayaraghavan and P.G. Tulpule. 1980. Nutritional and Toxicological evaluation of Karnal bunt affected wheat. *Indian J. Exptl. Biol.* **18**: 1333-1335.

Bhat, R.V., D.N. Rao, M. Vijayaraghavan and P.G. Tulpule. 1981. Toxicological evaluation of Karnal bunt in monkeys: A report. Bull. Food Drug Toxicological Research Centre, Hyderabad, India.

Bijral, J.S. and T.R. Sharma. 1995. *Triticum aestivum-Triticum araraticum* hybrids and their cytology. *Wheat Information Service* **81**: 20-21.

Bonde, M.R., G.L. Peterson and M.W. Schaad. 1997. Karnal bunt of wheat. *Pl. Disease* **81**: 1370-1377.

Bonde, M.R., J.M. Prescott, T.T. Matsumoto and G.L. Peterson. 1987. Possible dissemination of teliospores of *Tilletia indica* by the practice of burning wheat stubble. *Phytopathology* **77**: 639.

Brennan, J.P., E.J. Warham, J. Harnan, D. Byerlee and F. Coronel. 1990. Economic losses from Karnal bunt of wheat in Mexico. CIMMYT Economics Working Paper 90/2.

Cashion, N.L. 1984. Effect of fourteen fungicides on germination of secondary sporidia of the Karnal bunt pathogen *Neovossia indica. Phytopathology* **74**: 827.

Cashion, N.L. and E.S. Luttrell. 1988. Host parasitic relationship in Karnal bunt of wheat. *Phytopathology* **78**: 75-84.

Chona, B.L., R.L. Munjal and K.L. Adlakha. 1961. A method of screening wheat plants for resistance to *Neovossia indica. Indian Phytopath.* **14**: 99-101.

Commonwealth Mycological Institute. 1974. Distribution maps of plant diseases. No. 173, ed. 3.

Crous, P.W., Van Jaarsveld, A.B. Castlebury, L.A. Carries, and Z.B. Pretorius. 2000. Karnal Bunt found on wheat in South Africa. http:www.sassp.co.2a/x6/ newdiseases report/janurary/feature/karnalbunt. htm.

Davila, G.F. 1996. Karnal *Bunt.* ***In:*** *Bunt and Smut Diseases of Wheat: Concepts and Methods of Disease Management* (Eds. R.D. Wilcoxson and E.E. Saari) Mexico, D.F., CIMMYT, Mexico, 66 p.

Dhaliwal, H.S. and D.V. Singh. 1988. Up-to-date life cycle of *Neovossia indica* (Mitra) Mundkur. *Curr. Sci.* **57**: 675-677.

Dhaliwal, H.S. and D.V. Singh. 1989. Production and inter-relationship of two types of secondary sporidia of *Neovossia indica. Curr. Sci.* **58**: 614-618.

Dhiman, J.S. and P.S. Bedi. 1984. Studies on teliospore germination of *Neovossia indica*- the incitant of Karnal bunt of wheat. *Indian Phytopath.* **37**: 388.

Dhiman, J.S. and P.S. Bedi. 1989. Studies on the mode of perpetuation and pathogenic variability in *Neovossia indica* causing Karnal bunt of wheat. *Indian Phytopath.* **42**: 323.

Duran, R. 1972. Further aspects of teliospore germination in North American smut fungi. II. *Can. J. Bot.* **50**: 2569-2573.

Duran, R. and R. Cromarty. 1977. *Tilletia indica:* a hetrothallic wheat fungus with multiple alleles controlling incompatibility. *Phytopathology* **67**: 812-815.

Duran, R. and G.W. Fischer. 1961. *The genus Tilletia.* Washington State University Press, 138 p.

Fischer, G.W. 1953. *Manual of North American Smut Fungi.* Ronald Press. Co., New York.

Fuentes-Davila. 1984. *Biology of Tilletia indica* Mitra. M.S. Thesis, Deptt. of Plant Pathology, Washington State University, 66 p.

Gardner, J.S., J.W. Allen, W.M. Hess and R.K. Tripathi. 1983. Sheath structure of *Tilletia indica* teliospores. *Mycologia* **75**: 333-336.

Gill, K.S., I. Sharma and S.S. Aujla. 1993. *Karnal bunt of wheat.* Punjab Agricultural University, Ludhiana.

Goates, B.J. 1988. Histology of infection of wheat by *Tilletia indica*, the Karnal bunt pathogen. *Phytopathology* **78**: 1434-1441.

Gopal, S. and K.S. Sekhon. 1988. Effect of Karnal bunt disease on the milling, rheological and nutritional properties of wheat: Effect on the quality and rheological properties of wheat. *J. Food. Sci.* **53**: 1558-1559.

Gupta, R.P. and A. Singh. 1983. Effect of certain plant extracts and chemicals on teliospore germination of *Neovossia indica. Indian J. Mycol. Plant Pathol.* **13**: 116-117.

Hassan, S.F. 1973. Wheat diseases and their relevance to improvement and production in Pakistan. Proc. Fourth FAO/Rockefeller Foundation Wheat Seminar, pp. 84-86.

Holton, C.S. 1949. Observations on *Neovossia indica. Indian Phytopath.* **2**: 1-5.

Howard, A. and G.L.C. Howard. 1909. *Wheat in India. I. Production, Varieties and Improvement*. Thacker Sprink and Co., Calcutta, India.

Joshi, L.M., D.V. Singh and K.D. Srivastava. 1981. Meteorological conditions in relation to incidence of Karnal bunt of wheat in India. *Proc. 3rd Intern. Symp. Plant Pathol.*, IARI, New Delhi, 11 p.

Joshi, L.M., D.V. Singh, K.D. Srivastava and R.D. Wilcoxson. 1983. Karnal bunt: A minor disease that is now a threat to wheat. *Bot. Rev.* **49**: 309-330.

Kalha, C.S., S.N. Anand, R.S. Gupta and B.R. Gupta. 1987. Incidence of Karnal bunt of wheat in relation to different dates of sowing. *Research and Development Report* **4**: 100-101.

Khanna, A., M.M. Payak and S.C. Mehta. 1996. Teliospore morphology of some smut fungi. I. Electron microscopy. *Mycologia* **58**: 562-569.

Krishna, A. and R.A. Singh 1981. Aberrations in the teliospore germination of *Neovossia indica. Indian Phytopath.* **34**: 260-262.

Krishna, A. and R.A. Singh. 1982. Investigations on the disease cycle of Karnal bunt of wheat. *Indian J. Mycol. Plant Pathol.* **12**: 124.

Krishna, A. and R.A. Singh. 1982. Effect of physical factors and chemicals on the teliospores germination of *Neovossia indica. Indian Phytopath.* **35**:440-444.

Krishna, A. and R.A. Singh. 1983. Effect of some organic compounds on teliospore germination and screening of fungicides against *Neovossia indica. Indian Phytopath.* **6**: 233-236.

Krishna, A. and R.A. Singh. 1983. Multiple alleles controlling the incompatibility in *Neovossia indica. Indian Phytopath.* **36**:746-748.

Kumar, J. and S. Nagarajan. 1998. Role of flag leaf and spike emergence stage of wheat on severity of Karnal bunt. *Plant Dis.* **82**: 1368-1370.

Loke, C.M. and A.J. Watson. 1975. Foreign plant diseases intercepted in quarantine inspection. *Pl. Dis. Reptr.* **39**: 518.

Malick, V.S. and Mathre, D.E. 1998. Bunts and Smut of wheat: An international Symposium, NAPP, Ottawa, Canada.

Martin, P.M. 1986. Quarantine of Karnal bunt in Canada. *Proc. 5th Biennial Smut Workers Workshop*, CIMMYT, Mexico.

Mathur, R.S. 1968. The fungi and plant diseases. Published by the author. 34 pp.

Mathur, S.C. and S. Ram. 1963. Longivity of chlamydospores of *Neovossia indica (Mitra)* Mundkur. *Sci.& Cult.* **29**: 411-412.

Mc Rae, W. 1932. Report of the Imperial Mycologist. *Sci. Rep. Imper. Inst. Agric. Res.*, Pusa 1930-31, 73-86 pp.

McRae, W. 1933. Report of the Imperial Mycologist. *Sci. Rep. Imper. Inst. Agric. Res., Pusa* 1932-33. 134-160 pp.

Mehdi, V., L.M. Joshi and Y.P. Abrol. 1973. Studies on Chapati quality. VI. Effect of wheat grains with bunt on the quality of Chapaties. *Bull. Grain Technol.* **11**: 195-197.

Mitra, M. 1931. A new bunt of wheat in India. *Ann. Appl. Biol.* **18**: 178-179.

Mitra, M. 1935. Stinking smut (bunt) of wheat with special reference to *Tilletia indica* Mitra. *Indian J. Agric. Sci.* **5**: 51-74.

Mitra, M. 1937. Studies on the stinking smut or bunt of wheat in India. *Indian J. Agric. Sci.* **7**: 459-476.

Multani, D.S., H.S. Dhaliwal, P. Singh and K.S. Gill. 1988. Synthetic amphidiploids of wheat as a source of resistance to Karnal bunt (*Neovossia indica*). *Plant Breeding* **101**: 122-125.

Mundkur, B.B. 1940. A second contribution towards the knowledge of Indian Ustilaginales. *Trans. Brit. Mycol. Soc.* **24**: 312-336.

Mundkur, B.B. 1943. Karnal bunt, an air borne disease. *Curr. Sci.* **12**: 230-231.

Munjal, R.L. 1966. Bunt disease of wheat. *Sci. Reptr.* **3**: 33-36.

Munjal, R.L. 1970. *Studies on Karnal bunt of wheat*. Ph.D. thesis, Punjab University, Chandigarh.

Munjal, R.L. 1976. Studies on Karnal bunt (*Neovossia indica*) of wheat in northern India during 1968-69 and 1969-70. *Indian J. Mycol. Plant Pathol.* **5**: 185-187.

Nagarajan, S. 1991. Epidemiology of Karnal bunt of wheat incited by *Neovossia indica* and an attempt to develop a disease prediction system. Technical Report, Wheat-Programme, CIMMYT, Mexico., 1-62 pp.

Nagarajan, S. 2001. Pest risk analysis for shipping wheat from Karnal bunt (*Tilletia indica*) infected areas to disease free destination, DWR, Karnal, 114 p.

Nagarajan, S., S.S. Aujla, G.S. Nanda, I. Sharma, L.B. Goel, J. Kumar and D.V. Singh. 1997. Karnal bunt (*Tilletia indica*) of wheat – a review. *Rev. Plant Pathol.* **76**: 1207-1214.

Nanda, G.S., K. Chand, V.S. Sahu and I.Sharma. 1995. Genetic analysis of Karnal bunt resistance in wheat. *Crop Improvement* **22**: 189-193.

Nath, R., A.K. Lambat, P.M. Mukewar and I. Rani. 1981. Interceptions of pathogenic fungi in imported seed and planting material. *Indian Phytopath.* 34: 282-286.

Padwick, G.W. 1939. Report of the Imperial Mycologist. *Sci. Report. Agric. Inst.*, New Delhi 1937-38, 105-112 pp.

Prescott, J.M. 1984. Overview of CIMMYT's Karnal bunt programme, pp. 2-4. Proceedings of the Conference on Karnal bunt. April 16-18, 1984, Obregon, Sonora, Mexico.

Prescott, J.M. 1986. Sporidia trapping studies of Karnal bunt. Proc. 5th Biennial Smut Workers' Workshop held at Obregon, Sonora, Mexico, April 28-30, 1986, p. 22.

Peterson, G.L., M.R. Bonde, W.M. Dowler and M.H. Royer. 1984. Morphological comparisons of *Tilletia indica* from India and Mexico. *Phytopathology* **74**: 757.

Rai, R.C. and A. Singh. 1978. A note on the viability of wheat seed infected with Karnal bunt. *Seed Res.* **6**: 188-190.

Rai, R.C. and A. Singh. 1979. Effect of Chemical seed treatment on seed-borne teliospores of Karnal bunt of wheat. *Seed Res.* **7**: 186-189.

Rai, R.C., A. Singh and R.B. Rai. 1991. Haemotological and biochemical studies on albino rats fed on Karnal bunt infected wheat grains. *Narendra Dev. J. Agri. Res.* **6**: 1-6.

Rattan, G.S. and S.S. Aujla. 1990. Survival of Karnal bunt (*Neovossia indica*) teliospores in different types of soil at different depths. *Indian J. Agric Sci.* **60**: 616-618.

Roberson, R.W. and E.S. Luttrell. 1987. Ultrastructure of teliospore ontogeny in *Tilletia indica. Mycologia* **79**: 753-763.

Sawhney, R.N. and J.B. Sharma. 1996. Introgression of diverse genes for resistance to rusts into an improved variety, Kalyansona. *Genetics* **97**: 255-261.

Sekhon, K.S., N. Singh and P.P. Singh. 1992. Studies on the improvement of quality of wheat infected with Karnal bunt. I. Milling, rheological and baking properties. *Cereal Chemistry* (USA) **69**: 50-54.

Sidhartha, V.S. 1992. *Epidemiological studies on Karnal bunt of wheat*. Ph.D. Thesis, IARI, New Delhi, 105 p.

Sidhartha, V.S., D.V. Singh, K.D. Srivastava and R. Aggarwal. 1995. Some epidemiological aspects of Karnal bunt of wheat. *Indian Phytopath.* **48**: 419-426.

Sharma , A.K., K.S. Babu, R.K. Sharma and K. Kumar. 2007. Effect of tillage practices on *Tilletia indica* (Karnal bunt disease of wheat) in a rice wheat rotation of the Indo- Gangetic Plains. *Crop Protection* **26**: 818-821.

Sharma, I., N.S. Bains and R.C. Sharma. 2007. Karnal bunt resistance in wheat-an overview. *Journal of Wheat Research* **1**: 60-80.

Sharma, I., A. Sirari, B. Raj, N.S. Bains and R.C. Sharma. 2008. Heterothallism in *Tilletia indica:* Implications for physiological specialization. *Indian Phytopath.* **61**: 34-42.

Sharma, I., N.S. Bains, K. Singh and G.S. Nanda. 2005. Additive genes in nine loci governing Karnal bunt resistance in a set of common wheat cultivars. *Euphytica* **142**: 301-307.

Sharma, K.K., D.V. Singh, K.D. Srivastava, R. Aggarwal and S.K. Handa. 1994. Bio-efficacy and persistence of propiconazole against Karnal bunt of wheat. *Indian J. Plant Prot.* **22**: 93-95.

Shyama, G., K.S. Sekhon, K.L. Bajaj, R.P. Saigal and H. Singh. 1988. Effect of Karnal bunt disease on the milling rheological and nutritional properties of wheat: Studies on some chemical constituents and biological effects. *Food Sci.* **53**: 1560-1562.

Singh, A. 1994. *Epidemiology and management of Karnal bunt disease of wheat*. Res. Bull., 127, G.B. P.U.A. &T., Pantnagar, India, 165 p.

Singh, A and R.C. Rai. 1986. Chemical control of wheat diseases in India. **In**: *Problems and Progress of Wheat Pathology in South Asia*. (Eds. L.M. Joshi, D.V. Singh and K.D. Srivastava), pp. 374-394. Malhotra Publishing House, New Delhi, 401 p.

Singh, D.V., K.D. Srivastava, R. Aggarwal, S. Nagarajan and B.R. Verma. 1992. Mulching as a means of Karnal bunt management. *Plant Dis. Res.* **6**: 115-116.

Singh, D.V. 2005. Karnal bunt of wheat: A global perspective. *Indian Phytopath*. **58**: 1-9.

Singh, D.V. and K.D. Srivastava. 1992. Investigations on Karnal bunt (*Neovossia indica*) of wheat. Final Technical Report of PL-480 Project, IARI, New Delhi.

Singh, D.V. and K.D. Srivastava. 1997. Influence of cropping system on recurrence of Karnal bunt (*Neovossia indica*) of wheat. Final Report. ICAR Cess Fund Project, IARI, New Delhi, 29 p.

Singh, D.V. and K.D. Srivastava. 2000. Implications of Karnal bunt on grain trade. **In**: *Role of Resistance in Intensive Agriculture* (Eds. S. Nagarajan and D.P. Singh), pp. 78-85. Kalyani, Publishers, New Delhi.

Singh, D.V., K.D. Srivastava and R. Aggarwal. 1991. Results of coordinated experiments. AICWIP., 76-80 pp.

Singh, D.V., Aggarwal, R. and S. Nagarajan. 1991. Variability in *Neovossia indica* in the context of host resistance and disease management. Golden Jubilee

Symposium on Genetics Research and Education, Indian Society of Genetics and Plant Breeding, IARI, New Delhi (Feb. 12-15, 1991) 134-135 pp.

Singh, D.V., R. Aggarwal, J.K. Shreshtha, B.R. Thapa and H.J. Dubin. 1989. First report of *Tilletia indica* in Nepal. *Plant Dis.* **73**: 273.

Singh, D.V., K.D. Srivastava, L.M. Joshi and B.R. Verma. 1985. Evaluation of some fungicides for control of Karnal bunt of wheat. *Indian Phytopath.* **38**: 571-573.

Singh, D.V., P. Arora, K.D. Srivastava, S. Nagarajan and R. Aggarwal. 1990. Prediction of the frequency and severity of Karnal bunt occurrence in field using climatic variables. *Indian J. Plant Prot.* **18**: 225-229.

Singh, D.P., V.C. Sinha, L.B. Goel, Indu Sharma, S.S. Aujla, S.S. Karwasra, M.S. Beniwal, A. Singh, K.P. Singh, A.N. Tiwari, P.S. Bagga, B.K. Sharma, D.V. Singh, K.D. Srivastava, R. Aggarwal and B.R. Verma. 2003. Confirmed sources of resistance to Karnal bunt in wheat in India. *Pl. Dis. Res.* **18**: 37-38.

Singh, G., S. Rajaram, J. Montoya and G. Davila- Fuentes. 1995. Genetic analysis of Karnal bunt resistance in 14 Mexican bread- wheat genotypes. *Plant Breeding* **114**: 439-441.

Singh, P.J., H.S. Dhaliwal and K.S. Gill. 1989. Chemical control of Karnal bunt (*Neovossia indica*) of wheat (*Triticum aestivum*) by single spray of fungicides at heading. *Indian J. Agric. Sci.* **59**: 131-133.

Singh, N., K.S. Sekhon, A.K. Srivastava and P.P. Gupta. 1992. Studies on the improvement of quality of wheat infected with Karnal bunt II. Nutritional and biological effect. *Cereal Chemistry* (USA) **69**: 55-60.

Sirari, A., I. Sharma, N.S. Bains, B. Raj, S. Singh and R.L. Bowden. 2008. Genetics of Karnal bunt resistance in wheat: role of genetically homogenous *Tilletia indica* inoculum. *Indian J. Genet.* **68**: 10-14.

Smilanick, J.L., J.A. Hoffman and M.H. Royer. 1985. Effect of temperature, pH, light and desiccation on teliospore gemination of *Tilletia indica*. *Phytopathology* **75**: 1428-1431.

Tomar, S.M.S., K.D. Srivastava and D.V. Singh. 1991. Screening of wheat species and few amphidiploids for Karnal bunt resistance. *Annual Wheat Newsletter.* **37**: 64-65.

Torarbi, M., V. Mardoukhi and N. Jaliani. 1996. First report on the occurrence of partial bunt on wheat in the southern parts of Iran. *Seed and Plant* **12**: 8-9.

Warham, E.J. and P.A. Burnett. 1990. Influence of media on pathogenicity and morphology of secondary sporidia of *Tilletia indica. Plant Dis.* **74**: 525-527.

Williams, P.C. 1983. Incidence of stinking smut (*Tilletia spp.*) on commercial wheat samples in Northern Syria. *Rachis* **2**: 21.

Ykema, R.E., J.P. Floyd, M.E. Palm and G.L. Peterson. 1996. First report of Karnal bunt of wheat in the United States. *Plant Dis.* **80**: 1207.

Zadoks, J.C., T.T. Chang and C.F. Konzak. 1974. A decimal code for the growth stage of cereals. *Eucarpia Bulletin* **7**: 1-10.

Zhang, Z., L. Lange and S.B. Mathur. 1984. Teliospore survival and plant quarantine significance of *Tilletia indica* (causal agent of Karnal bunt of wheat) particularly in relation to China. *EPPO. Bull.* **14**: 119-128.

2.3.3 Hill Bunt

Hill bunt is one of the earliest known diseases of wheat. The disease is of historical importance as Tillet in 1755 demonstrated this disease to be contagious in nature. The pathogen is also the first fungus whose life cycle was discovered by Prevost in 1807. In recognition of Tillet's classic contribution, the genus of the bunt fungi was named *Tilletia.* This disease has several common names like European bunt, common bunt, stinking smut or complete bunt (Saari and Omar, 1996). In India, the disease is mentioned as hill bunt because it is principally restricted in the northern hilly region.

The disease is worldwide in distribution and regularly appears in the cooler hilly regions of Himachal Pradesh, Jammu and Kashmir, Uttaranchal and Punjab. It is essentially a disease of the hills but there are few records of the sporadic occurrence in the plains of India (Mundkur, 1944). However, source of infection of the disease to the plains is presumed to be contaminated seed carried from hills. Otherwise, the fungus cannot survive in the plains due to high temperature (Joshi *et al.,* 1986).

The yield losses caused by hill bunt are due to replacement of seed contents into fungal spores, the seed coat being left more or less intact. Mitra (1935) has reported 30–40% losses from Jubbal region of Himachal Pradesh. Bedi and Bhardwaj (1953) recorded 10–96% loss in yield in different areas in Kullu valley during 1948–49. Holton (1967) who estimated 25–50% and even 100% yield reduction in some fields, has considered hill bunt as a potentially dangerous disease under Indian conditions. However, hill surveys carried out since 1968 have not revealed such heavy losses in Himachal Pradesh except in few fields where farmers had sown same seed year after year without any fungicidal treatment (Singh, 1986). In Shimla hills 2–3% incidence was recorded by Goel *et al.* (1977).

In addition to reduction in yield, hill bunt can also deteriorate the quality of wheat flour. According to Mehdi *et al.* (1973) 3–5% infection with hill bunt results in the blackening of flour. It also imparts fishy odour to flour and chapaties are not acceptable in terms of palatibilty.

Symptoms: The symptoms of the disease are not usually evident until the heading stage although it is a systemic disease resulting from seedling infection. Diseased plants are moderately stunted but not readily distinguished. Infected plants ripen earlier than healthy ones. Bunted ears are narrower than disease free ears. At maturity ears remain erect, dark or olive green with expanded glumes with the tips of the bunt balls protruding. Awns of the beared varieties may be deformed or completely lost. Bunt balls are dull grey-brown, similar to normal seed in shape but usually more spherical (Plate 4C). Their fragile pericarps remain intact and the interior of the grain is filled with the greasy spore mass. The pericarp of bunted grains ruptures at harvest releasing spore mass, which emits foul, fishy odour due to the presence of volatile compound trimethylamine. It is because of this stinking smell that the disease is also referred to as stinking smut.

Hosts: In addition to *Triticum* species *T. caries* and *T. foetida* also occurs on numerous grasses including species of *Aegilops, Agropyron, Bromus, Dactylis, Elymus, Festuca, Hordeum, Poa, Lolium, Secale, x Tritcosecale* etc. (Duran and Fisher, 1961).

Causal Organism: Two species of the pathogen are involved with hill bunt disease of wheat. In recent past, accepted names for the hill bunt fungus were *Tilletia caries* (DC.) Tul. and *T. foetida* (Wall.) Liro. but under the international code of Botanical Nomenclature, the names have been reverted to *T. tritici* (Bjerk.) Wint. and *T. laevis* Kuhn, respectively. These two fungi are essentially identical except the differences in spore wall characteristics, which are controlled by very few genes (Fisher and Holton, 1957). Teliospores of *T. caries* are highly reticulate while they are smooth walled in case of *T. foetida*. The spores of both species are sometimes found in the same sorus.

The hyaline dikaryotic mycelium of the pathogen grows inter and intra-cellularly with in the terminal meristem of the host and then colonizes developing ovary. The tissues of seeds are converted into mass of teliospores. These spores are borne singly and are found at the ends of fertile swollen hyphae. Teliospores are light pale to dark olivaceous brown, globose, 14-25 µm in diameter. The exospore of *T. caries* spores usually has polygonal reticulations 0.5-1.5 µm deep. The exospore is smooth in *T. foetida*. Apiculus is absent. Among teliospores are mixed hyaline, smooth surface, thin walled, slightly smaller, spore like sterile cells that are formed during teliosporogenesis.

A teliospore germinates to form a short, non-septate promycelium or basidium, which bears apical whorl of 8–16 primary sporidia. Teliospores germinate most rapidly at 18–20°C, but most uniformly at 14-16°C after 4–5 days (Goates, 1996). The uninucleate sporidia are hyaline, filiform, slightly curved and less than 12 µm long. There are usually two mating types with in a whorl of primary sporidia designated by a (+) or a (-). The uninucleate sporidia fuse near the middle in compatible pairs by means of lateral conjugation hyphae and form H-shaped body and establish a dikaryon. The dikaryons germinate to produce either dikaryotic infection hyphae, which can directly infect the host or sickled shaped hyaline, dikaryotic secondary sporidia. The sporidia are forcibly discharged and on germination they give rise to infections hypha that penetrates host tissue intercellularly after coming in contact of the host cuticle (Fig. 42).

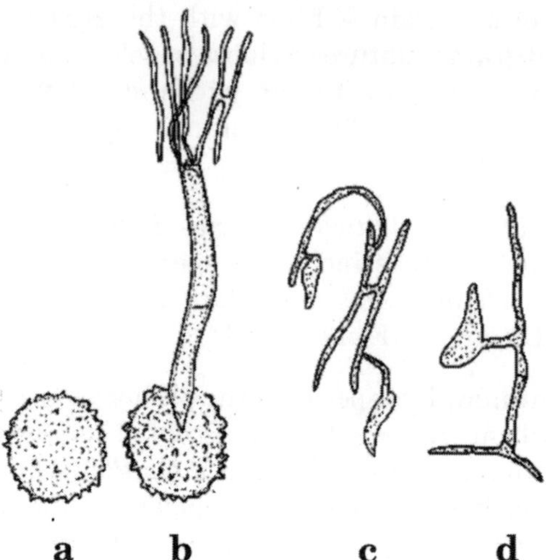

Figure 42: *Tilletia caries* (a) Teliospore (b) Germinated teliospore (c) Secondary sporidia, and (d) Germinating sporidia

Virulence of the pathogens is extremely variable. Genetic recombination occurs in each life cycle so the teliospores within a single bunt ball may differ greatly. Consequently, many pathogenic races have been recognized in both the species of the fungus. Hoffman and Matzger (1976) distinguished races of the pathogen on the basis of their virulence/avirulence responses by testing bunt isolates against 15 differential wheat cultivars (Table 11). These differential lines are monogenic for individual bunt resistant genes and each one is designated by "Bt". Races those of *T. caries* have been labeled as "T" and *T. foetida* as "L" (Table 12). In India, studies conducted have revealed the presence of races both in *T. caries* and *T. foetida* (Joshi, 1971). Sood and Singh (1984) identified four races *i.e.* L-1, L-3, L-A and L-B of *T. foetida* and T-1 and T-3 of *T. caries* from Himachal Pradesh and Almora.

The simultaneous occurrence of the bunt with loose smut, ear-cockle and ergot has been observed (Hanna, 1932; Mathur and Mishra, 1961).

Disease Cycle: Hill bunt is monocyclic and has one infection cycle per season. The perpetuation of the disease is by teliospores of the pathogens. At harvest, mature bunt balls are broken and release teliospores that contaminate soil and healthy seed. However, the infested soil is not the major source of inoculum in India. Teliospores are sticky in nature and easily adhere on the surface of dry seed during threshing. They can survive for many years and can spread from one region to another on the seed.

The teliospores germinate in response to moisture along with seed during crop season. At the time of germination diploid nucleus of the spore

Table 11. Differential cultivars used to determine pathogenic races of hill bunt.

Cultivars	Resistant gene	CI or PI number
Heines VII	*Bt-0*	PI 209794
Sel 2092	*Bt-1*	PI 554101
Sel 1102	*Bt-2*	PI 554097
Ridit	*Bt-3*	CI 6703
CI 1558	*Bt-4*	CI 1558
Hohenheimer	*Bt-5*	CI 11458
Rio	*Bt-6*	CI 10061
Sel 50077	*Bt-7*	PI 554100
PI 173438 x Elgin	*Bt-8*	PI 554120
Elgin x PI 178383	*Bt-9*	PI 554099
PI 178383 x Elgin	*Bt-10*	PI 554118
Elgin x PI 166910	*Bt-11*	PI 554119
PI 119333	*Bt-12*	PI 119333
Thule III	*Bt-13*	PI 181463
Doubbi	*Bt-14*	CI 13711
Carleton	*Bt-15*	CI 12064

under goes meiosis and then haploid nuclei pass into the promycelium and one nucleus migrates into each primary sporidium where it undergoes mitosis. The nuclei left in the promycelium pass into anucleate primary sporidia or degenerate (Goates and Hoffman, 1987). Thus primary sporidia are haploid. The dikaryophase is initiated when uninucleate primary sporidia of opposite mating types fuse to form H-bodies. At this stage H-shaped bodies lead to production of binucleate secondary sporidia. An infection hypha arises from the germinating sporidium which penetrates the coleoptile before the seedling emergence. Hyphae then grow intercellularly within the terminal meristem of the plant without materially affecting its normal health until the bunted ear-heads develop. The mycelium colonizes the developing ovary, displaces tissues within the pericarp and subsequently converts into mass of teliospores which are the source of inoculumn for next cycle (Fig. 43).

Infection occurs below the soil surface, shortly after the seed germination and prior to emergence. The incidence of the disease is enhanced by seedling depth of 7 cm as compared to 4 cm (Swinburne, 1963). Optimum infection depends upon the rate of seedling emergence. The optimum temperature for spore germination and penetration of seedlings ranges from 6–10°C (Purdy and Kendrick, 1963). The absence of the disease in Indian plains is due to the fact that temperatures ranging from 5–15°C are highly favourable to infection while lower and higher temperatures

Table 12. Virulence of hill bunt races against wheat bunt (*Bt*) resistance genes.

Race	*Bt 1*	*Bt 2*	*Bt 3*	*Bt 4*	*Bt 5*	*Bt 6*	*Bt 7*	*Bt 8*	*Bt 9*	*Bt 10*	*Bt 11*	*Bt 12*	*Bt 13*	*Bt 14*	*Bt 15*
T-1							x								
T-2	x						x								x
T-3		x					x								
T-4	x						x								
T-5	x	x					x								
T-6	x						x								
T-7	x						x								
T-8	x	x					x								
T-9					x		x								
T-10					x										
T-11		x	x												
T-12	x				x		x								
T-13	x	x	x										x		
T-14	x														
T-15	x	x			x		x								
T-16		x		x	x	x	x								
T-17				x		x	x								
T-18	x			x		x	x								
T-19	x	x	x				x								
T-20	x	x													
T-21	x	x		x		x	x								
T-22		x		x		x	x								
T-23	x	x		x		x	x		x						
T-24		x	x	x		x	x								
T-25	x						x			x					
T-26	x	x					x			x					
T-27	x	x		x		x	x			x					
T-28		x		x		x	x		x						
T-29	x	x					x		x	x					
T-30	x	x		x		x	x		x	x					
L-1							x								
L-2	x						x								x
L-3		x					x								
L-4	x						x								
L-5	x	x					x								
L-7	x	x					x							x	
L-8		x		x		x	x		x						
L-9	x	x	x												
L-10		x	x				x								
L-16	x	x		x		x	x								

x = indicates an infection level above 10%; T= race of *T. caries*; L= race of *T. foetida*

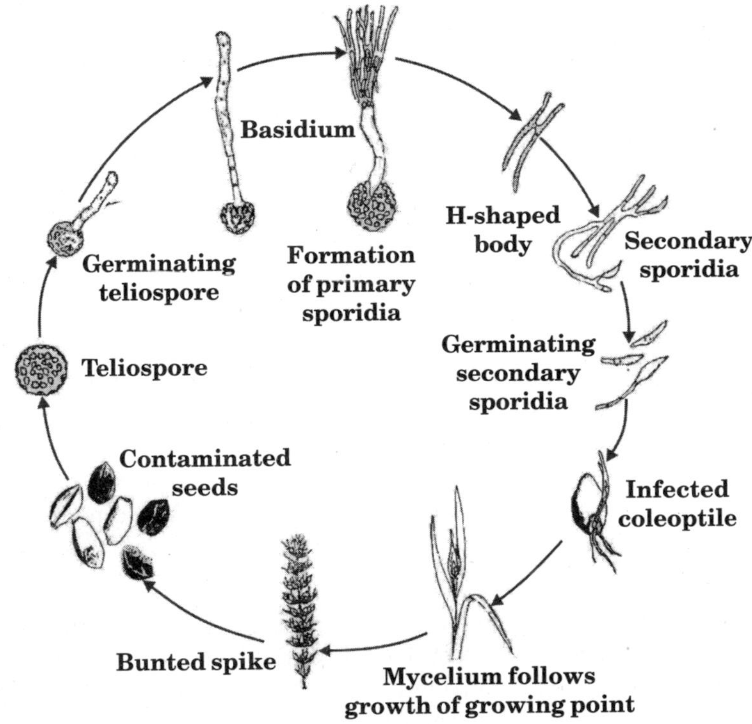

Figure 43: Disease cycle of hill bunt (*Tilletia caries / T. foetida*) of wheat.

are detrimental to infection (Goel *et al.,* 1977). If low temperature (approximately 10°C) prevails during early plant development, the resistance of wheat cultivars decreases (Zscheile, 1966). In general high sandy and humus rich soil favour hill bunt infection.

Management

Resistance: Genetics of bunt resistance in wheat has been studied extensively in USA. Beginning with the identification of four resistant genes M, H, T and R (now designated as *Bt 1*, *Bt 2*, *Bt 4* and *Bt 6*) based on the source cultivars Martin, Hussar, Turkey and Rio, respectively (Briggs, 1926; Metzger, 1970), 15 different heritable bunt resistant genes have been identified (Hoffman and Metzger, 1976) for their utility in breeding programme (Table 11). An additional gene, *Bt Z* that originated from a translocation into wheat from *Agropyron intermedium* has been useful for controlling bunt in Soviet Union (Varenitsa *et al.,* 1987).

There are very few reports of genetic analysis of bunt resistance in India. Kohli and Nambisen (1962) found one dominant gene for bunt resistance in cultivar Hussar. Upadhyay and Prakash (1979) concluded from their studies that most of the resistant Indian wheat including Kalyansona carry 'Brevor' factor for rust resistance. Sood (1983) observed that resistance in Federation 67 was controlled by a factor different from that in Kalyansona or Panjamo 62. Singh (1984) found that one dominant gene governed resistance in eight wheat cultivars. Singh and Chopra (1985) reported the association of bunt resistance gene *Bt 4* with cultivars Kalyansona, HP 1303, WL 1541, HB 501, HS 102 and IWP 502 and *Bt 8* gene with Kirac 66. Genes for bunt resistance have a weak or modifying effect on host resistance and their influence is slight so they have not been exploited in resistance breeding programmes.

Some wheat strains viz. HB 383, IWP 72, IWP 87, IWP 127, IWP 129, Kalyansona, WL 885, WL 1581 and UP 2002 possess high level of disease resistance (Bahadur *et al.,* 1984). Genotypes HB 501 and WL 1541 carry *Bt 4* gene for bunt resistance (Singh and Chopra, 1985). These strains may be a good source of resistance but significant headway in breeding programme will not be possible unless genotypes are evaluated against well defined pathogenic races.

Chemical Control: Seed treatment with fungicides for the control of hill bunt is highly effective. Copper fungicides and organo-mercurials as seed-dresser were used extensively considering the externally seed-borne nature of the pathogen (Prevost, 1807; Tisade and Cannon, 1929; Holton and Purdy, 1955). Kaul (1961) and Grewal *et al.* (1965) advocated the use of Ceresan and Agrosan GN @ 2 g/kg seed for bunt control in India. Since organo-mercurials are no longer desirable because they are hazardous to the environment, new systemic fungicides such as triadimol, carbendazim, carboxin, propiconazole etc. @ 2 g/kg seed very effectively control infection of the disease (Sharma *et al.,* 1971; Gaudet and Pachalski, 1992; Singh and Pandey, 2002). Rana and Banyal (2005) observed that wet seed treatment with Tilt 25 EC and dry seed treatment with Dividend 30WS were most effective in complete control of hill bunt. Besides hill bunt Tilt 25EC (propiconazole) is also highly effective against loose smut. (Aggrawal and Sood, 2006)

Cultural Control: Hill bunt incidence can be reduced by deep sowing (Bedi and Bhardawaj, 1953). Very early or late planting may be practiced to avoid the growth stages that are the most susceptible. But this cultural practice is not feasible and may reduce yields, and allow other diseases to be more vulnerable. Shallow sowing shortens the period of host susceptibility thus reducing disease incidence (Gaudet *et al.,* 1990).

REFERENCES

Aggrawal, P. and A.K. Sood. 2006. Management of hill bunt with fungicides effective against loose smut. *Seed Res.* **31**: 415-418.

Bahadur, P., S. Nagarajan, S.K. Nayar, L.B. Goel and S.K. Sharma. 1984. Evaluation of wheat and triticale against hill bunt of wheat. *Indian Phytopath.* **37**: 558.

Bedi, K.S. and C.R. Bhardwaj. 1953. Bunt or stinking smut of wheat and its control. *Punjab Farmer* **5**: 5-10.

Briggs, F.N. 1926. Inheritance of resistance to bunt, *Tilletia tritici* (Bjerk.) Winter in wheat. *J. Agric. Res.* **32**: 973-990.

Duran, R. and F.W. Fischer. 1961. *The genus Tilletia.* Wash. St. Univ. Press 138 p.

Fischer, G.W. and C.S. Holton. 1957. *Biology and Control of the Smut Fungi.* The Ronald Press Co., New York. 662 p.

Gaudet, D.A. and B.J. Puchalski. 1990. Influence of planting dates on the aggressiveness of common bunt races (*Tilletia tritici* and *T. laevis*) to Canadian spring wheat cultivars. *Can. J. Plant. Path.* **12**: 204-208.

Gaudet, D.A., B.J. Puchalski and T. Entz. 1992. Application methods influencing the effectiveness of carboxin for control of common bunt caused by *Tilletia trittci* and *T. laevis* in spring wheat. *Plant Dis.* **76**: 64-66.

Goates, B.J. 1996. Common bunt and dwarf bunt. **In**: *Bunt and Smut Diseases of Wheat: Concepts and Methods of Disease Management* (Eds. R.D. Wilcoxson and E.E. Saari), Mexico. D.F., CIMMYT., 12-25 pp.

Goates, B.J. and J.A. Hoffman. 1987. Nuclear behavior during teliospore germination and sporidial development in *Tilletia caries, T. foetida* and *T. controversa*. *Can. J. Bot.* **65**: 512-517.

Goel, L.B., D.V. Singh, K.D. Srivastava, L.M. Joshi and S. Nagarajan. 1977. *Smut and Bunts of Wheat in India.* Bull. Indian Agric. Res. Inst., New Delhi, 38 p.

Grewal, J.S., P.C. Joshi, K.D. Pathak and S.B. Mathur. 1965. Relative efficacy of some fungicides for control of hill bunt of wheat. *Indian Phytopath.* **18**: 94-96.

Hanna, W.F. 1932. Association of bunt of wheat with loose smut and ergot. *Phytopathology* **22**: 10-11.

Hoffman, J.A. and R.J. Metzger. 1976. Current status of virulence genes and pathogenic races of the wheat bunt fungi in the North western USA. *Phytopathology* **66**: 657-660.

Holton, C.S. 1967. Plant disease pitfalls for high yielding wheats in India. *Indian Phytopath.* **20**: 183-188.

Holton, C.S. and L.H. Purdy. 1955. Comparative effectiveness of hexachlorobenzene and mercury preparations in controlling soil borne common bunt in commercial field trials. *Pl. Dis. Reptr.* **39**: 842-843.

Joshi, L.M., D.V. Singh and K.D. Srivastava. 1986. Wheat and Wheat diseases in India. **In**: *Problems and Progress of wheat Pathology in South Asia* (Eds. L.M. Joshi, D.V. Singh and K.D. Srivastava), pp. 1-19. Malhotra Pub. House, New Delhi, 401 p.

Joshi, P.C. 1971. Race dynamics in the bunt species on wheat in India. *Proc. Sec. Int. Symp.,* IARI, New Delhi, 6 p.

Kohli, S.P. and P.N.N. Nambisan. 1962. Inheritance of resistance to bunt in hybrids of Hussar with Indian hill wheats. *Indian J. Genet.* **22**: 20-25.

Mathur, R.S. and M.P. Misra 1961. Simultaneous occurrence of *Tilletia foetida* (Liro and *Anguillulina tritici* (S.) G. Ben. in the same ear and grains of wheat in Pauri-Garhwal, Uttar Pradesh. *Curr. Sci.* **30**: 307.

Mehdi, V., L.M. Joshi and Y.P. Abrol. 1973. Studies on chapati quality VI. Effect of wheat grains with bunt on the quality of chapaties. *Bull. Grain Technol.* **11**: 195-197.

Metzger, R.J. 1970. Wheat genetics. *Wheat News Letter*. **17**: 122-125.

Mitra, M. 1935. Stinking smut (bunt) of wheat with special reference to *Tilletia indica* Mitra. *Indian J. Agric. Sci.* **5**: 1-24.

Mundkur, B.B. 1944. Some rare new smuts from India. *Indian J. Agric. Sci.* **14**: 49-52.

Prevost, I.B. 1807. Memoir on the immediate cause of bunt or smut of wheat and on preventives of bunt. *Phytopath. Classics* **6**: 1-96

Purdy, L.H. and Kendrick, E.L. 1963. Influence of environmental factors on the development of wheat bunt in the Pacific North West. IV. Effect of soil temperature and soil moisture on infection by soil-borne spores. *Phytopathology* **53**: 416-418.

Rana, S.K. and D.K. Banyal. 2005. Evolution of fungicides for the management of hill bunt of wheat. *Plant Dis. Res.* **20** : 38-39.

Saari, E.E. and F.M. Omar. 1996. Bunts and Smuts of wheat. **In**: *Bunt and Smut Diseases of wheat: Concepts and Methods of Disease Management* (Eds. R.D. Wilcoxson and E.E. Saari), Mexico D.F., CIMMYT., 1-11 pp.

Sharma, R.C., L.M. Joshi and K.D. Pathak 1971. Systemic fungicides for the control of hill bunt of wheat. *Indian Phytopath.* **24**: 604-605.

Singh, D.V. 1986. Bunts of wheat in India. **In**: *Progress and Problems of wheat Pathology in South Asia.* (Eds. L.M. Joshi, D.V. Singh and K.D. Srivastava), pp. 124-161. Malhotra. Publishing House, New Delhi, 401 p.

Singh, D.V. and V. Pandey. 2002. Fungal diseases of wheat and barley: Smuts and bunts. **In**: *Diseases of Field crops* (Eds. V.K. Gupta and Y.S. Paul), pp. 40-57. Indus Publ. Co., New Delhi, 464 p.

Singh, S.R. 1984. *Identification of genes for resistance to hill bunt in some varieties of Triticum aestivum* L., Ph. D. Thesis, IARI, New Delhi.

Singh, S.R. and V.L. Chopra. 1985. Identification and characteriation of genes for resistance to hill bunt in wheat-**In**: *Genetics and wheat improvement* (Eds. A.K. Gupta, S. Nagarajan and R.S. Rana), ICAR, New Delhi, 362 p.

Sood, A.K. 1983. *Studies on the genetics of resistance of wheat to bunt caused by Tilletia caries (D.C.) Tul. and Tilletia foetida (Wallr.) Liro,* Ph.D. Thesis HP Krishi Vishva Vidyalaya, 115 p.

Sood, A.K. and B.M. Singh. 1984. Occurrence and distribution of virulences of *Tilletia caries* and *T. foetida* causing bunt of wheat in Himachal Pradesh. *Indian Phytopath.* **37**: 388.

Swinburne, T.R. 1963. Infection of Wheat by *Tilletia caries,* the causal organism of bunt. *Trans. Brit. Mycol. Soc.* **46**: 145-156.

Tillet, M. 1755. Dissertation on the cause of the corruption and smutting of the kernels of wheat in the head, and on the means of preventing these untowards circumstances. Bordeux. 150 p.

Tisdale, W.H. and W.N. Cannon. 1929. Ethyl-mercury-chloride as a seed grain disinfectant. *Phytopathology* **19**: 80.

Upadhyay, M.K. and S. Prakash. 1979. Diverse genes for resistance to hill bunt of wheat. *Indian J. Genet.* **39**: 347-350.

Varenista, E.T., L.Y. Saakyan, A.F. Mozgovoi, G.V. Kochetygov and S.M. Gradskov. 1987. Using derivatives of the varieties Zarya as donors of resistance to bunt. *Lenina* **4**: 3-5.

Zscheile, F.P. 1966. Photoperiod-temperature-light relationship in development of wheat and bunt. *Phytopath. Z.* **57**: 329-374.

2.3.4 Black Point

Black point of wheat is identified as a disease with superficial necrotic lesions under the broad heading of seed discolouration. It was first described in USA by Bolley (1913) and the term black point was later used by Evans (1921). Seed discolouration of wheat grains is also known as smudge. The disease is termed as black point in USA and Kernel smudge in Canada. According to Hanson and Christensen (1953), discolouration should be used for a slightly bleached condition or loss of lustre of the grains while the term black point or kernel smudge connate distinct blackening of the grain due to some fungal infection.

The disease occurs in many parts of the world. It has been reported from Canada, USA, Argentina, Germany, Italy, Morocco, South Africa, Egypt, Algeria and Soviet Union (Gaur, 1986). The black point was reported for the first time in India by Dastur in 1932 from the Central Provinces (Dastur, 1932). Wheat crop in Punjab was severely damaged by this disease during 1964–65 (Parasher and Pracer, 1965). In 1967–68, black point had a wide spread occurrence due to prolonged wet weather that prevailed in Punjab, Haryana, Uttar Pradesh and Delhi (Dharmvir *et al.,* 1968). Rainfall at maturity of the crop caused 13 to 22% infection of black point in 1969 (Bhowmik, 1969). It is prevalent in almost all the states where wheat is grown (Joshi *et al.,* 1969; Singh *et al.,* 1989).

Generally, black point is not an economically important disease in India. Only when the disease percentage is high, it causes concern to trader and the consumer. Usually mycoflora associated with seed does not affect their germination and seedling emergence except pathogenic species of *Helminthosporium*. Chaudhary *et al.* (1984) recorded a reduction of 16% in germination of black point affected seeds. Infection of ear-heads with *Drechslera sorokiniana* at pre-anthesis stage or earlier is reported to reduce seed setting (Adlakha and Joshi, 1974). Aulakh *et al.* (1988) too noticed that *D. sorokiniana* caused seed rot and reduced seedling emergence and yield of the crop. Seed infected with *Alternaria* spp. are larger and heavier (Khetrapal and Agarwal, 1979) but the disease renders the affected grains unacceptable (King *et al.,* 1981).

Effect of the black point on quality of wheat grains ascertained by Gill and Tyagi (1972), showed negligible differences in protein, calcium, phosphorous and ash in diseased grains. However, Khetrapal and Agarwal (1979) reported increase in protein content. Chaudhary *et al.* (1984) recorded reduction in nitrogen, fatty acid, gluten, potassium, calcium, zinc and manganese.

Symptoms: As the name of the disease suggests, it appears as a black or dark brown tiny spot on the germinal end of the grain (Adlakha and

Joshi, 1974; Rana and Gupta, 1982). The spot or the point is normally confined around the embryo end but the necrotic lesion can extend up to middle half or rarely all over the grain surface (Plate 4D). According to Adlakha and Joshi (1974) grains infected with *Alternaria tenuis* have black discolouration of varied intensity and is restricted to embryonic region, whereas grains infected with *Helminthosporium sativum* have discolouration extended to other parts of the grain and show fine ridges on the surface which can be discernable with a hand lens. The grains due to infection of *Helminthosporium* are not only smaller but also lighter in weight and shrivelled.

Sometimes people confuse Karnal bunt with black point but these two diseases can be easily differentiated by their characteristic features (Table 13).

Causal Organism: Black point is mainly caused by species of *Alternaria* and *Helminthosporium*. Dastur (1932) recorded *H. sativum* as an incitant of the disease, where as Tandon (1946) isolated *Alternaria tenuis* (=*A. alternata*) from the black point of wheat (Fig. 44). Tandon was of the opinion that the *A. tenuis* acts as facultative parasite and causes blackening of the grains. Sometimes other fungi are also involved with the disease (Table 14).

Disease Development: Air borne inoculum of fungi is essential for the development of the disease. The spores of various fungi are present in

Table 13. Comparative account of black point and Karnal bunt diseases

	Black point	Karnal bunt
Pathogen	*Helminthosporium* spp. and *A. alternata*	*Nevossia indica* or *Tilletia indica*
Symptoms	The pericarp of grains become dark brown to black, usually restricted to germinal end.	The grains are only partially infected.
	No black powdery mass in infected portion of the grain.	Infected portion of the grain is converted into black sooty mass of spores.
	Infected grains do not give any foul smell.	Infected grains emit foul odour like a rotten fish due to production of trimethylamine.
Stage of infection	At any stage of grain formation.	At flowering during boot leaf stage.
Seed germination	Normal	Normal except badly infected grains.

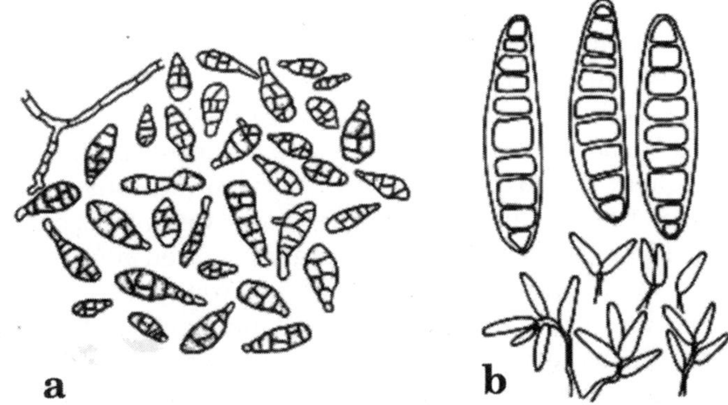

Figure 44: Principal causal agents of black point of wheat (a) *Alternaria alternata* (b) *Drecheslera sorokiniana*

Table 14. Fungi associated with black point of wheat in India

Fungi	Reference(s)
Alternaria sp.	Agarwal *et al.,* 1968
A. tenuis *A. alternata*	Tandon, 1946; Tandon and Srivastava, 1953; Parashar and Pracer, 1965; Dharmvir *et al.,* 1968; Bhowmik, 1969; Gill and Tyagi, 1970; Adlakha and Joshi, 1974; Agarwal *et al.,* 1972; Ayachi, 1974
A. triticinia	Prabhu and Prasada, 1967
Cladosporium sp.	Dastur, 1935
Cochliobolus tritici	Dastur, 1942
Curvularia lunata	Kumar and Nema, 1973
C. pallescens	Hasija, 1963
Drecheslera sorokiniana	Dastur, 1932; Parashar and Pracer, 1965; Parasher and Chohan, 1967; Dharamvir *et al.,* 1968; Sathe, 1972; Adlakha and Joshi, 1974
Fusarium sp.	Dastur, 1935; Dharamvir *et al.,* 1968
F. moniliformae	Ayachi, 1974
Helminthosporium sp.	Dastur, 1942; Agarwal *et al.,* 1968
H. sativum	Galloway,1935
Nigrospora sphaeria	Dastur, 1942
Penicillium sp.	Agarwal *et al.,* 1968
Rhizoctonia sp.	Dastur, 1942
Rhizopus sp.	Agarwal *et al.,* 1968
R. arrhizus	Ayachi, 1974
Sclerotium rolfsii	Dastur, 1942
Stemphylium sp.	Bhowmik, 1969
Ophiobolus sp.	Dastur, 1935

the field all the times but actual time of seed infection is a matter of conjecture. According to Scott and Sallans (1931) infection of black point takes place only during the anthesis stage of the crop. Adlakha and Joshi (1974) also produced disease symptoms with inoculation of ears with *A. alternata* and *D. sorokiniana* separately at post-anthesis stage under high humid atmosphere. Based on these observations they emphasized that black point appears in the field at dough or crop maturity if there is heavy rainfall. The prolonged wet weather favours spore germination of fungi. The fungal hyphae penetrate the integument and spread around the embryo to cause blackening of the germinal end of the developing grains. El-Helay (1947) could not find the invading hyphae in aleurone layer, the starch cells of the endosperm or the embryo.

Management

Resistance: Mexican germplasm used in breeding of wheat cultivars has shown susceptibility to black point. Some varieties like HD 1960, HD 2177, Lerma rojo, PV 18, S 308, Sonora 64 are quite susceptible (Ayachi, 1974; Goswami and Sehgal, 1968). In artificial inoculation tests varieties HD 1951, HD 1976 and E 5895 were found totally free from disease (Adlakha and Joshi, 1974).

Chemical Control: The black point grains sometimes show poor germination and cause seedling blight, root rot in the seedling (Kulkarni and Hedge, 1980). Seed treatment with organo-mercurial fungicides was previously reported to be highly effective in controlling seed-borne inoculum of black point fungi (Dharamvir, 1972; Mishra, 1974). Benlate, Bavistin, Emisan, Dithane M-45, Vitavax, Captan, Fytolan and Difolatan @ 2–2.5 g/kg seed have been found equally effective (Agarwal and Verma, 1975; Sinha and Thpliyal, 1984). Khetrapal and Agarwal (1979) recommended the use of Thiram on commercial scale for disinfecting the black point affected seeds.

Cultural Control: The severity of black point increases if seed is stored under high moisture conditions (more than 90% RH) and the seed moisture contents exceed 20%. Such conditions should be avoided (Mathur and Chahal, 1993).

REFERENCES

Adlakha, K.L. and L.M. Joshi. 1974. Black point of wheat. *Indian Phytopath.* **27**: 41-44.

Agarwal, N.S., S.P. Duggal and A.S. Yadav. 1968. Black tipped wheat kernels in the indigenously grown mexican varieties of wheat. *Bull. Grain Technol.* **6**: 113-116.

Agarwal, V.K. and H.S. Verma. 1975. Discolouration of seeds. *Seed Tech. News.* **5**: 1-2.

Agarwal, V.K., S.B. Mathur and P. Nergaard. 1972. Some aspects of seed health testing with respect to seed-borne fungi of rice, wheat, blackgram, greengram and soyabean grown in India. *Indian Phytopath.* **25**: 91-100.

Aulakh, K.S., S. Kaur, S.S. Chahal and H.S. Randhawa. 1988. Seed-borne *Drechslera* species in some important crops. *Plant Disease Research* **3**: 156-171.

Ayachi, S.S. 1974. Mycoflora of wheat II. A study on mycoflora associated with fresh wheat grains. *JNKVV. Res. J.* **8**: 148-149.

Bhowmik, T.P. 1969. *Alternaria* seed infection of wheat. *Pl. Dis. Reptr.* **53**: 77-80.

Bolley, H.L. 1913. Wheat soil troubles and deterioration. North Dakota Agr. Exp. Sta. Bull. 330.

Chaudhary, R.C., S.S. Aujla, I. Sharma. and R.P. Singh. 1984. Effect of black point on germination and quality characters of WL 711. *Indian Phytopath.* **37**: 351-353.

Dharamvir 1972. Efficacy of fungicides seed disinfection in relation to black point disease of wheat. *Pesticides* **6**: 15-16.

Dastur, J.F. 1932. Foot rot and black point disease of wheat in the Central Provinces. *Agric. Live Stk. India.* **3**: 275-282.

Dastur, J.F. 1935. Microscopic characters of the black point diseases of wheat in the Central Provinces. *Proc. Worlds Grain Exhib.* Conf., Reginia, Canada **2**: 253-55.

Dastur, J.F. 1942. Note on some fungi isolated from black point affected kernels in the Central Provinces. *Indian J. Agric. Sci.* **12**: 731-741.

Dharamvir, K.L. Adlakha, L.M. Joshi and K.D. Pathak. 1968. Preliminary note on the occurrence of black point disease of wheat in India. *Indian Phytopath.* **21**: 234-235.

El-Helaly, A.F. 1947. The black point disease of wheat. *Phytopathology* **37**: 773-780.

Evans, N.S. 1921. Black point of wheat. *Phytopathology* **11**: 515 p.

Golloway, L.D. 1935. New plant diseases recorded in 1935. *Bull. Pl. Prot.* **10**: 121-122.

Gaur, Ashok 1986. Black point of wheat. **In**: *Problems and Progress of Wheat Pathology in South Asia* (Eds. L.M. Joshi, D.V. Singh and K.D. Srivastava), pp. 230-241. Malhotra Publishing House, New Delhi, 401 p.

Gill, K.S. and P.D. Tyagi. 1970. Studies on some aspects of black point disease in wheat. *J. Res.* Ludhiana, **7**: 610-617.

Goswami, A.K. and K.L. Sehgal. 1968. Quality evaluation of wheat grains suffering from black point disease caused by *Helminthosporium sativum* and *Alternaria tenuis* Nees. *Jour. Food. Sci. Tech.* **6**: 35-37.

Hanson, E.W. and J.J. Christensen 1953. The black point disease of wheat in the United States. Tech. Bull. Univ. of Minnesota, 206 p.

Hasija, S.K. 1964. A new record of *Curvularia pallescens* Boed. on wheat grain. *Indian Phytopath.* **16**: 375-377.

Joshi, L.M., Dharamvir, K.L. Adlakha. 1969. Black point disease of wheat. *Proc. Indian Sci. Cong.* **56**, 12 p.

Khetarpal, R.K. and V.K. Agarwal. 1979. Studies on some aspects of black-point and Karnal bunt disease of triticale. *Indian Phytopath.* **32**: 292-94.

King, J.E. and A.D. Evens and B.A. Stewart. 1981. Black point of grain in spring wheats of the 1978 harvest. *Plant Pathology* **30**: 51-53.

Kulkarni, S. and R.K. Hedge. 1980. Black point disease of wheat in Karnataka. *Curr. Res.* **9**: 33.

Kumar, S.M. and K.G. Nema. 1973. Studies on internally seed-borne fungi of wheat. *Curvularia* spp., *Alternaria* spp. and *Helminthosporium rostratum* Drech.I. Morphological studies. *JNKVV Res. J.* **3**: 24-26.

Mathur, S.B. and S.S. Chahal. 1993. Black point. **In**: *Seed-borne Diseases and Seed health Testing of Wheat.* (Eds. S.B. Mathur and B.M. Cunfer), pp. 13-22. Danish Government Institute for Developing countries, Denmark, 167 p.

Mishra, R.P. 1974. Evaluation of seed dressing fungicides for effective control of black point disease of wheat. *JNKUV Res. J.* **8**: 92-94.

Parashar, R.D. 1970. Studies on the control of the black point disease of wheat caused by *Helminthosporium sativum* P.K. & B. and *Alternaria tenuis* Nees. *J. Res.,* Ludhiana **7**: 618-621.

Parashar, R.D. and Pracer, C.S. 1965. Black point or kernel smudge disease of wheat. *J. Res.,* Ludhiana **2**: 115-119.

Parashar, R.D. and J.S. Chohan. 1967. Effect of black point, both *Alternaria tenuis* Nees. and *Helminthosporium sativum* on seed germination under laboratory and field conditions and on yield. *J. Res.,* Ludhiana **4**: 73-75.

Prabhu, A.S. and R. Prasada 1967. Evaluation of seed infection caused by *Alternaria triticina* in wheat. *Proc. Int. Seed. Test. Assoc.* **32**: 647-654.

Rana, J.P. and P.K.S. Gupta 1982. Occurrence of black point disease on wheat in West Bengal. *Indian Phytopath.* **35**: 700-702.

Sathe, A.V. 1972. Black point of wheat grain. *FAO Plant Prot. Bull.* **20**: 9-11.

Scott, G.A. and B.J. Sallans. 1931. *Helminthosporium* Studies, Report of the Dominion Botanist for 1929. 97-100. Dominion Dept. Agric., Ottawa.

Singh, D.V., K.D. Srivastaa and L.M. Joshi. 1989. Occurrence and distribution of black point disease of wheat in India. *Seed Res.* **17**: 164-168.

Sinha, A.P. and P.N. Thapliyal. 1984. Seed disinfection in relation to black point disease of triticale. *Indian Phytopath.* **37**: 154: 156.

Tandon, R.N. 1946. Preliminary observations on a new disease of wheat at Allahabad. *Proc. Nat. Acad. Sci., Sec. B.* **16**: 1-5.

Tandon, R.N. and J.P. Srivastava 1953. Black point disease of wheat. *Proc. 40th Indian Sci. Congr.* Pt. III., 77 p.

2.3.5 Downy Mildew

Downy mildew was first reported on wheat in Italy in1900 (Traverso, 1902) and subsequently from USA, Japan, Mexico, Australia, Austria, France, Iran, Iraq, Pakistan and Turkey (Bhatti and Ilyas, 1986; Chahal and Singh, 1993; Rupent and Borlaugh, 1949; Tanaka, 1940; Weston, 1921). In India, the disease was first observed during 1968 by Tyagi and Anand (1969) from Punjab. It is now locally restricted in parts of Haryana, Delhi, Jammu and Kashmir and Uttar Pradesh (Joshi *et al.,* 1978; Goel *et al.,* 1986).

The disease is not economically important and is usually associated with water-logged or irrigated soil. It has a broad host range, including species of *Agropyron, Bromus, Cyperus, Festuca, Lophocloa, Phalaris, Poa* and *Polypogon* (Semenuik and Mankin, 1964; Singh and Bedi, 1991). Recently, Singh *et al.* (2009) have recorded symptoms of downy mildew in barley. The destructive potentialities of wheat downy mildew under favourable conditions are great (Goel *et al.,* 1986).

Symptoms: The symptoms of downy mildew most frequently appear during early growth stages of wheat plants. Diseased plants are stunted due to shortening of internodes and they become yellowish with excessive tillering. The leaves become twisted, thickened, leathery or warty, and usually fail to head. Leaves develop in close whorls around the culm giving a bushy appearance to the infected plants (Kawai and Hisanga, 1949; Zillinsky, 1983). Many affected plants die during the tillering to early elongation stages. Where heading occurs, the spikes are deformed and show 'crazy top' symptoms with several abnormalities like swollen and enlarged spikelets, curling of rachis, auxillary shoots growing from the nodes and excessive awn or loose ears due to rapid elongation of the rachis (Plate 5A). Diseased plants are typically scattered in or along standing water and many will be flagged with predominantly yellowed leaves (Weise, 1977). The diagnostic feature is the presence of numerous spherical yellowish-brown oospores in the hypertrophied and hyperplastic diseased parts.

Causal Organism: *Sclerophthora macrospora* (Sacc.) Thiram. Shaw & Naras. (Syn. *Sclerospora graminicola* (Sacc.) Schroet; *Sclerospora macrospora* Sacc.) is the most accepted nomenclature of downy mildew pathogen. However, Tanaka (1940) felt that the fungus has resemblance with *Phytophthora macrospora* (Sacc.) Ito & Tanaka.

The fungus produces coenocytic, thin-walled intercellular mycelium in the meristematic tissues of the host. Oospores are hyaline to yellowish, smooth walled, globose to sub-globose, multinucleate and 40–70 µm in diameter. They mature as host tissue senesce and are closely entangled with leaf residues. Oospores germinate in water or wet soil to produce thin walled sporangiophores of variable length which bear sporangium at the tip. Sporangia are lemon shaped, operculate and 60–100 × 30–65 µm on stalks (Shurtleff, 1980). Sporangia germinate within 1–2 hours after formation and liberate biflagellate motile zoospores that are 9–12 µm in size. Sporangia develop best under water between 10 and 25°C with limits near 7 and 31°C. Sometimes instead of zoospores, secondary sporidia are produced. The zoospores germinate within 2 hours of release and form slender germ tubes which measures 2–3 × 25–100 µm. Conidia and conidiophores are rare on wheat but frequent on other hosts giving them a downy appearance. Conidia are hyaline, 12–16 × 15–22 µm and borne on sterigmeta clustered at the terminus of conidiophore branches (Semeniuk and Mankin, 1964; Tanaka, 1940; Wiese, 1987).

Disease Cycle: *S. macrospora* is an obligate parasite. In India, both asexual and sexual stages of the pathogen have been observed (Tyagi, 1971) but the infection occurs by oospores persisting on leaf and ear residues. Zoospores are derived on germination of fertilized or unfertilized oospore. Infection usually takes place when lower leaves or leaf sheaths of wheat seedlings come in contact with water-borne zoospores. The germ tubes of the zoospores penetrate the seedling which appears to be more susceptible stage than adult plants. Four hours exposure of seedlings is sufficient to facilitate infection under water saturated conditions. Sporangia develop best under water between 10–20°C with optimum of 20°C (range 7–31°C). Sporangial production is maximum during mid night to early morning hours (Thirumalachar, 1953).

Bains and Jhooty (1986) showed downy mildew symptoms from seeds collected from diseased ear-head. But Chahal and Singh (1993) expressed that for confirmation of seed borne nature of the disease, detailed studies are required to detect and ascertain the location of the fungus in seed and its transmission.

Management

The disease is economically not very important in the country. It is because of this reason, not much work has been done on its control measures. Since the infection of downy mildew is more frequent in poorly drained and water logged fields, improved drainage and rotation of wheat with non-cereal crops can be beneficial.

Wheat downy mildew can be kept under control if proper agronomical practices like eradication of weed hosts, use of healthy seeds from a disease free crop, avoiding infested host debris etc. are followed.

Foliar spray of maneb followed by folpet application reduces disease incidence partially (Akhtar and Lil-Haq. 1987). Wheat genotypes Arjun, HD 2364, HD 2384, WG 357 and KSML 3 were found resistant to this disease.

REFERENCES

Akhtar, M.A. and E. Lil-Haq. 1987. Effect of fungicides on downy mildew of wheat. *Rachis* **6**: 49-50.

Bains, S.S. and J.S. Jhooty. 1986. Seed transmission of *Sclerophthora macrospora* in wheat. *Seed Research* **13**: 154-156.

Bhatti, M.A.R. and M.B. Ilyas. 1986. Wheat diseases in Pakistan. **In**: *Problems and Progress of Wheat Pathology in South Asia* (Eds. L.M. Joshi., D.V. Singh and K.D. Srivastava), pp. 20-30. Malhotra Publishing House, New Delhi, 401 p.

Chahal, S.S. and P.P. Singh 1993. Downy midew. **In**: *Seed –Borne Diseases and Seed Health Testing of Wheat*. (Eds. S.B. Mathur and B.M. Cunfer), pp. 69-72. Danish Government Institute of Seed Pathology, Denmark, 168 p.

Goel, L.B., S.D. Singh and V.C. Sinha. 1986. The mildews of wheat. **In**: *Problems and Progress of Wheat Pathology in South Asia* (Eds. L.M. Joshi, D.V. Singh and K.D. Srivastava), pp. 176-190. Malhotra Publishing House, New Delhi, 401 p.

Joshi, L.M., K.D. Srivastava, D.V. Singh, L.B. Goel and S. Nagarajan. 1978. *Annotated Compendium of Wheat Diseases in India. ICAR*, New Delhi, 331 p.

Kawai, I. and M. Hisanga. 1949. Downy mildew of cereals and control. *Agric. Hort.*, Tokyo. **24**: 656.

Semeniuk, G. and C.J. Mankin. 1964. Occurrence and development of *Sclerophthora macrospora* on cereals and grasses in South Dakota. *Phytopathology* **54**: 409-416.

Shurtleff, M.C. 1980. Crazy top. pp. 28.-29. *Compendium of Corn Diseases*, 2nd Edition, APS Press, St. Paul, Minnesota, 105 p.

Singh, D.P., S.S. Karwasra, M.S. Beniwal and R.S. Beniwal. 2009. Downy mildew of barley in India - a new report. *Indian Phytopath*. **62** : 134.

Rupent, J.A. and N.E. Borlaugh. 1949. Downy mildew of wheat in Mexico. *Plant Dis. Reptr*. **33**: 382-383.

Tanaka, I. 1940. *Phytophthora macrospora* (Sacc.) S. Ito & Tanaka on wheat plant. *Ann. Phytopath. Soc. Japan.* **10**: 127-138.

Thirmulachar, M.J., C.G. Shaw and M.J. Narashimhan. 1953. The sporangial phase of the downy mildew on *Eleusine coracana* with a discussion on identity of *Selerospora macrospora* Sacc. *Bull. Torrey Bot. Club* **80**: 299-307.

Traverso, G.B. 1902. Note critiche sopra II. *Sclerospora parasitica* di graminaceae. M'pighi **16**: 280-290.

Tyagi, P.D. 1971. Downy mildew of wheat. *Proc. Second Int. Symp. Pl. Pathol.,* IARI, New Delhi, 76 p .

Tyagi, P.D. and H.C. Anand. 1968. Downy mildew of wheat. *Plant Dis. Reptr.* **52**: 569.

Wiese, M.V. 1987. *Compendium of Wheat Diseases*. Amercan Phytopathological Society, APS Press, St. Paul, Minnesota, 106 p.

Weston, W.H. 1921. The occurrence of wheat downy mildew in United States. U.S. Dept. Agric. Circ. 186, 6 p.

Zillinsky, F.J. 1983. *Common Diseases of Small Grain Cereals : A guide to identification*. CIMMYT, Mexico, 141 p.

2.4 SOIL BORNE DISEASES

A number of useful and harmful fungi live in the soil as members of soil microflora or they colonize in the soil for longer or shorter durations. Some of these soil borne organisms are responsible for causing serious diseases in plants. The spores of *Urocystis agropyri* present in the soil provide the inoculum for infection of flag smut of wheat. *Drechslera sorokiniana, D. tetramera, Sclerotium rolfsii, Curvularia specifera, C. verruciformis, Fusarium* species, *Pythium grarminicolum* and *Rhizoctonia solani* that are soil borne in nature, have been associated with wheat and cause seedling blight and root rot diseases (Joshi *et al.,* 1978). However, Rhizoctonia root rot and Pythium root rot are comparatively more important than others (Kulkarni and Nargund, 1986). On the other hand Sclerotium foot rot (*Sclerotium rolfsii*) is important in central-peninsular India. *Geaumannomyces graminis* var. *tritici* (syn. *Ophiobolus graminis*) causing take-all disease has been recorded in wheat but it is not serious disease in the country.

2.4.1 Flag Smut

Flag smut is distributed in many parts of world's wheat growing countries (Purdy, 1965). It was first reported in South Australia in 1868 (McAlpine, 1905). However, the disease was earlier recorded on *Agropyron* sp. in Europe and was attributed to *Uredo agropyri* (Preuss, 1848). It has now been reported from Australia, Chile, China, Egypt, Japan, Pakistan, South Africa, Spain, USA and Italy (Ballantyne, 1906). In India, Sydow and Butler (1906) reported the occurrence of flag smut for the first time from Layallpur in Punjab (now in Pakistan). The introduction of this disease in Punjab is believed to be through exchange of wheat material from Australia (Bedi, 1957). The disease is prevalent in the states of Uttar Pradesh, Madhya Pradesh, Bihar, Rajasthan, Punjab, Haryana and Himachal Pradesh (Mundkur, 1944; Bedi, 1957) but it is less significant.

The disease leads to considerable loss of productivity of infected plants. Kothari and Dange (1968) recorded heavy incidence of flag smut in 1964–65 in few fields near Jobner in Rajasthan resulting in tremendous yield loss. In 1968 mild to severe incidence was observed in a number of fields in Chittor (Rajasthan) and as high as 80% loss was estimated in one of the fields. Sethi and Singh (1971) noticed severe infection of the disease in isolated areas in Kullu valley of Himachal Pradesh where 30–35% plants were affected with about 80% tillers becoming unproductive. Bhatnagar *et al.* (1975) estimated 7.23 to 20.85% loss in grain weight of wheat varieties Arjun, Lalbahadur and D134.

Symptoms: The disease is evident from the late seedling stage until maturity of the crop. It is characterized by distortion of stem, culm and leaf. The plants become stunted showing conspicuous twisting of plant parts. The first symptom of flag smut appears in the leaf blade, mainly in the flag leaf in the form of black streaks running parallel to the veins. The smut sori are formed in the mesophyll tissues between the veins. In early stages, the streaks are covered with epidermis and are grayish-black in colour but later, the epidermis ruptures and black powdery mass of spores is released (Plate 5B). Soon infected leaves are shed away and often the whole plant is dead. In case of severe infection the number of tillers is reduced and the infected plants do not produce ears and if ears are formed, the grains produced are shrivelled and are not normally viable. Very frequently, the entire stool bears no grain, the ear being replaced by twisted mass of leaves. As a rule, every shoot of the plants is infected (Fisher and Holton, 1957; Joshi *et al.,* 1978).

Causal Organism: Wolff (1873) was the first to identify the fungus causing flag smut of wheat and designated it as *Urocystis occulata* (Wallr.) Rab. Fisher (1953) regarded the wheat and grass attacking forms of the pathogen as the same species and named the causal agent *Urocystis agropyri* (Preuss) Schroter. However, Vanky (1985) considered the flag smut fungus on wheat to be distinct from the one on grasses and suggested the name *Urocystis tritici.* The *U. tritici* is still used even when agreeing that two species, *U. agropyri* and *U. tritici* are synonymus.

The fungus produces linear black erumpent streaks in host leaves. Spore balls occur in the streaks which are globose, ellipsoidal to oblong, dusty, golden brown to black and 18–52 μm in diameter. The spore ball consists of 1–6 teliospores (chalmydospores), normally 3 spores within a sheath of hyaline, flattened sterile cell. Teliospores are globose to subglobose, dark reddish brown, smooth walled, 8–18 μm in diameter. Teliospores germinate *in situ* to produce a short promycelium which is generally without any septa but sometimes may have 1–2 septa. The promycelium terminally bears a cluster of 1–4 primary sporidia. Mature sporidia are cylindrical, translucent, smooth, thin walled, aseptate, 12–15 × 3.0 μm in size. *U. agropyrii* has shown less physiologic specialization than any other cereal smuts. Some races of the fungus have been reported from USA, China, Japan and South Africa (Ballantyne, 1996) but no work is reported from India (Pal and Mundkur, 1941).

Disease Cycle: The disease is soil as well as seed borne. The fungus gains access to seed at the time of harvest and soil is infested by smut spores released from crop residue during tillage. Teliospores can survive in soil for 4 to 7 years (Sattar and Hafiz, 1952). After a pre-wetting period, the teliospores germinate in the soil. At this stage meiosis takes place in the diploid teliospores (2n) followed by mitosis. Haploid nuclei migrate to developing primary sporidia, generally one nucleus per sporidium. Plasmogamy occurs *in situ* between mononucleate sporidia and H-body (n + n) is formed by sporidial fusion and infection hypha. The infection (n + n) arising from H-body penetrates the cuticle of the coleoptile before emergence of seedlings. After infection the fungus grows both inter and intracellularly until it begins to sporulate, generally first in the leaf blade and then in the leaf sheath and other parts of the plant (Fig. 45).

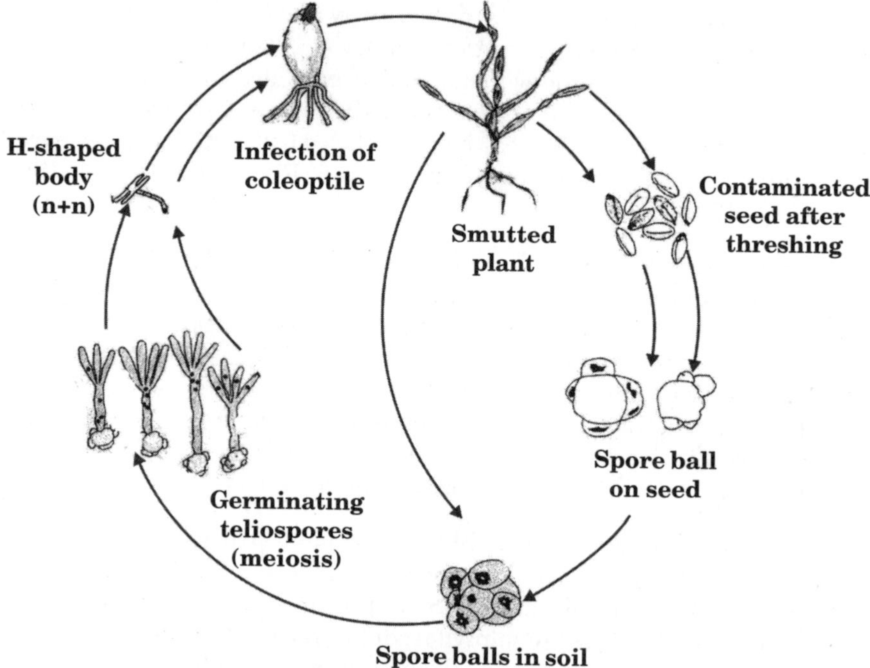

Figure 45: Disease cycle of Flag smut (*Urocystis agropyri*) of wheat

Infection in the seedling stage of plant growth is necessarily influenced by environment temperature, soil pH as well as cultural practices. Soil temperature from 20–22°C and a moderately moist soil with pH range of 5.5 to 8.7, are conducive to infection. It has been found that the disease is less common in calcium deficient soil. Deep sowing increases chances of infection probably due to the fact that coleoptile remains below the soil for a longer period and it is exposed for greater length of time to allow infection by the smut spores (El-Helaly, 1948).

The co-existence of flag smut with common bunt (McAlpine, 1905), loose smut (Aujla and Sharma, 1977), dwarf bunt (Holton and Jackson, 1951) and stripe rust (Purdy and Holton, 1963) in the same plant has been reported under field conditions.

Management

The control of flag smut depends on varietal resistance, chemical seed treatment, cultural practices and seed certification.

Resistance: Resistant cultivars are effective in managing the disease but most of the wheat and triticale varieties are susceptible to the disease in India. Durum varieties PDW 237, Raj 1555 and WH 896 have been found

Table 15. Flag smut (*Urocystis agropyri*) resistant lines/ accessions of wheat.

	Genotype	Pedigree
1	A-9-30-1 (D)	A 206/GAZA
2	BIJGA YELLOW	DHARWAD LOCAL/GAZA
3	DT 46	JNIT/140/DTS/1209
4	DWR 1006 (D)	SULA/CREX//AAZ
5	GW 1139	MACS 2340/IW 5070
6	HD 2189	HD 1963/HD 1931
7	HD 2329	HD 1962//E 4870/K 65/3/HD 1553/UP 262
8	HD 4502 (D)	PI 'S'/TAC 125//4* TC/2B//LAK–OTA
9	HI 1384	MRL 'S'/SVC 'C'/VA–1
10	HI 8381 (D)	JO "S"–AA "S"/PG "S"
11	HP 1744	CNO/PRI//CHILENO/GERUDA
12	HPW 42	VEE 'S'/4/PVN 'S'/CBB/CNO 'S'/3/JAR/ORZ 'S'
13	HS 240	AV/KA/Bb/WOP "S"/PVN "S"
14	HS 365	HS 207/SONALIKA
15	HS 295	CQT/AZ//IAS 55/ALD "S"/3/ALD "S"/NAFN/4/PIN "S" /PEN ZSL–127
16	MACS 1967	GULAB/CPAN 1471
17	MACS 2846	CPAN 6079/MACS 2340
18	NIDW 15	DOM 50
19	PBW 299	BB/KAL/WL 711/PBW 65
20	PBW 435	HD 2160/CALIDAD
21	PBW 34	AA "S"/FG "S"
22	PBW 343	ND/VG 9144//KAL//BB/3/VACO 'S'/4/VEE # 5 'S'
23	PBW 396	CNO 67/MFD//MON 'S'/3/SERI
24	PDW 215 (D)	DWL 5031/DWL 5002
25	PDW 237 (D)	YAV 'S'/TEZ 'S'
26	RAJ 1555 (D)	COCORIT 'S'/RAJ 911
27	WH 896 (D)	STN 'S'/WH 852

to be resistant (Singh, 2008). Resistance to flag smut has also been observed in a number of wheat genotypes evaluated under multi-location tests (Table 15). These stocks are valuable source for disease resistance breeding (Sharma *et al.,* 2002).

Chemical Control: Seed dressing with systemic fungicides such as carboxin @ 2 g/kg seed or fenpropimorp, triadimefon, benomyl, butrizol, oxycarboxin or penconazole @ 1.5 g/kg seed reduces the inoculum and results in good control of the disease (Bhatnagar *et al.,* 1975, 1978; Singh and Pandey, 2002).

Cultural Control: Cultural practices such as crop rotation, early and shallow sowing (less than 1.5 cm) deep ploughing and burning of the stubbles reduce infection and severity of flag smut (Hafiz, 1948; Bedi, 1957).

REFERENCES

Aujla, S.S. and Y.R. Sharma. 1977. Simultaneous occurrence of *Ustilago nuda tritici* and *Urocystis agropyri*. *Indian Phytopath.* **30**: 262.

Ballantyne, B. 1996. Flag smut. **In**: *Bunt and Smnt Diseases of wheat: Concept and Methods of Disease Management,* (Eds. R.D. Wilcoxson and E.E. Saari), CIMMYT, Mexico, 66 p.

Bedi, K.S. 1957. Fighting the flag smut. *Indian Farming* **7(6)**: 25-27.

Bhatnagar, G.C. 1979. Screening of wheat varieties against flag smut. *Proc. 18th All Indian Wheat Research Workers Workshop*. Ranchi, 1979.

Bhatnagar, G.C., R.B.L. Gupta and V.L. Mishra. 1978. Relative efficacy of different fungicides in controlling flag smut *(Urocystis agropyri)* of wheat in field conditions. *Pesticides* **12**: 27-28.

Bhatnagar, G.C., R.B.L. Gupta, V.L. Mishra and G. Singh. 1975. Assessment of losses and chemical control of flag smut of wheat caused by *Urocystis tritici. Indian J. Mycol. Pl. Pathol.* **5**: 35-36.

El-Helaly, A.F. 1948. Influence of cultural conditions on the flag smut of wheat. *Phytopathology* **38**: 688-697.

Fischer, G.W. and C.S. Holton. 1957. *Biology and Control of the Smut Fungi*. The Ronald Press Co., New York.

Fischer, G.W. 1953. *Manual of North American Smut Fungi*. The Ronald Press Co., New York.

Hafiz, A. 1948. Smut diseases and their control by cultural methods. *Sci. J.R. Coll. Sci.* **18**: 77-80.

Holton, C.S. and T.L. Jackson. 1951. Varietal reaction to dwarf bunt and flag smut and the occurrence of both in the same wheat plant. *Phytopathology* **41**: 1035-1036.

Joshi, L.M., K.D. Srivastava, D.V. Singh, L.B. Goel and S. Nagarajan. 1978. *Annotated Compendium of Wheat Diseases in India*. ICAR, New Delhi, 332 p.

Kothari, K.L. and S.R.S. Dange. 1968. Flag smut, a potential threat to wheat cultivation in India. *Farmer & Parliament*. June, 1968.

McAlpine, D. 1905. Flag smut of wheat. *J. Dept. Agri.,* Victoria. **3**: 168-169.

Mundkur, B.B. 1944. Some rare and new smuts from India. *Indian J. Agric. Sci.* **14**: 49-52.

Pal, B.P. and B.B. Mundkur. 1941. Studies in Indian cereal smuts. III. Varietal resistance of Indian and other wheats to flag smut. *Indian J. Agric. Sci.* **21**: 687-694.

Preuss, C.G.T. 1848 **In**: Stum J. Deutschlands Flora 6(25-26), Table 1 (1848); Linneae 24: 99-153. Quoted by Purdy (1965).

Purdy, L.H. 1965. Flag smut of wheat. *Bot. Rev.* **31**: 565-606.

Purdy, L.H. and C.S. Hoton. 1963. Flag smut of wheat, its distribution and co-existence with stripe rust in Pacific North West. *Plant Dis. Reptr.* **47**: 516-518.

Sattar, A. and A. Hafiz. 1952. Some factors influencing the incidence of flag smut of wheat. *Pakistan J. Sci.* **4**: 25-30.

Sethi, G.S. and H.B. Singh. 1971. Flag smut threatens Kalyansona. *Intensive Agric.* **9**: 13.

Sharma, A.K., D.P. Singh, J. Kumar, M.S. Saharan, K.S. Babu, A.K. Singh and S. Nagarajan. 2002. *Disease and Insect Pest Resistant Genotypes of wheat and triticale.* Res. Bull. No. 15, Directorate of wheat Research, Karnal, India. 18 p.

Singh, D.P. 2008. Disease problems of wheat and management approaches. **In**: *A Compendium of Lectures on Integrated Pest Management in Wheat Based Cropping Systems* (Eds. A.K. Sharma and D. P. Singh), DRW Compendium No. 2. Directorate of Wheat Research, Karnal, India, 258 p.

Singh, D.V. and V. Pandey. 2002. Smuts and Bunts. **In**: *Diseases of Field Crops* (Eds. V.K. Gupta and Y.S. Paul), Indus Publishing Co., New Delhi, 464 p.

Sydow, H.P. and E.J. Butler. 1906. Fungi Indiae Orientalis. *Ann. Mycol.* **4**: 427.

Vanky, K. 1985. Ustilaginales. Symbolae Botanical Upsalienses **24**(2). Tegelbruksvagen, Gagnef, Sweden, 309 p.

Wolff, R. 1873. Beitrag Zur Kenntnis der Ustilagineen. *Bot. Ztg.* **31**: 657-661.

2.4.2 Root Rots

Rhizoctonia Root Rot

This disease mostly occurs in Madhya Pradesh (Sharma and Jain, 1967) and Uttar Pradesh (Saksena and Kumar, 1971) in dry regions. The importance of the disease varies from region to region. Since root rot caused by *R. solani* is not prevalent on large scale, actual yield losses are not fully known.

Symptoms: The affected seedlings of wheat remain stunted with yellowing of leaves that later dry up and curl. The injury to roots ranges from necrosis of root tips to rotting of primary and secondary root system. The roots are poorly developed; discoloured, blackish brown and the rootlets are either not properly developed or destroyed. Sclerotia of the fungus are observed on infected roots. The disease in the field is noticed by no growth patches by rapid successional killing of seedlings. Normally *R. solani* does not favour sheath blight symptoms in wheat but 6 rice isolates of the fungus induced sheath blight under artificial inoculated conditions. However, sheath blight lesions remained restricted to the inoculated sheath and did not spread to other sheaths or lamina (Biswas and Samajpati, 2006).

Causal Organism: *Rhizoctonia solani* Kuhn is associated with root rot disease of wheat. The perfect stage of the fungus *Thanetephorus cucumeris* (Frank) Dark. exists in nature. Other synonyms of this stage are *Pellicularia filamentosa* (Pat.) Rogers and *Corticium solani* Bourd & Galzin.

The mycelium of the fungus is septate, hyaline and sparsely branched at right angle with a constriction at the point of origin. The hyphal cells are multinucleate (2–25 nuclei) and 5–12 µm wide, and 250 µm long with dolipore septum. Sclerotia are borne on the surface of mycelial cushion, grey to brown in colour, spherical and variable in size. Holobasidia present in the hymenium are arranged in recemose or cyme. They are light brown, barrel-shaped or sub-cylinderical and 15–25 × 5–19 µm in size. Basidiospores are hyaline, oblong or ellipsoidal, unilateral, flattened, apiculate, soft, smooth walled and 6–14 × 4–8 µm in size.

Disease Cycle: *R. solani* is soil borne in nature. It survives in the form of sclerotia on plant debris over a long period and remains in dormant state in the dry soil. The survival of sclerotial bodies is largely influenced by such soil factors as organic matter and nutrients. During crop season the sclerotia serve as primary source of inoculum and when conditions are favourable, they absorb moisture and germinate to produce basidiospores. The infection hyphae of germinating spores penetrates the tissue of the roots, spread inside the host system and make the roots of the seedlings weak. Dry conditions and temperature around 30°C are condusive for infection of root rot (Sharma and Jain, 1967). The secondary infection usually takes place through mycelium produced by pathogen in the vicinity of infected roots or by sclerotia which spread from one site to another by means of agricultural implements and irrigating water.

Management

Resistance: Wheat varieties resistant to *R. solani* are very rare. Some genotypes namely Kalyansona, HY 5, HY 34, HY 38, HY 172, HY 173, HY 277, HY 278, HY 633, S-331, PKD 9,MP 4, MP14, MPO 212, NP 404, NP 839, HI 316, HI 346, HD 1739, HD 1944, HD 1977, NI 5439 and J-1 had very less mortality due to *R. solani* (Chaurasia *et al.,* 1973). Saxena and Kumar (1971) also recorded least susceptibility of Lalbahadur and UP 301 against pathogen.

Chemical Control: Seed treatment with PCNB, carboxin and carbendazim @ 2 g/kg proved inhibitory to *R. solani* (Shukla *et al.,* 1972). Application of fungicide rhizoctol @ 0.05% as soil drench also reduced disease incidence in Madhya Pradesh (Agarwal and Singh, 1969).

Biological Control: *Trichoderma harzianum, Gliocladium virens, Bacillus subtilis, Pseudomonas fluorescens* strains applied to wheat seed reduced the effect of *R. solani* and increased seedling growth (Marshal, 1982). *T. harzianum* is produced in powder formulations that are sprayed onto foliage or soil. *G. virens* products are incorporated as granules. *P. fluorescens* strains produce fungicidal secondary metabolites, including phenazine-1-carboxylate, that inhibit the growth of *Rhizoctonia* (Hewitt, 1988).

Cultural Control: All the diseased material should be destroyed and crop rotation practiced. Application of potash @ 60 kg/ha reduces Rhizoctonia root rot in lentil (Prasada and Chaudhary, 1984).

Pythium Root Rot

This is not widespread disease and has been found to occur in certain tracts in India. It was first reported from Dharwad and Bombay in 1928 (Subramanian, 1928). Later mention of this disease has been made from Madhya Pradesh, Maharashtra and Andhra Pradesh. It is also reported from USA, Canada, Italy and Soviet Union (Mishra and Singh, 1967; Wiese, 1977).

Symptoms: The disease symptoms appear soon after emergence of wheat seedlings. The disease is difficult to diagnose with certainty at early stages as young seedlings are stunted and foliage turns light green to yellow as if suffering from nutritional disorders. However, at later stage of disease development a light brown discolouration and soft rot of the roots and rootlets appear followed by wilting. If the pathogen attacks older plants then cortical tissues of collar region become discloured and soft leading to collar rot. The leaf sheaths turn blackish brown and split into shreds. Adequate moisture and favourable temperature (15–20°C) lead to damping off of the infected plants. Root rot affected plants are predisposed to other microbial attack.

Causal Organism: *Pythium graminicolum* Subram. is associated with root rot and foot rot of wheat. Some morphologically similar species of *Pythium* namely, *P. arrhenomanes* Drech., *P. tradicresceus* Vanterpool, *P. arestoraporum* Vanterpool, *P. volutum* Vanterpool & Trnscott, *P. aphanidermatum* (Edson) Fitz. and *P. myrlotylum* Drechs. also occur as parasites on graminaceous hosts (Dickson, 1956).

P. graminicolum is a saprophytic fungus of family Pythiaceae. It has thin, filamentous, branched and non-septate mycelium. The fungus reproduces both asexually and sexually. Asexual reproduction is by sporangia which arise terminally on the aerial sporangiophores. Sporangia are spherical, lobate with prominent exit tubes 3.5 µm wide and 15 µm long. Germinating sporangium put forths a protuberance which swells up to form a vesicle or zoosporangium. Each zoosporangium contains 15–48 biflagellate zoospores, about 8–12 µm in diameter. Sexual reproduction of the fungus is oogamous. The oogonium arises as intercalary or terminal by swelling of mother hypha. Similarly, antheridium also arises just below the oogonium. Oogonium is smooth, spherical and 17–36 µm (Av. 25 µm) in diameter and the antheridium is monoclinous. Fertilized oospores are single, spherical, smooth and thick walled (2 µm thick), 15–35 µm in diameter and germinate only by germ tubes.

Disease Cycle: The fungus inhabits the soil as sporangia, oospores or dormant mycelia in host residues. When the crop is sown the sporangia germinate under conditions of abundant moisture and produce a vesicle where they are organized into biflagellate zoospores. Liberation of zoospores takes place by rupture of the vesicle. After swimming in water for some time zoospores come to rest, withdraw the flagella; get rounded up and encysted, and germinate by forming a germ tube. Later the germ tube invades the tissues of the roots and cause infection. Under adverse climatic conditions sporangium directly germinates and attacks roots of seedlings.

During sexual phase of infection all nuclei of oogonium except one nucleus are disorganized and then surviving nucleus moves towards peripheral region. Same process takes place in antheridium. Soon after fertilization tube develops from the antheridium which facilitates transfer of protoplasmic contents to oogonium. The male functional nucleus fuses with female nucleus to give rise to oospore with thick wall. After overwintering oospore germinates, produces zoospores to infect wheat crop. Oospore can also germinate directly to cause infection without forming zoospore.

In Madhya Pradesh farmers preserve water in field during rainy season by raising mud bunds so that adequate soil moisture is available at sowing period. This cultural practice is called Haveli system. The incidence of Pythium root rot is higher in Haveli sowing because of high prevailing temperature and soil moisture content. Rainfall after primary infection promotes rapid spread of the disease. Disease development is more in compact soil as compare to loose sandy conditions.

Management

Resistance: Cultivars resistant to Pythium root rot are not available but wheat genotype RR 21 was reported to be tolerant to *P.graminicolum* (Kauraw and Singh, 1975).

Chemical Control: Treating seed with fungicides like thiram, oxadixyl, metalaxyl, hymexazol have shown inhibitory effect against *Pythium* species (Hewitt, 1998).

Cultural Control: *P. graminicolum* inhabits the soil as dormant oospore/ sporangia and mycelia in the left over host residues. The collection and destruction of these structures reduces the potential inoculum. A suitable crop rotation for few years starves out the pathogen. Balanced application of fertilizers also reduces the severity of the disease. According to Wiese (1977) restoring phosphate levels, or at least balancing them with nitrogen, give wheat a marked advantage over the pathogen by promoting root growth.

2.4.3 Foot Rot

Sclerotium Foot Rot

The disease was first reported in India and Burma by Subramanian (1928). It was later observed in some areas of Madhya Pradesh, (Agarwal and Singh, 1969), Maharashtra and Karantaka (Reddy *et al.,* 1971; Kulkarni, 1977) and West Bengal (Chattopadhyay, 1953).

In an experiment Chattopadhyay (1953) noticed that wheat seedlings became infected within 15 days of germination in inoculated soil and gradually died, there being some 50–60% mortality after three weeks.

Symptoms: The pathogen attacks young seedlings at the collar region. Plants affected by foot rot are characterized by yellowing of leaves followed by loss of vigour and premature death. They are easily pulled out from the soil as the root system is poorly developed and side roots are badly destroyed. White mycelial growth of the fungus appears on crown and root region. Dirty white to brown mustard seed like sclerotia are found adhering around the collar region. The affected plants in advance stage of infection collapse due to weakening of basal region of stem.

Causal Organism: *Sclerotium rolfsii* Sacc. causes foot rot disease in wheat. The fungus is omnivorous, attacking a large number of plant species. The perfect stage of the fungus *Corticium rolfsii* Curzi (Syn. *Pellicularia rolfsii* West.) sometimes develops on dead host tissues. *P. rolfsii* in India was recorded as a pathogen of root and stem rot of wheat by Upadhyaya and Pavgi (1967). The mycelium of *S. rolfsii* is hyaline, branched, septate, selender and thin walled with clamp connection at the septum. The mycelial growth is white and feathery. Sclerotia are spherical, mustard seed-like, 0.5–2 mm in diameter and light brown initially, becoming dark brown with age. The hymenium is a dull white radiating mass, profusely branched and forming a network of short, stout, broad basidial initials. The basidia are thin walled, hyaline, broadly clavate. Each basidium gives rise to 2–4 divergent sterigmeta on which hyaline, pyriform basidiospores are borne.

Kulkarni and Ahmed (1967) produced whitish, button like growth, consisting of hymenia, basidia and basidiospores on a new culture medium consisted of 2% agar, 1% uric acid and a combination of micro elements.

Disease Cycle: The pathogen has been found associated with the seed but usually survives in the soil in the presence of sufficient organic matter (Nargund, 1981). It has a wide range of hosts on which it can survive during off-season. Sclerotia buried in the soil and inside the dead tissues germinate under congenial environment. Sclerotial germination is favoured by a temperature range of 27–30°C with 2–5 soil pH at 2 cm depth. On germination the sclerotia produce basidiospores in the basidia. In the root zone the basidiospore penetrates the tissues of roots by producing infection hypha. Once inside the host the organism becomes virulent and develops disease symptoms. Sclerotia produced in the vicinity of infected roots are transported by irrigating water and agricultural implements to cause secondary infection.

Nargund *et al.* (1982) noticed that disease severity increases with the increase in inoculum density in infested soil. Minimum of 2% inoculum of *S. rolfsii* in infested soil is necessary to cause the disease. The growth of the pathogen is good at pH 5-8 (Chattopadhayay and Mustafa, 1977). Ramarao and Raju (1980) observed that mortality of wheat seedlings to *S. rolfsii* occurred at 25 to 50% moisture holding capacity and decreased with the increase of soil moisture. Reddy and Kulkarni (1972) recorded more damage to seedlings at 25 than 30, 35 or 40% moisture level in black clay soil. Sclerotial production is high in nitrogen deficient soil or when the ratio of nitrogen and carbon in the soil is more.

Management

Resistance: Agarwal and Singh (1969) tested several wheat varieties and reported HY-65 as highly tolerant to *S. rolfsii*. Some cultivars/lines, namely Lerma Rojo 64A, UP 301, HD 2181, BR 2104, BW 46, CPAN 1796, CPAN 2190, CPAN 1827, DL 153-7, DWR 43, HD 2735, HD 2189, HD 2190, HD 2320, HD 2323, HD 2325, HD 2328, HD 2329, HI 784, HI 936, HUW 12, HUW 100, HUW 154, HUW 187, HW 741, IWP 72, K 7906, NI 8188, PBW 11, Raj 1948, Raj 1987, Raj 2005, UP 115, VL 421, WH 291, WH 320 and WH 322 have been reported as tolerant or resistant to *S. rolfsii* (Reddy *et al.*, 1971; Kulkarni *et al.*, 1978; Nargund *et al.*, 1982).

Chemical Control: Seed treatment with Bisdithane, Arasan, Tritisan, Captan, Thiram and Brassicol was found effective in checking early stage seedling mortality of wheat caused by *S. rolfsii* (Agarwal and Singh, 1969; Mishra and Chand, 1970). Bahadur *et al.* (1974) also reported effectiveness of BAS 3191F, Vitavax, Brassicol and Rhizoctol against the pathogen. Baytan, Banodanil, Vitavax and Brassicol have been found to be good seed-dressing and soil-drenching fungicides (Kulkarni,1980).

Biological Control: *Bacillus subtilis* showed antagonism against wheat isolates of *S. rolfsii in vitro* (Hegde *et al.*, 1980). Manjappa (1979) and Nargund (1981) reported that *Trichoderma viride* and *Streptomyces* sp. were also antagonistic to Sclerotium foot root of wheat.

Cultural Control: Clean cultivation of crop may be of help. Diseased material in the field should be destroyed and crop rotation practiced. Deep ploughing of the field is useful in destroying the Sclerotia buried in the soil, thereby reducing the inoculum potential. Application of nitrogenous fertilizers and gypsum as calcium at higher rate may be effective in reducing disease incidence.

REFERENCES

Agarwal, S.C. and S.P. Singh. 1969. Efficacy of some fungicides controlling foot rot of wheat. *Indian Phytopath.* **22**: 476-477.

Agarwal, S.C. and S.P. Singh. 1969. Varietal reaction of wheat against *Selerotium rolfsii. Indian Phytopath.* **22**: 511-513.

Bahadur, P., V.C. Sinha and Y.M. Upadhyay. 1974. Evaluation of some systemic and non systemic fungicides as seed dressing against foot rot of wheat caused by *Sclerotium rolfsii* Sacc. *Pesticides* **8**: 42-44.

Biswas, A. and N. Samajpati. 2006. Behaviour of *Rhizoctonia solani* Khun in rice - wheat cropping system in West Bengal. *J. Mycopathol. Res.* **45**: 195-199.

Chattopadhya, S.B. and T.P. Mustafa. 1977. Behaviour of *Macrophomina phaseoli* and *Sclerotium rolfsii* with relation to soil fernore and soil pH. *Curr. Sci.* **46**: 226-228.

Chattopadhyay, S.B. 1953. Root rot and foot rot of wheat caused by *Sclerotium rolfsii* Sacc. and *Curvularia* specifera (Bain) Boed = *Helminthosporium tetramera* McKinney. *Science and Culture* **19**: 101-102.

Chaurasia, R.K., A.B. Mishra and S.P. Singh. 1973. Screening of wheat varieties against *Rhizoctonia solani. JNKYV Res. Jour.* **7**: 178-179.

Dickson, J.G. 1956. Diseases of Field Crops (2nd ed.), Mc Graw Hill Book Co., Inc., London.

Hedge, R.K., S. Kulkarni, A.L. Siddaramaiah and K.S. Krishna Prasad. 1980. Biological control of *Sclerotium rolfsii* Sacc. causal agent of foot rot of wheat. *Curr. Res.* **9**: 67-69.

Hewitt, H.G. 1998. *Fungicides in Crop Protection*, CAB International, U.K, 221 p.

Kauraw, L.P. and R.S. Singh. 1975. Studies on the root rot complex of wheat in India-I. Varietal reaction. *Indian J. Mycol. Plant Path.* **5**: 96-97.

Kulkarni, N.B. 1959. Foot rot disease of wheat. *Poona Agric. College Mag.* **49**: 240-242.

Kulkarni, N.B. and L. Ahmed. 1967. Studies on the basidial formation by *Sclerotium rolfsii* Sacc. VII. Modified medium including basidial stage by wheat isolate of *S. rolfsii*. *Science and Culture* **33**: 127-128.

Kulkarni, S. 1977. Soil borne, seed borne and seedling diseases of wheat. *Kisan World* **4**: 36-37.

Kulkarni, S. 1980. Chemical control of foot rot of wheat in Karnataka. *Pesticides* **14**: 29-30.

Kulkarni, S., A.L. Siddaramaiah and K.S. Krishan Prasad. 1978. Varietal resistance of promosing wheat cultivars to foot root disease of wheat. *Curr. Res.* **7**: 208-209.

Manjappa, B.H. 1979. *Studies on survival and variation in Sclerotium rolfsii* Sacc. M.Sc. Thesis, UAS, Bangalore, 86 p.

Marshall, D.S. 1982. Effect of *Trichoderma harzianum* seed treatment and *Rhizoctonia solani* inoculum concentration on damping – off in snapbean in acidic soils. *Plant Disease* **66**: 788.

Mishra, A.B. and S.P. Singh. 1969. Wheat disease situation in Madhya Pradesh, India. *PANS* **15**: 71-73.

Misra, R.P. and J.N. Chand. 1970. Efficacy of different fungicides against foot– and root–rots caused by *Sclerotium rolfsii* of wheat. *PANS* **16**: 327-330.

Nargund, V.B. 1981. *Studies on foot-rot of wheat caused by Sclerotium rolfsii* Sacc. in Karnataka. M.Sc. Thesis, UAS, Bangalore, 99 p.

Nargund, V.B., S. Kulkarni and R.K. Hegde. 1982. Studies on foot rot of wheat caused by *Sclerotium rolfsii* Sacc. in Karnataka. *Proc. 21st All India Wheat workers workshop*, UAS., Bangalore.

Prasad, B. and K.C.B. Chaudhary. 1984. Influence of potash and phosphorus on mortality of lentil due to *Rhizoctonia solani* and *Sclerotium rolfsii. Agricultural Science Digest* **4**: 59-61.

Ramarao, P. and U. Raja. 1980. Effect of soil moisture on development of foot and root rot of wheat and other soil microflora. *Indian J. Mycol. Plant Path.* **10**: 17-22.

Reddy, H.R. and B.G. Kulkarni. 1972. Studies on the influence of soil moisture on foot rot of wheat caused by *Sclerotium rolfsii* Sacc. *Mysore J. Agric. Sci.* **6**: 10-13.

Reddy, H.R., K. Fazalnoor and R.K. Hegde. 1971. Screening of commercial varieties and genetic stocks of wheat against foot rot caused by *Sclerotium rolfsii Sacc. Mysore J. Agric. Sci.* **5**: 252-256.

Saksena, H.K. and K. Kumar. 1971. Rhizoctonia root rot of cereals in India. *Proc. Second Inter. Symp. Pl. Pathol.*, IARI, New Delhi, 72 p.

Sharma, O.P. and A.C. Jain. 1967. Root rots of wheat in Madhya Pradesh. *Indian Phytopath.* **20**: 267-269.

Shukla, T.N., Z.U. Ahmad and S.K. Garg. 1972. Root rot of lentil. *Indian Phytopath.* **25**: 584-585.

Subramanian, L.S. 1928. Root rot and sclerotial disease of wheat. *Bull. Agric. Inst., Pusa* **17**: 7.

Upadhyaya, R. and M.S. Pavgi. 1967. Some new hosts for *Pellicularia rolfsii* (Sacc.) West. from India. *Science and Culture* **33**: 71-73.

Wiese, M.V. 1977. *Compandium of Wheat Diseases*. American Phytopathological Society, APS Press, St. Paul, Minnesota. 106 p.

2.4.4 Take-all Disease

The name 'Take-all' disease of cereals was applied to devastating seedling blight in Australia about 150 years ago. The disease is widely distributed and recognized to be an important determinant of yield in western Australia, Europe, South Africa, Japan and North-south America (Weise, 1977). In India, take-all disease was first recorded by Padwick (1940) from Pusa (Bihar) though its presence had been suspected before (Dastur, 1928). Ghurde (1963) isolated the fungus from Madhya Pradesh.

Damage to wheat due to take-all is related to the extent of foot (basal stem) and root colonization by the pathogen. The disease in India is more of academic interest but in other countries it causes significant reduction in yield. Losses in the Pacific North-west of America are estimated to be 10–50%. In high risk areas of western Australia, take-all disease accounts for yield losses up to 40%. In England and Wales, estimates are for losses between 1 and 4% (Hornby and Bateman, 1991; Hewitt, 1998).

Symptoms: The infection occurs from the seedling stage onward. The chief symptoms of the disease are most often observed soon after heading. Infected plants are stunted, foliage turns to pale green and tillering is reduced. The ear-heads become bleached and sterile. The disease, therefore is also known as 'white heads' and 'dead-heads'. Diseased plants have dark brown to black dry rot or decay on roots and basal stems. Mats of dark mycelial strands develop superficially under the lowest leaf sheath. Plants ripen prematurely and are prone to lodging. They break easily at their crown when pulled from soil.

Causal Organism: The take-all fungus was originally isolated from a grass and the name was described *Rhaphidophora graminis*. The pathogen involved with cereals was included under *Ophiobolus graminis* Sacc. Later it was renamed as *Gaeumannomyces graminis* (Sacc.) Arx. & Olive. Isolates from wheat are designated *G. graminis* var. *tritici* Walker.

G.graminis var. *tritici* develops perithecial initials among the mats of dark mycelia usually found at the base of stem under the leaf sheath. Perithecia erupt through leaf sheaths and produce mature ascospores in asci. The perithecia are black spherical to oblong, 200–400 µm in diameter with beaks 150 to 300 µm long. Asci are 10–15 × 80–130 µm, and each has a distinct apical ring 2–3 µm in diameter. The ascospores are 8 per ascus, narrow, thread like, hyaline, septate, taper towards the end and about 70–90 µm × 2.5–3.5 µm in size.

Disease Cycle: The fungus persists in the soil as mycelia and perithecia on plant debris. Under favourable environment during crop season, the perithecia become active and release ascospores in the soil. Ascospores are released after wet periods and transported from place to place by rain splash or wind. Most plant to plant spread of take-all takes place through runner hyphae (Fig. 46). In general, the infection mainly occurs by direct mycelial penetration of root and stem tissues (Zillinsky, 1983). Weise (1977) felt that hyphae infection is more important epidemiologically. Roots contacting hyphae are colonized superficially and penetrated through infection pegs beneath hyphopodia.

Root rot results from vascular occlusion. Infection of take-all disease is favoured at 12–18°C temperature. The density of inoculum in the soil is influenced by organic matter, alkaline, compacted, nitrogen and phosphorous deficient soil (Garret, 1940, 1970). There is also effect of soil texture on microbial abbreviation of saprophytic survival by take-all fungus of wheat. Ascopsores are reported to produce sugars in sandy-clay soil which help in cellulolysis of wheat straw, favouring multiplication of fungal hyphae (Garret, 1985).

Management

Resistance: Cultivars highly resistant to take-all disease are not available and many wheat genotypes differ in tolerance in USA/ Australia and Europe.

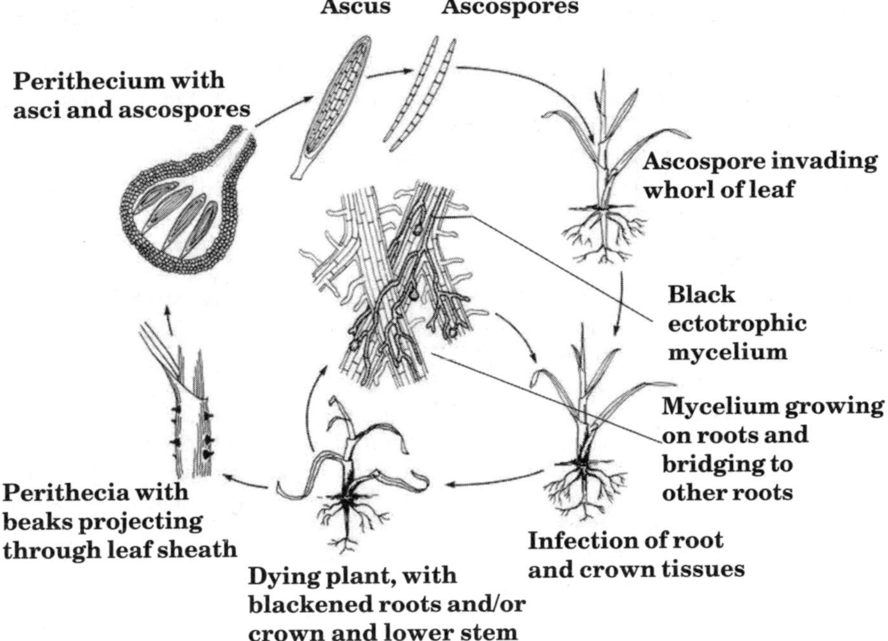

Figure 46: Disease cycle of take-all (*Gaeumannomyces graminis)* on graminaceous hosts.

Cultural Control: *G. graminis* var. *tritici* is a weak saprophyte and is reduced to insignificant levels with in 1 or 2 years if susceptible hosts are not available. Therefore, cultivation of non-susceptible crops like legumes as a rotational break is recommended to lower the inoculum potential.

Chemical Control: Seed treatments with triazole fungicides (triadimenol, flutriafol) provides some protection to roots in USA and Europe (Hewitt, 1998).

Biological Control: The biocontrol of *G. graminis* var. *tritici* has been successful. *Pseudomonas fluorescens* (Cook, 1988) and *Bacillus* spp. (Copper and Campbell, 1986) can be applied to the rhizosphere directly or through seed treatment. Most claims for the success of biological control come from the USA and Australia and several patents have been filed (Hewitt, 1998).

REFERENCES

Copper, A.L. and R. Campbell. 1986. The effect of artificially inoculated antagonistic bacteria on the prevalence of take- all disease in field experiments. *Jour. Appl. Bacteriology* **60**: 150-160.

Cook, R.J. 1988. Management of the environment for the control of pathogens. **In**: *Biological control of Pests, Pathogens and Weeds. Developments and Prospects*. (Eds. R.K.S. Wood and M.J. May), Royal Society, London, 61-72 p.

Dastur, J.F. 1928. Annual Report of the mycological section for the year ending March, 1927. Report Dept. Agric. Central Provinces and Berar for the year 1926-27, 7 p.

Garrett, S.D. 1940. Soil condition and take all disease of wheat. V. Further experiments on the survival of *Ophiobolus graminis* in infected stubble buried in soil. *Ann. App. Biol.* **25**: 199-204.

Garrett, S.D. 1970. *Pathogenic Root Infecting Fungi.* Combridge Univ. Press, London, 294 p.

Garrett, S.D. 1985. Effect of soil texture on microbial abbreviation of saprophytic survival by take all fungus of wheat. *Proc. Ind. Acad.* (Pl. Science) **94**: 85-90.

Ghurde, V.R. 1963. On the occurrence of *Ophiobolus graminis* Sacc. in India. *Curr. Sci.* **32**: 282.

Hewitt, H.G. 1998. *Fungicides in Crop Protection*, CAB International U.K. 221 p.

Hornby, D. and G.L. Bateman. 1991. Take-all disease of cereals. Home Grown Cereals Authority Review No. 20, 147 p.

Joshi, L.M., K.D. Srivastava, D.V. Singh, L.B. Goel and S. Nagarajan. 1978. *Annotated Compendium of Wheat Diseases in India*, ICAR, New Delhi, 332 p.

Padwick, G.W. 1940. A new disease of wheat in India. *Curr. Sci.* **9**: 179-180.

Kulkarni, Srikant and V.B. Nargund. 1986. Soil borne diseases of wheat. **In**: *Problems and Progress of Wheat Pathology in South Asia*, Malhotra Publishing House, New Delhi, 401 p.

Rangaswami, G. 1993. *Diseases of Crop Plants in India*. Prentice – Hall of India Pvt. Ltd., New Delhi, 498 p.

Wiese, M.V. 1977. *Compandium of Wheat Diseases.* American Phytopathological Society, APS Press, St. Paul, Minnesota, 106 p.

Zillinsky, F.J. 1983. *Diseases of Small Grain Cereals: A Guide to Identification*, CIMMYT, Mexico, 141 p.

Chapter

Viral Diseases

Viruses in wheat crop are not of major economic importance, though a few of them have been reported from India. As far as viral diseases are concerned, no wide-spread epidemic of any virus in the country has been on record. It is possible that either viral problems may be really absent or they do occur but go unreported due to lack of detection (Basu and Niazi, 1986). Secondly, symptoms of viruses may often be confused with similar to those caused by mineral deficiency diseases.

Information on the viral diseases in India is very limited. Generally, it deals with localized incidence of wheat viruses. In the recent past, the following viruses of wheat have been recorded:

1. Barley yellow dwarf virus : Aphid transmitted: *Rhopalosiphum maidis, R. padi, Sitobion miscanthi.*
2. Maize streak virus : Leaf hopper transmitted: *Cicadulina mbila*
3. Eastern wheat strait virus : Leaf hopper transmitted: *C. mbila*
4. Wheat spindle streak mosaic virus : Sap transmitted and soil borne

The viruses which have been found to infect wheat under experimental conditions are as under:

1. Cardamom mosaic streak virus : Sap and aphid transmitted: *R. maidis* and *R. padi, Brachycaudus helichrysi,*

		S. miscanthi, Micromyzus kalimpongensis transmitted
2. Barley mosaic virus	:	Sap and aphid transmitted: *R. maidis*
3. Maize vein enation virus	:	Leaf hopper transmitted: *C. mbila*

The above mentioned virus diseases are extremely limited in their prevalence, distribution and intensity and are of academic interest at present.

3.1 BARLEY YELLOW DWARF

Barley yellow dwarf (BYD) is the most wide-spread cereal virus attacking wheat, barley, oats, triticale, maize, rice and nearly 100 graminaceous grasses. Although the disease is believed to be present as early as 1890, the first report of its occurrence in USA dates back to 1951 (Oswald and Houston, 1951). Currently, barley yellow dwarf is recognized in several countries as a serious constraint to wheat cultivation (Bruehl, 1961; Plumb, 1983). In India, wheat is apparently free from this virus. In the course of survey Slykhuis (1962, 1963) recorded typical symptoms of the disease on barley, wheat and oats in the states of Uttar Pradesh, Punjab, Haryana, Madras and Bombay. The first authentic report of disease occurrence in the country was made by Nagaich and Vashistha (1963) from Shimla hills in 1958. The prevalence of BYD on large scale in the hilly regions of Mukteshwar and Bhowali (Uttaranchal) was subsequently recorded by Singh *et al.* (1979). The disease attacks wheat in Karnataka also (Kulkarni and Hegde, 1980).

Since the disease is sporadic in northern hilly region and it causes no serious damage to wheat crop, estimation of yield loss has not been attempted in the country. In other countries like USA, the annual loss may range from 1–3 percent. According to Burnett (1984) the disease can implict upto 40% loss under favourable conditions. Gill (1970) recorded 7% yield loss to wheat crop in Canada.

Symptoms: The symptoms of BYD are more pronounced on barley than wheat. Leaves of affected plants begin to turn yellow in about two to three weeks after infection. The yellowing progresses towards the base of lamina and a characteristic yellow colour replaces the normal green pigment of the leaves. The diseased plants are stunted and bushy in appearance (Plate 5C). Earhead formation is severely affected in infected plants (Nagaich and Vashistha, 1963).

Causal Virus: Barley yellow dwarf virus belongs to Luteovirus group (Randles and Rathjen, 1995). The virus particles are isometric - icosahedral measuring 25–30 nm in diameter consisting of single stranded RNA.

There exist different isolates of barley yellow dwarf virus. According to Rochow (1970) the isolates can be classified into five groups on the basis of vector specificity namely, RPV, RMV, MAV, SGV and PAV. Based on serological test these five isolates are categorized into two groups viz., BYDV I consisting of PAV, MAU and SGV isolates and BYDV II with RPV and RMV isolates of the virus (Rochow and Carmichael, 1979).

Transmission: BYD virus is not transmissible by sap inoculation. The isolates of this virus are transmitted by different insect vectors (Table 16).

Table 16. Vectors involved with transmission of BYDV isolates

Isolate	Vector
RPV	*Rhopalosiphum padi*
RMV	*R. maidis*
MAV	*Macrosiphum granarium or Sitobion avenae*
SGV	*Schizaphis graminum*
PAV	*R. padi and S. avenae*

Nymphs of the aphids can also effectively transmit the disease. Aphid vectors (*Sitobion gramineum* and *S. avenae*) have been shown to favour long distance dispersal of the virus in northern America. Similarly winged aphids transmit the disease in large scale in Canada. PAV isolate of the virus is wide spread and causes most severe symptoms (Miller and Raschova, 1997).

Host Range: Besides cereal hosts *i.e.* Wheat, barley, oats, triticale, more than 100 graminaceous hosts including *Avena, Agropyron, Agrostis, Lolium, Bromus, Dactylis, Panicum, Paspalum, Phleum, Poa* etc. get infection of barley yellow dwarf virus (Sharma *et al.* 2002). Latch (1977) noted that 50–80 % incidence of barley yellow dwarf virus is in grass-pastures.

Disease Cycle: There has not been any indepth study on the disease so far in India. Since the virus has a wide host range and insect vector, these sources are supposed to favour the survival of BYDV and transmit the virus during the wheat growing season. A report of BYDV indicates that October-sown wheat crop is prone to viral infection in hilly regions of Kumaon whereas December-sown crop remain free from barley yellow dwarf virus apparently due to lack of vector population (Singh *et al.,* 1979).

Management

BYD tolerant wheat genotypes like NS 879/4, Arjun, DWR 16, DWR 32, HD 2189, HD 2278, HW 657 an H 10–5–7 hold promise for breeding resistant varieties (Singh *et al.* 1979; Kulkarni and Hegde, 1983). Gene *BdV2* detected in *Agropyron intermedium* and *Aegilops binucialis* which has its origin in Ethiopia, may be useful for incorporating resistance in wheat.

Destruction of wheat hosts and volunteer plants may help in reduction of initial inoculum and population of insect vectors. Adjustment in planting dates may also be a good cultural practice.

3.2 MAIZE STREAK VIRUS

The maize streak virus (MSV) is prevalent in Africa and Asia. The disease in India was first recorded in 1972 at the Indian Agricultural Research Institute, New Delhi on wheat variety Hira (Seth *et al.*, 1972). Kulkarni *et al.* (1980) noticed the incidence of the disease on wheat variety Bijaga yellow at Dharwad (Karnatka) and Mali *et al.* (1983) reported this disease on wheat in Maharashtra. In Burma and Mauritius it appears in variable proportions (Ricaud and Felix, 1976). Maize streak virus has wide host range and is considered economically important in a number of hosts in some counties (Knoke *et al.*, 1992).

Symptoms: The virus affected plants are stunted and chlorotic. The younger leaves of the infected plants first show small stripes or streaks on one side of the mid ribs. Subsequently developed leaves are completely chlorotic with green area appearing as small islands. Leaf sheaths are also chlorotic. Long stripes run parallel to the mid ribs (Plate 5D). When the plants get infection at the seedling stage all the tillers simultaneously develop the disease symptoms. In case of late infections the existing tillers also develop sings of virus in due course of time. The earheads of the affected plants are smaller in size.

Causal Virus: On the basis of symptomatology, host range and vector, the virus has been identified as Pennisetum strain of maize streak virus. It is a Gemini virus. The virus particles are 30 × 20 nm in size.

Transmission: The virus is not transmitted by sap inoculation. The transmission on MSV is by the viruliferous jassid, namely *Cicadulina mbila*. Transmission may be possible through even a single nymph and adult. The transmission of this virus by the vector requires minimum acquisition feeding to be 10–15 minutes and the incubation period 7–9 days in the host at 25–30°C. The incubation period of MSV in *C. mbila* is stated to be 2–3 weeks during October-November and 4–6 weeks during December to February (Basu and Niazi, 1986).

Host Range: The maize streak virus occurs naturally on maize, pearl-millet, sorghum and barley.

Disease Cycle: The virus has several cereal hosts to survive during off season. Obviously, the incidence of the disease in the field depends upon the cropping sequence, availability of vector and susceptible host. Since the virus is transmitted by the vector (*C. mbila*) to regular wheat crop, the rate of spread of the disease in field is proportional to the vector population and its transmission efficiency.

Management

Destruction of volunteer plants and weeds can help in reduction of source of inoculum to be transmitted by the surviving vectors. For vector control application of insecticides like carbaryl, dimethote, phorate, aldicard, carbofuran etc. has been advocated (Sharma *et al.*, 2002). Host resistance is the better option for managing the disease. Choudhary *et al.* (1978) rated wheat strains CE 403, CE 407, Ce 408, HD 804, DH 2058, HD 2060, HD 2062, HD 2063, HD 2067, DH 2068, DH 2070, K 818, K 834, K 843, Raj 847, Raj 849, Raj 853, Raj 856, Raj 858, UP 252, UP 304, UP 325, UP 329, UP 331, UP 335, UP 336, WH 111, WH 114, WH 117, WH 118, WH 119, WH 120, WH 121, WG 325, WG 844, WL 350, WL 360, WL 368, WL 371, WL 377, WL 386, WL 387, WL 395, WL 396, WL 397 and WL 1002 as resistant (0.1-2.0% infection) under artificial epiphytotic conditions.

3.3 EASTERN WHEAT STRAIT VIRUS

The strait symptoms of disease in India were first observed on wheat and barley in Shimla hills in 1962 by Nagaich and Sinha (1974). As high as 35% of the plants were found to be affected in some fields causing about 22% loss in local wheat.

Symptoms: The main symptoms of wheat strait virus consist of fine chlorotic stripes developing on the leaves and also sometimes on leaf-sheaths which become more pronounced with aging of the plants.

Infected plants are invariably stunted and generally produce completely or partially sterile earhead with shrivelled grains of poor quality.

Causal Virus: The particles of the virus are polyhedral, 40 nm in diameter and scattered in the cytoplasm of the phloem cells of the host. The strain of the virus existing in India is claimed to be distinct from those causing strait disease in Europe, Australia and North America on the basis of its vector.

Transmission: The disease is not transmitted through sap of infected plants, seed, soil and aphids. The transmission of the virus is taken place through the leaf hopper, *Cicadulina mbila.*

Host Range: Besides wheat, barley is severely affected by strait virus. Symptoms as chlorotic stripes have also been noticed on *Narenga* grass around infested fields in Shimla hills (Nagaich and Sinha, 1974). However, the symptoms are milder as compare to wheat and barley.

Disease Cycle: The eastern wheat strait disease has not assumed major role in damaging the crop in the main wheat belt. Hence, a precise account of its disease cycle is lacking. However, surveys of wheat fields in Shimla and Kumaon indicate that the crop is infected before the advent of severe winter and disease appears all of sudden during March-April (Basu and Niazi, 1986).

Management

No specific control measures are recommended but protective measures are likely to be rewarding if adopted at early stage of growth to prevent primary spread of the virus.

3.4 WHEAT SPINDLE STREAK MOSAIC

The natural occurrence of wheat spindle streak mosaic virus (WSSMV) was recorded on wheat by Ahlawat *et al.* (1976) from Kalimpong region of Darjeeling in West Bengal. The disease is extremely limited in prevalence, distribution and intensity, and it is only of academic interest. Survey data reveal that disease incidence varies from 2–10% in different fields in the hilly tract of Kalimpong.

Symptoms: The main symptoms of the disease are consisted of distinct chlorotic and necrotic dashes and discontinuous streaks on wheat leaves (Plate 6A). Leaves of virus affected plants are abnormally erect and may roll adoxially.

Causal Virus: Slykhuis (1970) described wheat spindle streak mosaic as a soil-borne virus from Ontario. The virus particles are filamentous measuring 2000–3000 × 10 nm.

Transmission: The disease is transmissible on wheat by sap and soil. About 60% transmission of the virus is caused by sap and 30% by soil (Basu and Niazi, 1986). Being soil-borne its transmission is reported to be associated with fungus, *Polymyxa graminis.* Tests have failed to transmit the virus by insect vectors namely, *Myzus persicae* and *Cicadulina mbila.*

Host Range: WSSMV has been found to be associated mainly with wheat but it could also develop symptoms on other cereals like barley, oats and maize.

Disease Cycle: There is very little information regarding the occurrence and spread of the disease in the field. The infection of the virus is more pronounced near irrigation channels. The virus takes about 6–8 weeks to develop symptoms in the glasshouse when maximum temperature is below 14°C.

Management

Since the virus is not a serious threat to wheat cultivation in the plain, no specific control methods are advocated.

3.5 WHEAT STREAK MOSAIC

A virus disease which was originally described as 'Chirke' disease of large cardamom (*Ammomum subulatum*) by Raychaudhuri and Chatterjee (1961) was later referred to as wheat streak mosaic by Ganguli *et al.* (1968) because it was found to produce typical symptoms on wheat under natural and experimental conditions of infection. By and large the incidence of this disease is confined to Darjeeling district of West Bengal. The surveys have shown that disease sometimes makes its appearance in West Bengal, Bihar, Uttar Pradesh, Haryana, Himachal Pradesh and Madhya Pradesh also (Ganguli *et al.,* 1968). However, the prevalence of this virus in the field is negligible and it does not seem to pose a threat to cultivation of wheat in the main belt of the country.

Symptoms: The disease is characterized by mosaic streak symptoms on wheat leaves. These streaks with the passage of time coalesce gradually and eventually cause browning and drying up of the leaves. By using fluorescent microscopy, Mayee and Ganguli (1974) observed modification of nuclei as well as formation of inclusion body-like structure in the cytoplasm of epidermal cells of infected leaves.

Causal Virus: Ganguli and Raychaudhuri (1971) purified the virus by extraction in phosphate buffer followed by high and low speed centrifugation which revealed isometric polyhedral particles of 40 nm in diameter under election microscopic observation.

Transmission: The virus is sap transmissible and also spreads by the agency of aphids, *Rhopalosiphum maidis, R. padi* and *Brachycaudus helichrysi.* Chakraborty and Sardar (1984) identified *Micromyzus kalimpongensis* Basu as a new aphid vector of the disease. The transmission of wheat streak mosaic has been shown to be stylet-borne (Basu and Raychaudhuri, 1972).

Disease Cycle: The investigation undertaken by Basu and Raychaudhuri (1972) shows that *R. padi* and *S. avenae* play more important role in the spread of the virus in non-persistant manner. These seem to be more important than *R. padi* as vector in the Indian plains while *B. helichrysi* spreads the disease in the hills of Darjeeling region. According to Basu and Niazi (1986) there is lack of inoculum of the virus in the plains and the chances of the transmission of the virus through migratory vectors from long distance are very remote, so the possibility of disease appearance in the major wheat growing regions is rare.

Management

No control is recommended. Wheat genotypes *viz.*, Ridley, E 4647, E 6003 and E6831 proved to be resistant, whereas NP 717, N 745, NP 803 and NP 809 showed moderate resistance to the virus (Ganguli *et al.*, 1970).

3.6 BARLEY MOSAIC VIRUS

In the host range studies, the barley mosaic virus was found to cause infection in wheat plants mechanically (Dhanraj and Raychandhuri, 1969). But there has not been any report of its occurrence in the country since the disease was described (Basu and Niazi, 1986).

Symptoms: The affected plants are stunted, chlorotic and develop typical mosaic symptoms on leaves.

Causal virus: Electron microscopy reveals the virus particles to be spherical and 40 nm in diameter. The thermal inactivation point ranges from 53 to 55°C and dilution end point is between 1:1000 and 1:500. The virus gives negative results with barley stripe mosaic virus antiserum (Rayhandhuri and Nariani, 1977).

Transmission: The virus is readily transmitted mechanically to barley, wheat and oats. The disease is also aphid transmissible by vector *Rhopalosiphum maidis*. Dhanraj and Raychandhuri (1969) reported 2–45% seed transmission of the virus in different barley varieties.

Host Range: Besides barley, the virus can also infect wheat and oats.

Management

No specific control measures are recommended.

3.7 MAIZE VEIN ENATION

The maize vein enation virus was described from Kalimpong (Ahlawat and Raychaudhuri, 1976). The disease transmitted by *Cicadulina mibla* was also found to infect wheat plants under artificial conditions of inoculation. Symptoms of the disease in glasshouse infected plants were similar as that of maize showing leaf chlorosis, severe swelling and several white spindle shaped enations 13-15 days after inoculation.

REFERENCES

Ahlawat, Y.S., A. Majumdar and V.V. Chenulu. 1976. First record of wheat spindle streak mosaic in India. *Plant Dis. Reptr.* **60**: 782-783.

Ahlawat, Y.S. and S.P. Raychandhuri. 1976. Vein enation: A new virus disease of maize in India: *Curr. Sci.* **45**: 273-274.

Basu, A.N. and S.P. Raychadhuri. 1972. Further studies on transmission of wheat mosaic streak virus with notes on epidemiology of the aphid vectors. *India. J. Ent.* **34**: 115-117.

Basu, A.N and F.R. Niazi. 1986. Viral problems of wheat in India. **In**: *Problems and Progress of Wheat Pathology in South Asia* (Eds. L.M. Joshi, D.V. Singh and K.D. Srivastava), pp. 242-253. Malhotra Publishing House, New Delhi, 401 p.

Bruehl, G.W. 1961. Barley yellow dwarf, a virus disease of cereals. *Phytopathology Monograph* No. 1, 52 p.

Burnett, P.A. 1984. Barley yellow dwarf. *A Proceedings of the Workshop*, CIMMYT, Mexico, 6-12 pp.

Chakraborty, N.K. and K.K. Sardar. 1984. *Annual Report*, Div. of Mycology and Plant Pathology, IARI, New Delhi, 56 p.

Chaudhary, G.G., Singh, G. and G.C. Bhatnagar. 1978. Reaction of wheat varieties to maize streak virus in Rajasthan. *India Phytopath.* **31**: 403-404.

Dhanraj, K.S. and S.P. Raychaudhuri. 1969. A note on barley mosaic in India. *Plant Dis. Reptr.* **53**: 766-767.

Ganguli, B. and S.P. Raychaudhuri. 1971. Purification of wheat streak mosaic virus. *Phytopath. Z.* **70**: 11-14.

Ganguli, B., S.P. Raychaudhuri and M.R. Nimbalkar. 1968. Mosaic streak disease of wheat and its control. *Proc. 1st Summer Inst. on Pl. Dis. Cont.*, IARI, New Delhi, 78 p.

Ganguli, B., S.P. Raychandhuri, B.C. Sharma and M.R. Nimbalkar. 1970. Varietal resistance of wheat *Triticum aestivum* to mosaic streak. *Indian J. Agric. Sci.* **40**: 653-656.

Gill, C.C. 1970. Aphid nymphs transmit an isolate of barley yellow dwarf more efficiently than do adults. *Phytopathology* **60**: 1747-1752.

Knoke, J.K., R.E. Gingery and R. Louie 1992. Maize dwarf mosaic, maize chlorotic dwarf and maize streak. ***In**: Plant Diseases of International Importance – Diseases of Cereals and Pulses* (Vol. 1) (Eds. U.S. Singh, A.N. Mukhopadhyay, J. Kumar and H.S. Chaube), pp. 235-281.

Kulkarni, S. and R.K. Hegde 1980. Barley yellow dwarf of wheat – a new disease to Karnataka. *Curr. Res.* **9**: 119.

Kulkarni, S. and R.K. Hegde. 1983. Reaction of wheat varieties to barley yellow dwarf disease. *Plant Pathology Newsletter* **1**: 11-12.

Kulkarni, S., R.K. Hegde and A.B. Basavarajaiah. 1980. Occurrence of wheat streak in Karnataka. *Curr. Res.* **9**: 135.

Latch, G.C.M. 1977. Incidence of barley yellow dwarf virus in rye-grass pastures in

Newzealand. *Newzealand J. Agril. Res.* **20**: 87-89.

Mali, V.R., N.T. Vyanjana and A.U. Ekbote. 1978. Studies on a streak disease of wheat in Maharashtra. *J. Maharashtra Agric. Univ.* **3**: 76-77.

Mayee, C.D. and B. Ganguli. 1974. Fluorescence microscopy in relation to wheat mosaic streak virus disease. *Indian Phytopath.* **27**: 608-609.

Miller, W.A. and Raschova. 1997. Barley yellow dwarf viruses. *Ann. Rev. Phytopathol.* **35**: 167-190.

Nagaich, B.B. and R.C. Sinha. 1974. Eastern wheat striate: A new viral disease. *Plant Dis. Reptr.* **58**: 968-970.

Nagaich, B.B. and K.S. Vashistha. 1963. Barley yellow dwarf. A new viral disease for India. *Indian Phytopath.* **15**: 318-319.

Oswald, J.W. and B.R. Houston. 1951. A new virus disease of cereals transmitted by aphids. *Plant Dis. Reptr.* **35**: 471-475.

Plumb, R.T. 1983. Barley yellow dwarf virus- a global problem. **In**: *Plant Virus Epidemiology* (Eds. R.T. Plumb and J.M. Thresh), pp. 185-198. Blackwell Scientific Publications, Oxford.

Randles, J.W. and J.P. Rathjen 1995. Luteroviruses. **In**: *Virus Taxonomy. Sixth Rept. Int. Comm. Taxon. Viruses* (Eds. F.A. Murphy, C.M. Fauquet, D.H.L. Bishop, G.P. Ghabrial, A. W. Jarwis and G.P. Martelli), pp. 379-383. Springer Verlag, New York.

Raychaudhuri, S.P. and S.N. Chatterjee. 1961. 'Chirke' a new threat to cardamom. *Indian Fmg.* **11**: 11-12.

Raychaudhuri, S.P. and T.K. Nariani. 1977. *Virus and Mycoplasma Disease of Plants in India,* Oxford & IBH Publ. Co., New Delhi, 102 p.

Ricaud, C. and S. Felix. 1976. Identification and relative importance of virus diseases of maize in Mauritius. **In**: *Proc. 2nd South African Maize Breeding Symposium. Tech. Commu.* Dept. Agric. Tech. Serv., S. Africa. No. 142, 105 p.

Rochow, W.F. 1970. Barley yellow dwarf virus. *Description of plant viruses.* No. 32 Commonwealth. Mycol. Inst., Assoc. Appl. Biologists, Kew, Surrey, UK.

Rochow, W.F. and L.E. Carmichael. 1979. Specificity among barley yellow dwarf viruses in enzyme immunosorbent assays. *Virology* **95**: 655-660.

Seth, M.L., S.P. Raychandhuri and D.V. Singh. 1972. Occurrence of maize streak virus on wheat in India. *Curr. Sci.* **41**: 684.

Sharma, P.N., O.P. Sharma and S.K. Sharma. 2002. Virus diseases of cereal crops **In**: *Diseases of Field Crops* (Eds. V.K. Gupta and Y.S. Paul), pp. 353-371. Indus Publ. Co., New Delhi, 464 p.

Singh, D.V., H.C. Joshi, M.N. Singh, B.B. Nagaich and L.M. Joshi. 1974. Prevalence of barley yellow dwarf virus on cereals and occurrence of smut in triticales in Kumaon hills. *Indian Phytopath.* **32**: 98-99.

Slykhuis, J.T. 1970. Factors determining the development of wheat spindle mosaic caused by a soil brone virus in Ontario. *Phytopathology* **60**: 319-331.

Slykhuis, J.T. 1962. An international survey for virus diseases of grasses. *FAO Plant Prot. Bull.* **10**: 1-16.

Chapter

Nematode Diseases

Although the first plant parasitic nematode *Anguina tritici,* the causal agent of ear cockle disease of wheat was reported as early as 1743 (Needham, 1743), the importance of nematodes as limiting factors in agricultureal production was realized after World War II (Raski, 1958, Singh, 1963). Of late, a number of nemic species are being encountered in the rhizosphere of wheat crop. At present the wheat rhizosphere abounds in having over 25 species of parasitic nematode in India (Table 17), but only two nemic species are of major economic importance *i.e. Anguina tritici,* the ear cockle nematode and *Heterodera avenae*, the cereal cyst nematode (Seshadri and Sethi, 1978, Sethi and Dhawan, 1986).

Among these nematode species, six are endoparasitic and the rest are ectoparasitic. Most of the species are not of economic importance but with changing agricultural technology, they have a potential to become serious pest on wheat. For example, wheat gets moderate infection of *Meloidogyne incognita* under artificial inoculation (Pathak and Yadav, 1978). Wheat cultivated after tobacco-bajra rotation develops signs of yellowing and paling of the seedlings due to root knot (*M. incognita*) nematode in Gujarat. Under bajra-wheat rotation, the population of *Tylenchorhynchus* increases when wheat is sown after bajra (Singh, 2008). While the importance of other species cannot be underestimated, much of the work done in the country on nematode problems in wheat in mainly confined to ear cockle (ECN) and cereal cyst (CCN) nematodes.

Table 17. Plant parasitic nematodes recorded on wheat in India

Nematode species	Recorded by
Anguina tritici (Ear-cockle nematode)	Milne, 1919
Heterodera avenae ('Molya' disease)	Vasudeva, 1958
Meloidogyne javanica (Root-knot nematode)	Prasad *et al.*, 1964
M. incognita	Patel, 1984
Basiria nasikensis	Darekar & Khan, 1978
Tylenchorhynchus vulgaris	Upadhyaya & Swarup, 1981
T. eramicolus	Chabra and Bindra (1971)
T. (Merlinius) brevidens	Sethi & Swarup, 1968
T. ewingi	Sethi & Swarup, 1968
T. capitatus	Sethi & Swarup, 1968
T. mashhoodi	Sethi & Swarup, 1968
Pratylenchus thornei	Sethi & Swarup, 1971
P. brachyurus	Sethi & Swarup, 1971
P. neglectus	Sethi & Swarup, 1971
P. penetrans	Sethi & Swarup, 1971
P. coffeae	Sethi & Swarup, 1971
P. delattrei	Yadav & Verma, 1971
Hirschmanniella mucronata	Rashid *et al.*, 1973
H. oryzae	Rashid *et al.*, 1973
Hoplolaimus indicus	Yadav & Verma, 1971
Longidorus brevicaudatus	Khan & Khan, 1972
Merlinius brevidens	Upadhyaya and Swarup, 1981
Pseudhalenchus indicus	Bajaj *et al.*, 1981
Psilenchus graminis	Bajaj *et al.*, 1981
Rotylenchulus reniformis	Prasad *et al.*, 1964
Xiphinema americanum	Khan & Khan, 1972
Heterodera zeae	Srivastava & Swarup 1975
Helicotylenchus dihystera	Sethi, 1965

4.1 EAR COCKLE DISEASE

The ear cockle nematode enjoys a unique position in the nematological history as the first plant parasitic nematode ever reported (Needham, 1743). In India, ear cockle nematode in wheat was first recorded by Milne (Swarup and Gokte, 1986). Monitoring of the disease over past several years shows the prevalence of this pest in all wheat growing tracts of the country but the incidence of the disease is by and large low (Koshy and Swarup, 1971). However, in some fields the prevalence even goes upto 80 per cent (Seshadri and Sethi, 1978). At present ear cockle disease occurs in Bihar, eastern Uttar Pradesh, Rajasthan, Madhya Pradesh and Chattisgarh (Paruthi and Bhatti, 1985; Paruthi and Gupta, 1987; Nath and Pathak, 1993). Now the disease has almost vanished from Punjab, parts of Haryana and Jammu and Kashmir (Kaur and Sharma, 2003; Singh, 2008). Similarly, ear cockle which existed in all five continents of the world has been eradicated from Western Europe, North America, Australia and USSR by adopting modern improved cleaning procedures. However, ECN is damaging pest in 3rd world countries causing 20–50% loss. It is a pest of regulatory significance in developed countries (Singh, 2008).

In some parts of India, sporadic incidences with high intensity have been recorded from time to time. The disease appeared in Patna, Gaya, Munger, Darbhanga, Madhubani in Bihar in 1992 and 1997 and in Panna district of Madhya Pradesh in 1999, and damaged wheat crop seriously owing to severe intensity (Singh *et al.*, 2000). Annual loss on a country-wide basis is estimated to be 2 to 3 crores in rupees (Sethi and Dhawan, 1986). According to Gaur (2008) the ear cockle nematode with less than 1% level of incidence may cause huge losses amounting to Rs. 750 million at current price.

Nematode galls have been found non-toxic to rats (Midha and Swarup, 1974). Galls contaminate the wheat grain which ultimately leads to deterioration in the quality of flour. Flour milled from the grains with more than 5% galls has a dark colour and it is also not tasty (Swarup and Gokte, 1986).

Symptoms: 'Ear cockle' and 'Tundu' are two distinct phases of the same disease complex which are determined by environmental conditions. In such case both the nematode (*Anguina tritici*) and bacterium (*Corynebacterium tritici*) are associated. Under low humidity and high temperature the earcockle phase appears whereas Tundu phase is most conspicuous under high humidity and low temperature.

The first symptom of ear cockle infection can be observed with in 30 days of germination of the seedlings, in the form of slight swelling at the base which is followed by twisting and crinkling of the leaves (Plate 6B). The infested seedlings exhibit prostate condition with profuse tillering, giving the false impression of better crop stand. Severely infested seedling are stunted and may even die prematurely. Diseased plants produce ear heads earlier than healthy ones, usually shorter in size with spreading glumes bearing either very short or no awns at all (Gupta and Swarup, 1968).

The most characteristic symptom of the disease is the transformation of wheat grains in a ear into cockel or seed-galls (Fig. 47). Either all or a few grains are converted into gall. Normally, the number of galls produced in a spikelet varies from 1 to 5. Rarely galls can also be seen on awns. Seed-galls are green during early growth stage but become hard, dark brown to black bodies on maturity.

Figure 47: Ear cockle - (a) healthy wheat seeds (b) nematode galls.

Host range: *Anguina tritici* has a very limited host range, which includes *Triticum aestivum, T. durum, T. compactum, T. dicoccum, T. spelta,* rye, triticale and barley. Barley has been found to be poor host since fewer galls are formed. *Aegilops* sp. is highly susceptible (Dukel, 1957). Oat is not a host of ECN (Paruthi and Gupta, 1987; Dhawan and Pankaj, 1998).

Life Cycle: *Anguina tritici*, the causal agent of earcockle is an amphimictic species with only one generation per year (Fig. 48). Being highly host-specific, the nematode requires a wheat crop to complete its life cycle. The main source of inoculum is the nematode galls sown along

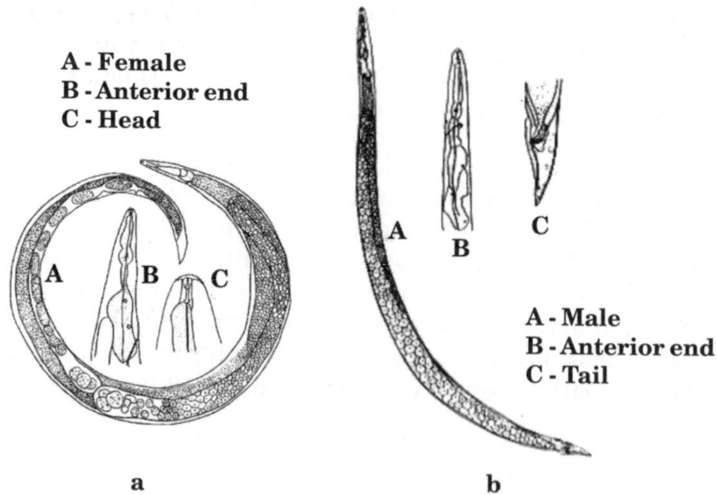

Figure 48: *Anguina tritici*. (a) Female nematode (b) Male nematode.

with seed material. The galls on reaching the soil, become moist and under suitable temperature and moisture, the larvae emerge out of the soften walls of the galls in 10–15 days. The optimum soil temperature and moisture for juvenile emergence have been found to be 15°C and 20% moisture condition respectively (Midha and Swarup, 1972). The larvae then migrate to the growing point of the young seedlings, while it is still below the soil base. The entry of the larvae into seed takes place both at brush and embryonic ends (Pathak and Swarup, 1983). These larvae feed ectoparasitically onto the growing point and are carried to developing ear heads along with lengthening of the clums. With the initiation of flower primodia, they invade the floral tissues, aborting further development of staminate as well as overial tissues. When the ovary is aborted, grain formation ceases and the larvae get enclosed in cavities formed by overgrowth of floral initials which eventually turn into galls.

After invasion of floral tissues, the larvae in the developing galls undergoes three successive moults with in 3–5 days before they become adult males and females, generally in the ratio of 1:2. By the time of galls turn brown by the middle of March, the adults die and the next generation of second-stage larvae in quiescent stage fill the entire cavity of the gall. Each nematode gall contains as many as 3000–12000 juveniles in quiescence or anhydrobiotic state. (Leukel, 1924). The re-infestation of the wheat crop takes place when the contaminated seed material with nematode galls is sown in the next crop season. The galls can maintain viability even after 32 years of storage (Limber, 1973).

The disease development is favoured in the sowing made in October and November when the soil temperature ranges between 20–30°C and 11–27°C respectively (Saxena and Khan, 1964). Maximum infection is reported to occur when the galls are placed in the soil at 2 cm depth and the larval concentration is 10^4/1000 g soil (Midha and Swarup, 1972).

Concomitant occurrence of ear cockle with hill bunt (*Tilletia foetida*), loose smut (*Ustilago tritici*) and Karnal bunt (*Neovossia indica*) in earheads has been observed (Mathur and Mishra, 1961, Paruthi and Bhatti, 1980, 1982) in wheat crop.

Management

Ear cockle disease is the seed contaminant. Hence, the use of clean and gall free seed is of utmost importance. Galls are smaller than the healthy wheat grains and by sieving, they can be separated. Water floatation method can also be applied for removal of galls from seed lot. For removal of galls, seed lots should be floated in 2–5 per cent brine solution. The light weight galls, which float on the water surface, can easily be separated and destroyed away. Seed, thus obtained should be washed 2–3 times in plain water and shade dried before sowing. The use of well controlled hot water treatment at 40–55°C for 60 min also kills nematodes in the planting material.

So for, none of the varieties have shown high degree of resistance to ear cockle except cultivar Saber-beg found to be resistant in Iraq (Al-Baldawi *et al.*, 1977). Dhawan *et al.* (1979) made an attempt to use a nematophagus fungus, *Arthrobotrys oligospora* for biological control of the nematode.

It is possible to control the disease in the field with the use of systemic nematicides like aldicarb and thionazin. However, the chemical control of the disease is not considered eco-friendly and economical because cleaning procedure is cheapest method of disease control.

4.2 CEREAL CYST NEMATODE

The cereal cyst nematode (CCN) was first recorded from Germany in 1874 by Khun (1874) and has been since reported from several countries including India. Its association with wheat as '*molya*' disease was established in 1958 by Vasudeva (1958). The disease now occurs practically in all wheat growing regions of the country including Rajasthan, Punjab, Haryana, Delhi, Jammu & Kashmir and Himachal Pradesh. The incidence of CCN is more in the dry and warmer areas of Rajasthan, Punjab and Haryana, and less in cooler climate (Mishra *et al.*, 2005).

A European origin of cereal cyst nematode, *Heterodera avenae* has been suggested by Meaghar (1977). The theory of European origin indicates that instead of wheat, oat and rye must be the original hosts since these crops have originated in Europe while wheat and barley are from Asia. It is speculated that the nematode cysts must have been carried from Europe to Asia with the Europeans engaged in overseas expeditions and later got adopted to different conditions, suitable for wheat growth (Swarup and Gokte, 1986).

A number of attempts have been made to estimate the loss caused by CCN. Singh and Swarup (1964) suggested that level of nematode population in soil is directly correlated with damage to crops and the causal agent of the disease is capable to implicate losses between 50–100 per cent. Annual monetary loss to the tune of 80 million rupees in Rajasthan was estimated by Berkum and Seshadri (1970). Gaur (2008) noted that *molya* disease causes losses worth Rs. 400 million in Rajasthan alone.

Symptoms : In nematode-infested wheat fields, usually patchy growth of plants is visible. The number of patches being one or more depends on degree of infestation. Each patch seldom exceeds 1 meter in diameter. These patches tend to spread with monoculture and may cover the entire field within a span of 3–4 years. Besides patchy growth of stunted plants and general chlorosis, the affected plants are stiffer, thinner with fewer leaves and narrower leaf blades. Tillering is greatly reduced having thin and weak clums. Emergence of the earheads, if formed, is delayed, have very few grains. If the infestation is uniformly heavy, there is practically no grain formation.

The disease is characterised by the symptoms on roots. The main root of the diseased plant is elongated with excessive branching of rootlets at the extreme end, giving it a bunchy and twiggy appearance. These rootlets show slight pearl like swelling at the point of cyst attachment (Plate 6 C&D). As many as 15–20 lemon shaped cysts can be found on a single plant root. The size of the cyst is about 600–70 µ long and 400–50 µ wide. Such plants can easily be pulled out of the soil (Seshadri and Sethi, 1978; Swarup and Gokte, 1986). Roots of infected plants become predisposed to soil-borne diseases like root rot (Megher and Chamber, 1971).

Host Range: The host range of *H. avenae* is mainly restricted to graminaceous hosts. In addition to wheat, barley and oats, the nematode also infects *Alopecurus geniculatus, Arrhenatherum elatius, Avena brevis, A. byzantina, A. sterilis, Agrostis castellana, Bromus tectorum, B. arvensis, Cenchrus ciliaris, Dactylis glomerata, Elymus caninus, Echinocloa frumantacea, Festuca arundinacea, F. kashmiriana, F. pratensis, Holcus lanatus, Lolium multiflorum, L. perenne L. temulentum, Melica ciliata, Poa bulbosa, P. pretensis, P. trivialis, Phalaris canariensis, P. paradoxa, Polypogon monspeliensis, Stipa lagascae, Secale cereale, Vulpia ciliata, V. muralis, V. myuros* and *Zea mays*. The only non-graminaceous host-recorded is *Senebiera pinnatifida,* belonging to family cruciferae (Gill and Swarup, 1971; Valdeolivas and Romero, 1983). Maize is considered to be an inefficient host since there is post larval mortality and the females do not break out through root cortex, resulting in hampered fertilization. Sorghum is also considered a weak host for the nematode (Swarup and Gokle, 1986).

Life Cycle: *H. avenae* exhibits sexual dimorphism. Males are vermiform and residents of soil while females are lemon shaped bodies turning to brown cysts after death. Under Indian conditions, the white females protuding from the infested roots can be seen clearly. These cysts require a dormancy which extends from April to October, though availability of congineal temperature, may shorten dormancy period by at least two months (Gill and Swarup, 1971). During sowing season, the second-stage larvae emerge from the cyst in late October after the embryo undergoes the first moult within the egg. The emergence of second-stage larvae continue upto March. The optimum temperature for hatching and larval emergence is about 20°C. But the emergence process can take place at wide range of 10–25°C. Alternating or fluctuating temperature seem to stimulate larval emergence (Swarup and Gill, 1972).

The free living second-stage larvae of *H. avenae* hatched out of egg constitute the infective stage of the nematode. Soon after emergence, they migrate through the soil in search of suitable host plant for penetration. The penetration takes place usually in the meristematic tissues behind the root cap, and occasionally in other parts of the root as well. A small elbow develops which is the first sing of the presence of the nematode in the plant. The larvae are attached to the root mostly by their neck while the most of the body remain outside the root. Immediately after penetration the nematode moves through the cortex towards the steal where they established themselves for feeding. Their movement in the host has been observed to be both inter and intra-cellular. Eventually, the larvae position themselves parallel to the stealer region with head embeding in the parenchymatous cells of vascular tissues. The movemnt of the larvae in the cortex results in some damage to cortical cells. The nematode at this stage become sedentary and under goes three moults over a period of 4-5 weeks and then white swellon females protrude from the roots in late December- February.

The sexes of the nematode can be distinguished in the third stage larvae. The female developed paired ovaries while in the male a single testis is formed. The matured third-stage larva is slender in appearance, slightly tapering at the anterior end and has a short round tail and 20–30 µ long stylets. It is stouter than second-stage female larva. Within a short span of time fourth moult occurs and the fully developed fifth stage adult male larva leaves the roots, wanders in the soil and then dies. At the same time, adult female larva formed after the fourth moult, grows into typical lemon shaped structure which retain about 200–500 eggs within the body. Soon after female dies, its cuticle hardens into tough resistant brown cyst. The cysts help the nematode to survive in the soil to cause infection in the next crop season (Fig. 49). The nematode can complete only one generation in an year. Male larva is essential for normal reproduction but parthenogenetic reproduction may occur at a very low frequency under abnormal conditions (Dhawan and Swarup, 1984).

Pathotypes: Biological races or biotypes of *H. avenae* have been known to occur in the country. The existence of two different biotypes in Ambala and Mahendrgarh was reported in Haryana. Jaipur and Udaipur populations also proved to be different in their reaction on same differential hosts. Later on, it was established that Sirsa population was different than Mahendregarh. Punjab population also exhibited different reactions as compared to Rajasthan and Southern Haryana (Singh, 2008).

Management

Cultural Practices: Good crop husbandry practices improve the plant tolerance of nematode attack. Fallowing of field and cultivating non-host

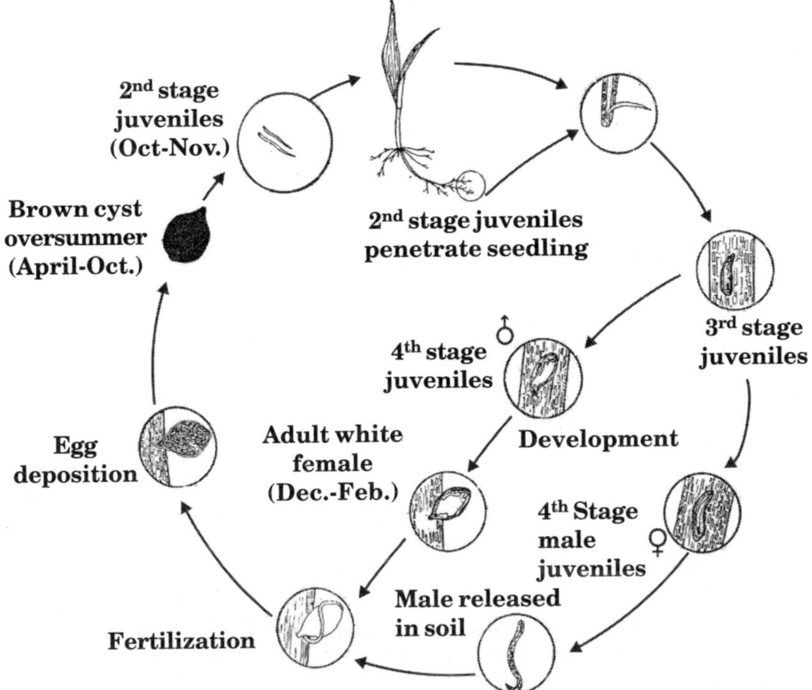

Figure 49: Life cycle of *Heterodera avenae*

crops like carrot, onion, mustard and chickpea reduces the population of *H. avenae* considerably. Deep ploughing at weekly intervals during May-June results in decrease of *molya* nematode population (Sethi and Dhawan, 1986). Since *H. avenae* causes more damage to wheat in soils deficient in Mn, proper application of this micro-nutrient may help in overcoming nematode damage (Gaur, 2008).

Resistance: None of the wheat cultivars exhibits any resistance to cereal cyst nematode in the country. CCN resistant variety CCNRV-1 is advocated for cultivation in disease prone areas (Mishra *et al.*, 2005). Besides CCNRV-1, resistance to *H. avenae* has also been detected in wheat strains DWR174, CCNRV-3 and CCNRV-7 (Sharma *et al.*, 2002).

Chemical Control: Earlier the use of soil fumigants like DD @ 300 litre/ha and DBCP @ 45 litre/ha were used (Jain and Sehgal, 1982; Swarup *et al.*, 1976). Now granular nematicides like Aldicarb 10G and Carbofuran 3G @ 1.5 kg a.i. /ha, respectively are recommended (Handa *et al.*, 1978, 1980; Singh, 2008).

Biocontrol Agents: *Catenaria vermicola* and *Cylindrocarpon uniseptum* were found infecting juveniles and egg of *H. avenae* (Chowdhary and Dhawan, 1984; Chowdhary and Kaushal, 1984).

Alternative Approach: Alternative to exclusive chemical management strategy of molya disease through combination of *Trichoderma viride* + FYM + Carbofuran 3G @ 1.0 kg a.i/ha is being exploited for managing the disease (Singh, 2008).

REFERENCES

Al-baldawi, A.S., Z.A. Stephan and A.H. Alwan. 1977. Studies on the wheat gall nematode in Iraq. **In**: *Year Book of Plant Protection Research* 1, 1974-76, Iraq.

Chowdhry, P.N. and S.C. Dhawan. 1984. *Catenaria vermicola in Heterodera avenae* nematode- a new record. *Curr. Sci.* **53**: 96.

Chowdhry, P.N. and K.K. Kaushal. 1984. *Cylindrocarpon uniseptum* sp. Nov.- A new fungus from India: *Curr. Sci.* **53**: 205-207.

Dhawan, S.C. and Pankaj. 1998. Nematode pests of wheat and barley. **In**: *Nematode Diseases in Plants* (Ed. P.C. Trivedi), pp. 12-25. CBS Publishers and Distributors, New Delhi.

Dhawan, S.C. and G. Swarup. 1984. Some observations on reproduction in *Heterodera avenae* Woll. *Indian J. Nematol.* **14**: 206-207.

Dhawan, S.C., K.K. Jain and G. Swarup 1979. A nematode trapping fungus of *Anguina tritici* larvae. *India J. Nematol.* **9**: 180-182.

Gaur, H.S. 2008. Root health and nematode problems of agricultural crops with special reference to wheat based cropping systems. **In**: *Compendium of Lectures on Integrated Pest Management in Wheat based Cropping Systems* (Eds. A.K. Sharma and D.P. Singh), pp. 143-154. Directorate of Wheat Research, Karnal, 257 p.

Gill, J.S. and G. Swarup. 1971. On the host range of the cereal cyst nematode, *Heterodera avenae* Woll. 24, the causal organism of molya disease of wheat and barley in Rajasthan, India. *Indian J. Nematol.* **1**: 63-67.

Gupta, P. and G. Swarup. 1968. On the ear cockle and tundu diseases of wheat. I. Symptoms and histopathology. *Indian Phytopath.* **21**: 318-322.

Handa, D.K., R.L. Mathur and B.N. Mathur. 1978. Studies on the integrated control of 'molya' disease of wheat and barley caused by *Heterodera avenae. Indian J. Mycol. Plant Pathol.* **8**: 70.

Handa, D.K., R.L. Mathur, B.N. Mathur and G.L. Sharma. 1984. Survey and assessment of losses in wheat and barley by 'molya' disease caused by *Heterodera avenae* in the light soils of Rajasthan. *Indian J. Mycol. Plant Pathol.* **8**: 73.

Jain, R.K. and S.P. Sehgal. 1982. Efficacy of nematicides in controlling the cereal cyst nematode, *Heterodera avenae* on wheat in light sandy soil in Rajasthan. *Indian J. Mycol. Plant Pathol.* **12**: 37-39.

Kaur, D. and R.C. Sharma. 2003. Effect of seed replacement on the infestation of ear cockle nematode of wheat in Punjab. *Seed Res.* **31**: 228-233.

Khun, J. 1874. Uber das Vorkommen von Rubennematoden an den Wurzein der halmfruchte. Z. Wiss. Landw. *Arch. Kgl Preuss.* Landes-Okon-Kolleg **3**: 47-50.

Koshy, P.K. and G. Swarup 1971. Distribution of *Heterodera avenae, H. zeae, H. cajani* and *Anguina tritici* in India. *Indian J. Nematol* **1**: 106-111.

Leukel, R.W. 1924. Investigation on nematode disease of cereals caused by *Tylenchus tritici. J. Agric. Res.* **27**: 925-926.

Limber, D.P. 1973. Notes on the longevity of *Anguina tritici* (Steinbuch 1799) Filipjev 1936 and its ability to invade wheat seedlings after thirty two years of dormancy. *Proc. Helm. Soc.*, Washington (USA). **40**: 272-274.

Mathur, R.S. and M.P. Misra. 1961. Simultaneous occurrence of *Tilletia foetida* (Wall.) Liro and *Anguillulina tritici* (S)G. Ben. in the same ear and grains of wheat in Pauri Garhwal, Uttar Pradesh. *Curr. Sci.* **30**: 307.

Meagher, J.W. 1977. World dissemination of the cereal cyst nematode (*Heterodera avenae*) and its potential as a pathogen of wheat. *J. Nematol.* **9**: 9-15.

Meagher, J.W. and S.C. Chamber. 1971. Pathogenic effects of *Heterodera avenae* and *Rhizoctonia solani* and their interaction on wheat. *Australian J. Agric. Res.* **22**: 189-194.

Midha, S.K. and G. Swarup. 1972. Factors affecting development of ear-cockle and tundu disease of wheat. *Indian J. Nematol.* **2**: 97-104.

Midha, S.K. and G. Swarup. 1974. Studies on the wheat seed gall caused by *Anguina tritici. Indian J. Nematol.* **4**: 53-63.

Mishra, B., Jag Shoran, R. Chatrath, A.K. Sharma, R.K. Gupta, R.K. Sharma, Randhir Singh, J. Rane and A. Kumar. 2005. *Cost effective and Sustainable Wheat Production Technologies*. Directorate of Wheat Research, Karnal, Technical Bulletin No. 8, 36 p.

Nath, R.P. and K.N. Pathak. 1993. Widespread occurrence of ear cockle disease of wheat in Bihar. *Indian J. Nematol.* **23**: 124-130.

Needham, T. 1743. A letter concerning certain chalky tubulous concretions called malm; with some microscopial observations on the farina of the red lily, and the worms discovered in smutty corn. *Phil. Trans. R. Soc.* **42**: 173; 174; 634-641.

Paruthi, I.J. and D.S. Bhatti. 1980. Concomitant occurrence of *Anguina tritici* and Karnal bunt fungus (*Neovossia indica*) in wheat seed galls. *Indian J. Nematol.* **10**: 256-257.

Paruthi, I.J. and D.S. Bhatti. 1985. Estimation of loss in yield and incidence of *Anguina tritici. International Nematology Network Newsletter* **2**: 13-16.

Paruthi, I.J. and D.S. Bhatti. 1982. Simultaneous occurrence of tundu and loose smut on wheat. *Indian J. Nematol.* **12**: 106.

Paruthi, I.J. and D.C. Gupta. 1987. Incidence of tundu in barley and Kanki in wheat field infested with *Anguina tritici. HAU Jour. Res.* **17**: 78-79.

Pathak, A.K. and B.S. Yadav. 1978. Host range of root-knot nematode *Meloidogynae incognita* on selected graminaceous field crops. *Indian J. Mycol. Plant Pathol.* **8**: 74.

Pathak, K.N. and G. Swarup. 1983. Migration of *Anguina tritici* juveniles to germinating wheat and barley seed and the mode of entry. *Indian J. Plant Pathol.* **1**: 147-152.

Raski, D.J. 1958. Plant parasitic nematodes. **In**: *Plant Pathology Problmes and Progress,* 1908-1959, Dept. Agric., W. Australia, 384-394 pp.

Saxena, S.K. and A.M. Khan. 1964. Effect of temperature on the development of ear-cockle disease of wheat and reaction on 16 varieties to *Anguina tritici* (Steinbuch, 1799) Filipjev, 1936. *Labdev J. Sci. Technol.* **2**: 238-239.

Seshadri, A.R. and C.L. Sethi. 1978. Nematode problems. ***In**: Wheat Research in India*, ICAR, New Delhi, 244 p.

Sethi, C.L. and S.C. Dhawan. 1986. Nematode diseases of wheat in India. **In**: *Problems and Progress* of *Wheat Pathology in South Asia* (Eds. L.M. Joshi, D.V. Singh and K.D. Srivastava), pp. 254-276. Malhotra Publ. House, New Delhi, 401 p.

Sharma, A.K., D.P. Singh, J. Kumar, M.S. Saharan, K.S. Babu, A.K. Singh and S. Nagarajan. 2002. *Disease and Pest Resistant Genotypes of Wheat and Triticale*. Research Bull. No. 15, Directorate of Wheat Research, Karnal, 18 p.

Singh, A.K. 2008. Nematode problems in wheat-based system. **In**: *Compendium of Lectures on Integrated Pest Management in Wheat based Cropping System* (Eds. A.K. Sharma and D.P. Singh), pp. 155-161. Directorate of Wheat Research, Karnal, 257 p.

Singh, A.K., A.K. Sharma and S. Nagarajan 2000. Ear cockle disease of Wheat-Management at ease. *Wheat Extension Bulletin 10*, Directorate of Wheat Research, Karnal.

Singh K. and G. Swarup. 1964. Studies on the population of *Heterodera avenae* Wollen. causing molya disease of wheat and barley in Rajasthan. *Indian Phytopath.* **17**: 212-215.

Singh, R.S. 1963. *Plant Diseases*. Oxford & IBH Publishing Co. New Delhi, 512 p. (Revised, 1973).

Swarup, G. and Gill, J.S. 1972. Factors influencing emergence of larvae from cysts of *Heterodera avenae* Woll. 1924. *Indian J. Exptt. Biol.* **10**: 219-223.

Swarup, G. and N. Gokte 1986. Nematode diseases in wheat **In**: *Plant Parasitic Nematodes of India- Problems and Progress* (Eds. G. Swarup and D.R. Dasgupta), IARI, New Delhi, 497 p.

Swarup, G., R.L. Mathur, A.R. Seshadri, D.J. Raski and B.N. Mathur. 1976. Response of wheat and barley to soil fumigation by DD and DBCP against 'molya' disease caused by *Heterodera avenae. Indian J. Nematol.* **6**: 150-155.

Valdeolivas, A. and M.D. Romero. 1983. New hosts of *Heterodera avenae. Nematologica Mediterrnea* **11**: 205-207.

Van Berkum, J. and A.R. Seshadri. 1970. Xth International Symposium, Italy, 136-137 pp.

Vasudeva, R.S. 1958. *Annual Report*, Indian Agricultural Research Institute, New Delhi.

Vasudeva, R.S. and M.K. Hingorani. 1952. Bacterial disease of wheat caused by *Corynebacterium tritici.* Phytopathlogy **42**: 291-292.

Chapter

Bacterial Disease

Wheat hardly suffers much from any bacterial disease in India except yellow ear rot or 'tundu' disease caused by *Corynebacterium tritici*. However, the reports of the following pathogenic bacteria associated with wheat in other countries, are available in literature (Zillinsky, 1983):

Bacterial stripe/black chaff	:	*Xanthomonas translucens*
Basal glume rot	:	*Pseudomanas atrofaciens*
Bacterial leaf blight	:	*Pseudomonas syringae*
Yellow ear rot/spike blight	:	*Corynebacterium tritici*

The occurrence of these diseases in India has not been reported so far, although a bacterial disease of wheat caused by *Pseudomonas tritici* was recorded in Punjab by Chaudhary (1935) and *Xanthomonas translucens* was noticed in experimental plot containing wheat varieties from USA (Vasudeva, 1954) but these two pathogens have not appeared any where in the country. The only bacterial disease which is worth mentioning is yellow ear rot.

5.1 YELLOW EAR ROT

Yellow ear rot of wheat is also referred as spike blight and is commonly known as 'tundu' disease. The disease occurs in North Africa, Australia and Asia. The disease in India occurs in Punjab, Haryana, parts of Uttar Pradesh and Rajasthan. It was first reported by Hutchinson (1911), as a bacterial disease. However, it was demonstrated that presence of ear cockle nematode (*Anguina tritici*) is essential for disease development (Vasudeva and Hingorani, 1952). All attempts to reproduce the disease with bacterium alone have failed. The bacterium itself could not induce 'tundu' symptoms, though a faint yellow streak was produced by artificial inoculation. Based on these observations Gupta and Swarup (1968, 1972) conclusively proved that the nematode larvae act as vector and carry the bacterium to the plant.

Symptoms: Yellow ear rot disease is generally confined to the ears. It is characterised by the production of sticky, yellow slime or gum on abortive spikes. Spikes emerging from the sheath are covered with a mass of sticky bacterial exudate. This yellow slime can be seen trickling down the infected ears under humid conditions. Under dry conditions the slime becomes hard, brittle and yellowish brown in colour. Clums, which are infected, die in young stage or grow until heading stage. The flag leaves are distorted. The emerging spikes are narrow and smaller than the healthy ones (Plate 7A). Diseased ears produce little or no viable seed.

Disease Cycle: The disease is dispersed through contaminated seed, or soil in association with nematode galls. The bacterium has been found to be present both inside the gall as well as on its surface (Swarup and Singh, 1962, Gupta, 1966). In the soil the larvae of the nematode are liberated, which carry the bacterial inoculum upto developing ears to produce aborted spikes having seed galls.

A mixture of $10^4 \times 5$ larvae of the nematode and bacterial suspension of 0.40 optical density is reported to give significantly high infection. In general the expression of 'tundu' disease is more under low temperature and high humidity. Seshadri and Sethi (1978) observed that maximum disease symptoms in 25 days old seedlings develop at 100% relative humidity and average temperature of 25°C. Under such conditions the multiplication of *C. tritici* is rapid and yellow bacterial slime is produced profusely.

Management

Since the source of inoculum of 'tundu' disease is contaminated seed material with nematode galls and the nematode is the vector, the effective method of its control is the use of clean seed free from the galls as advised for the management of ear-cockle disease.

REFERENCES

Chandhuri, H. 1935. A bacterial disease of wheat in the Punjab. *Proc. Indian Acad. Sci.* **1**: 579-585.

Gupta, P. 1966. *Studies on 'Ear-cockle' and 'Tundu' Diseases of Wheat*. Ph.D. thesis, IARI, New Delhi.

Gupta, P. and G. Swarup. 1968. On the ear-cockle and tundu diseases of wheat. I. Symptoms and histopathology. *Indian Phytopath.* **21**: 381-383.

Gupta, P. and G. Swarup. 1972. Ear-cockle and yellow ear-rot diseases of wheat. II. Nematode-bacterial association. *Nematologica* **18**: 320-324.

Hutchinson, C.M. 1917. A bacterial disease of wheat in Punjab. Mem. *Dept. Agric. India, Bact. Ser.* **1**: 169-175.

Seshadri, A.R. and C.L. Sethi. 1978. Nematode Problems. **In**: *Wheat Research in India,* ICAR, New Delhi, 244 p.

Swarup, G. and N.J. Singh. 1962. A note on the nematode- bacterium complex in 'tundu' disease of wheat. *Indian Phytopath.* **15**: 294-295.

Vasudeva, R.S. 1954. Report of the Division of Mycology and Plant Pathology. *Sci. Rep. Agric. Res. Inst.,* New Delhi, 75-87 pp.

Vasudeva, R.S. and M.K. Hingorani. 1952. Bacterial disease of wheat caused by *Corynebacterium tritici. Phytopathology* **42**: 291-292.

Zillinsky, F.J. 1983. *Common Disease of Small Grain Cereals: A Guide to Idetification,* CIMMYT, Mexico, 141 p.

Chapter

Phanerogamic Parasites

Phanerogams are flowering, seed producing angiospermic plants. Some of the flowering plants are parasitic on agriculturally important crops and cause considerable loss (Kumar, 1940). The common parasitic flowering plants may belong to following classes-

Holoparasite: These types of phanerogams completely lack chlorophyll and hence are fully dependent of their host for elaborate nutrition. Examples of holoparasites or complete parasites are *Orobanche* spp. and *Cuscuta* spp.

Partial Parasite: Partial parasites are green chlorophyllous plants and are dependent on the host only for water and mineral nutrients. Important parasites in this category are Striga *spp.* and *Loranthus* spp.

In India, wheat crop suffers very little from the attack of phanerogams, although the parasitism of *Orobanche* sp. (Sharma, 1953) and *Striga densiflora* (Chaudhari and Dhake, 1965) with wheat has been recorded. These weedy parasites pose no threat to wheat and are not of any economic significance.

6.1 *OROBANCHE*

This is an angiospermic annual, monocot plant, parasitic on wheat roots. *Orobanche* is commonly known as broomrape. It was found to attack wheat variety NP52 which was growing alongwith *Brassica campestris* var. *sarson* in 1951 by Sharma (1953).

Symptoms: The tissues of the parasite, *Orobanche* merge with those of the roots of wheat to form a swelling from which aerial flowering shoot arise. When the host is carefully up rooted, the parasitic roots are seen intertwined with host's root system. As a result of the drain of nutrient and mineral supply, the growth of the host plants is checked. Affected wheat plants loose vigour and exhibit poor growth. In severe attack the plant may even die.

Parasite: The parasite (*Orobanche sp.*) appears as clumps of whitish, yellowish, brownish or purplish annual stems arising from the ground at or near the base of the host plants. The stem of the parasite is stout, fleshy, 15 cm or more in height and is covered by small, thin and brown scaly leaves. The flowers appear in to axil of leaves and are tubular. A large number of tiny seeds are produced in capsular fruits (Plate 7C).

Disease Cycle: Broomrape survives as seed which remains dormant in the soil for a number of years. Seeds of the parasite burried in the field germinate when the sowing of host is done. The germination of seed is stimulated by the root exudates of host plants. On germination the primary root of the parasite forms a nodule of tissues in contact with the host root. New roots and stem of the parasite make new contacts with other host roots, giving rise to a large number of parasitic plants all over the field near the host plants. On maturity the seeds are released from the capsular fruit in the soil which start fresh infestation during next crop season.

Eradication: Spraying of soil with 25% copper sulphate ($CuSO_4$) has been reported to be successful in destroying the parasite (Singh, 1963). Linke and Saxena (1989) reported effective control of broomrape in the lentil field by applying 120 g a.i./ha of glyphosate at the vegetative stage.

Orobanche spp. can be controlled by soil-solarization (Sauerborn and Saxena, 1987). Effective solarization is achieved by using polythene sheets on soil during summer hot season for 30-50 days.

The best protection against broomerape is to separate the seeds mechanically from the host before sowing. The emerging shoots of the weed-parasite should be pulled out by hand at the time of browning of flowers but prior to maturity of the capsules and seeds. The shoots should be destroyed or can be fed to animals. *Orobanche* infestation can be reduced through long crop rotation with non-wheat and mustard crops.

6.2 *STRIGA*

Striga is commonly known as witch-weed. The species of *Striga* namely, *S. asiatica, S. lutea* and *S. hermonthica* parasitise wheat, barley, oats, maize, sorghum, sugarcane and millets in Australia, South eastern Asia, South Africa and South eastern coastal region of USA (Weise, 1977). Roots of *S. densifolia* showing distinct haustorial connections with root system of wheat plants were collected by Chaudhuri and Dakhe (1965) for the first time in India. The naturally infested field at Badnapur (Maharashtra) showed an average of 13.68% atttack of the parasite. Wheat variety PW5 was severely affected showing attack of 27.7 percent.

Symptoms: Wheat plants infested with *Striga* are slightly chlorotic and generally poor in vigour. The roots of the parasite form haustoria which penetrate the host roots to draw water and nutrients. The heavy infestation of the weed outcompetes and kills host plants.

Parasite: Although the *Striga densifolia* is an obligate parasite but it is self-sufficient chlorophyllous flowering weed plant. The plants are erect, annual, fairly high (20–30 cm) at maturity. The leaves of the parasite are bright green with alternate leaf-arrangement. Flowers are tubular with a lipped corolla of yellow colour (Plate 7D). The seeds of *Striga* are very minute and are produced in great abundance in capsules.

Disease Cycle: Witch weed survives as seed which remain viable and dormant in the soil for a number of years. Seeds on germination develop into infections gerlings in response to root exudates of host plants. After germination the parasite grows below the soil surface for about 1–2 months and produces underground stem and roots. The roots of the parasite form haustoria which penetrate the host roots drawing water and minerals from the host. Once established on the host the parasite emerges in 4 to 6 weeks and produce mature seed with in 8 to 10 weeks to repeat the disease cycle.

Striga weed is more common in light sand and sandy loam soil and its growth in different soil types is influenced by temperature. Short distance dissemination of seeds takes place through rain and irrigation water while wind is the chief agent for long distance dissemination.

Eradication: In early stage of growth of the parasite, manual weeding can be useful. Flooding of infested field for some time and then draining out the water also helps in eradicating the weedy parasite. Amine salt formulations of 2, 4-D in rotation with hosts of *Striga* are preferable. Soil injection of methyl bromide under polythene cover has been successfully used to eliminate witch-weed (Singh, 1963).

REFERENCES

Chaudhari, B.B. and S.B. Dakhe 1965. Occurrence of *Striga densiflora* on wheat. *Sci & Cult.* **31**: 381.

Kumar, L.S. 1940. Flowering plants which attack economic plant I: *Striga. Indian Farming* **1**: 596.

Linke, K.H. and M.C. Saxena 1989. *Orobanche* studies: Chemical control. **In**: *Food Legume Improvement Programme*, Annual Report 1989, pp. 243-249, ICARDA., Syria.

Sauerborn, J. and M.C. Saxena 1987. Effect of soil solarization on *Orabanche* spp. infestation and other pests in faba bean and lentil. **In**: *Parasitic Flowering Plants*, (Eds. H.C. Weber and W. Forstreuter), Marberg, FRG, 733-744 pp.

Sharma, S.L. 1953. *Orobanche* on wheat 'NP 52' *Curr. Sci.* **22**: 56.

Singh, R.S. 1963. *Plant Diseases*. Oxford and IBH Publication, New Delhi, 564 p. (revised edition, 1978).

Wiese, M.V. 1977. *Compendium of Wheat Diseases*. American Phytopathological Society, APS Press, St. Paul Minnesota, 106 p.

Chapter

Nutritional Disorders

Mineral nutrients support soil fertility and plant growth. If the soil is nutrient deficient then the hunger sign of nutrient deficiency results in the crop from impaired metabolism within the plant. The role of nutrients for nourishment of plants and physiological process was recognised by Justus Von Leibig (1803–1873) who laid the foundation of mineral nutrition of plant as a scientific discipline (Lauchhi and Bieleski, 1983). Based on Leibig's observations and speculation 'mineral element theory' was proposed by Arnon (1950) which emphasises that essentiality of mineral nutrients should be referred to following criteria:

1. The element is required for the plant growth and reproduction.
2. It can not be replaced by another element.
3. It has a direct or indirect role in the metabolism of the plant.

Those elements which do not fulfil the above criteria but stimulate plant growth, are termed as beneficial elements.

The chemical analysis has revealed the involvement of more than 60 elements with plant. Among these, only 16 mineral elements *i.e.* carbon (C), hydrogen (H), oxygen (O), nitrogen (N), phosphorus (P), potassium (K), calcium (Ca), magnesium (Mg), sulphur (S), iron (Fe), manganese (Mn), zinc (Zn), copper (Cu), boron (B), molybdenum (Mo) and chlorine (Cl) are considered essential elements. The availability of carbon, hydrogen and oxygen by the plant is made from air and water where as rest of the elements are compensated by the soil in the form of oxides, salts, nitrates, phosphates, sulphates, sulphite etc. (Steward, 1963).

In terms of quantum of nutritional requirement, the essential elements are classified as micro-nutrient where the nutrient is required in trace amount and macro-nutrient where nutrients are consumed by the plant in large quantity. Currently seven micro-nutrients (B, Cu, Fe, Mn, Mo, Cl and Zn) and six macronutrients (N, P, K, S, Ca and Mg) are considered essential for plant growth (Singh and Bana, 1986).

7.1 FUNCTIONS OF ESSENTIAL ELEMENTS IN PLANTS

Essential mineral elements have specific and non-specific functions in plant growth. Their role in physiological processes like translocation, stomatal closing and opening, membrane integrity, protein synthesis, auxin metabolism, nucleotide synthesis, nitrogen fixation, nitrate reduction etc. is well known. An element may be an integral constituent of essential metabolic complex or macro-molecular assembly. They take important part in cellular organization, energy transfer, activation of enzymes in a enzyme-mediated process, electron transport etc. (Epstein, 1972; Mengel and Kirkby, 1982; Marschner, 1986). Tolerance to disease stress is enhanced in the presence of essential elements.

It is clear that a nutrient may have several functions within the plant. Symptoms that are most characteristic for a particular nutrient are those in which one specific function has a much greater requirement than other functions. However, according to Snowball and Robson (1991) confusion may occur when several elements play a role in the formation of a key component in physiological process (*e.g.* nitrogen and magnesium play role in chlorophyll formation) or when nutrient is involved in assimilation or metabolism of another nutrient (*e.g.* molybdenum deficiency in wheat promotes symptoms of nitrate toxicity). Thus the nutrient function varies for the essential elements as elaborated below:

Nitrogen (N): Nitrogen is the most abundant element in plants. It is a constituent of amino acids, proteins, nucleotides, co-enzymes and chlorophyll. It plays vital role in various metabolic processes of the plant. Nitrogen is available to plants as nitrate which is reduced to ammonia and then incorporated into organic compounds (Beevers and Hageman, 1969). N-deficient plants lack essential amino acids, thereby affect protein synthesis. Its deficiency also inhibits chlorophyll synthesis and causes chlorosis. Deposition of anthocyanin pigment in the epidermal cells of stem and leaves of wheat is caused due to N-deficiency (Blank, 1947).

Phosphorus (P): The major role of phosphorus is in energy metabolism. It is involved in synthesis of high energy bonding – adonosine triphosphate (ATP) and adonosine diphophate (ADP) which are needed during phosphorylation of compounds (Fuijino, 1967). Phosphorus occurs in phospholipids including those of membranes, in sugar phosphate, nucleotides and co-enzymes. It also activates some enzymes. Besides these physiological functions, phosphorus is an important element for cell division and constituent of cell membrane, chloroplast and mitochondria This element is essential in the timely differentiation and maturation of plant tissues. It is absorbed by the plants mainly as dehydrogen phosphate ion ($H_2P0_4^-$).

Potassium (K): The main role of potassium in plants is of an activator of several enzymes. It also stimulates amino acids which initiate protein synthesis through peptide bonds. Stomatal opening and closing is regulated by action of potassium (Fujino, 1967). Thus it affects the photosynthesis by regulating the exchange of CO_2 in leaves (People and Koch, 1979). Potassium also has a role in promoting the translocation of the photosynthates from leaves.

Sulphur (S): Sulphur is constituent of amino acids, proteins and co-enzymes *viz.*, cystine, cysticine and methionine, thiamine, biotin, ferredoxin and co-enzyme A. It has a role in protein synthesis and cellular organization (Singh and Bana, 1986).

Calcium (Ca): Calcium is an activator of enzymes. Amylase is the only enzyme known to have Ca as co-factor. Calcium pectate, the principal constituent of middle lamella of cell wall provides mechanical strength to plant tissues. It is immobile in the phloem of plants and plays an important role in the growth of meristematic tissue. (It and Fujiwara, 1967; Rasmussan, 1967). Thus Ca has a role in morphological resistance against fungal infections.

Magnesium (Mg): It is metabolic constituent of chlorophyll. Magnesium is an activator of almost all enzymes which act on phosphorylated substrates. Therefore, the role of Mg in energy metabolism and CO_2 assimilation is of paramount importance (Stocking and Ongun, 1962). Magnesium shows antagonism with calcium and potassium so absorption of one element in excess may cause deficiency of another element.

Boron (B): Boron plays key role in the transport of sugars in plants (Gauch and Dugger, 1954). It is also involved in nucleotide synthesis (Snowball and Robson, 1991).

Copper (Cu): This element is the constituent of ascorbic acid, oxidase, phenolases, plastocyanin and Cu-proteins (Vallee and Wacker, 1970; Snowball and Robson, 1991). Its essential metabolic role is in the biosynthesis of lignins and activity of polyphenol- oxidase (Graham, 1976). These roles of copper explain the correlation in copper nutrition with shoot growth and resistance to lodging as an indicator of stem strength as a result of lignin accumulation.

Iron (Fe): Iron is the constituent of ferrodoxin and iron porphyrins. It is actively associated in electron transport during physiological processes. Heme-proteins are the primary carriers of iron in plants. This element is essential for maintenance of chlorophyll and plays an important role as chelating agent in plants (Carell and Price, 1965).

Manganese (Mn): It is an activator of enzymes. Mn is essential for functioning of oxidative as well as carboxylating enzymes (Scrutton *et al.*, 1966). It is directly involved in electron transportation in chloroplast in photosystem II of photosynthesis (Chenial, 1970).

Molybdenum (Mo): It is an essenial element for the functioning of nitrogenase and nitrate reductase enzyme activity which is the first step in reduction of nitrate to ammonium. Thus Mo plays very important role in nitrogen metabolism (Evans, 1956).

Chlorine (Cl): This element is required for oxygen evolution in photosystem II in photosynthesis (Bove *et al.*, 1963). Chlorine deficient plant tend to accumulate free amino acids.

Zinc (Zn): It is activator of enzymes. Zinc is an essential constituent of enzyme carbonic dehydrogenase (Vellee and Wacker, 1972). It has a role in nucleotide synthesis and auxin metabolism (Skoog, 1940). The RNA and ribosomes contents sharply decline in plant cells deficient in zinc.

7.2 NUTRIENT DEFICIENCY SYMPTOMS

Visual symptoms, stripe plots, soil and plant tissue analysis are methods of identifying nutrient deficiencies. Many symptoms of deficiency are sufficiently characteristics and are adequate to permit visual identification (Fig. 71). Although the visual symptoms may be the first clue to nutrient deficiency but analysis of an appropriate tissue is required to confirm whether the diagnosis of a nutrient is correct. Stripe plots method is conducted to confirm the speculation regarding the presence of a deficiency. In this method comparisons are made between untreated and treated plots with desired nutrient. In these comparisons, the mineral element can be applied either to the soil or on the foliage to overcome the effect of the deficiency. For some nutrients, symptoms are less characteristic and their presence could indicate one of several possible stresses. In this situation plant and soil analysis is required to distinguish between such stresses (Chapmans, 1986; Snowball and Robson, 1983). Since the visual symptoms give an indication of nutrient deficiency, these are described as under.

Nitrogen: Nitrogen deficient wheat plants are pale in comparison to healthy ones. The yellowing is due to break down in chlorophyll production. The symptoms first appear on the oldest leaves which become yellowish with the new leaves remaining relatively green (Fig. 71a). The chlorosis begins at the tip and gradually merges into light green area further down the leaf. Ultimately oldest leaves become totally chlorotic, changing from yellow to almost white in colour (Jones, 1966).

In the field pale green or yellow patches are seen, which may spread rapidly in a short period. Affected plants show restricted growth of root and shoot. Shoots are short, thin with poor tillering and spikes are very small in size.

Phosphorus: The most characteristic feature of phosphorus deficiency in wheat during early stage of vegetative growth is the reduced growth and vigour of the plant. P-deficient plants appear dull dark green with slight mottling and coiling of the oldest leaf. Plant maturity is delayed and size of the earhead is reduced.

Specific symptoms are on the old leaves. Chlorosis appears at the tip of the leaf which gradually moves down (Fig. 71b). In contrast to N-deficiency, necrosis of chlorotic area is fairly rapid covering whole leaf. The base of the leaf stays dark green and the tip becomes purple to dark brown (Chapman and Mason, 1969).

Potassium: Deficiency symptoms of potassium are mostly visible in the oldest leaves. Severe K-deficiency causes necrosis in the old leaves which begins as a necrotic speckling along the length of the leaf blade. Bluish-green leaves may show tip and marginal scorching (Fig. 71 c). As a result of spread of necrotic tissue, an arrow of green tissue from the base upwards towards the center of leaf blade can be seen (Chapman, 1966).

K-deficiency in wheat may lead to stunting of plant with reduced internodes and excessive tillering. The tillers die off and surviving shoots are susceptible to lodging.

Magnesium: The symptoms of magnesium resemble those of potassium and iron deficiencies. However, in case of Mg-deficiency new leaves of wheat plants develop yellowing in contrast to the old leaves affected by K-deficiency. The new leaves show marked chlorosis and rolling of leaf blade (Fig. 71 e). Over times, the chlorosis of the leaves becomes mottled and necrotic. Wheat plants in the field may appear as patchy yellowing with reduced growth and having folded new leaves (Gauch, 1972).

Iron: Deficiency symptoms of iron and magnesium are more or less identical. In both deficiencies, new leaves are affected first and become chlorotic. However, the contrast between old and new leaves of the Fe-deficient plants is more marked than Mg-deficiency. Deficiency of iron causes longitudinal interveinal chlorosis during early stage of growth of wheat plants (Fig. 71 f). This characteristic feature is more regular for Fe-deficiency than for Mg or Mn deficiencies. Moreover, Fe-deficient plants remain relatively erect, while Mn deficient plants are generally floopy. If iron deficiency is severe, the new growth may appear without chlorophyll and turn white. The leaf apices of the chlorotic leaves become necrotic and wither away (Agarwal, 1971; Randhawa, 1967; Prince, 1968).

Sulphur: S-deficiency symptoms in wheat are similar to nitrogen chlorosis. But the deficiency of sulphur differs from N-deficiency in that overall yellowing in young leaves (instead of old leaves) appears with greater degree in whole plant. Leaf tip necrosis may appear (Fig. 71 d). Under extreme deficient situation of sulphur, earhead formation may also be influenced (Singh and Bana, 1986).

Manganese: In wheat, symptom of Mn-deficiency appears first as formation of light gray flecks or streak at the base of fully opened youngest leaf. This effect in new growth spreads from base towards the apex covering entire length of the leaf and the interveinal stripes develop in middle region of foliage. Although the stripe is interveinal, it differs from Fe in its irregularity and in the flecking associated with it (Fig. 71g). Mn deficiency exhibits yellow-patchy growth of the plants in the field (Rammamoorty and Desai, 1946; Agarwal, 1971; Keisling *et al.*, 1984).

Zinc: Generally, the symptoms of zinc deficiency in wheat appear on middle-aged leaves. Leaves develop interveinal chlorotic mottling and necrotic lesions. The necrotic areas soon get intensified and coalesce resulting in the collapse of the affected leaves near the middle. Consequently, the apical part of these leaves turn dry, papery and droop.

Severe Zn-deficiency in the field can result in stunted chlorotic plants with many collapsed leaves due to necrosis in the middle of the leaves. Whole field may be affected but in general there are chlorotic patches within the crop.

Calcium: Symptoms of calcium deficiency always appear in newly emerging leaves as necrotic spotting in the middle of the leaf. The affected area quickly expands and the leaf collapses midway before unrolling. Ca-deficient roots of wheat show stunted laterals (Chapman, 1966).

Copper: Wither tip is the characteristic feature of copper deficiency. Young wheat leaves show sudden dying and curling of the tip end of the leaf blade upto half of the length. In case of severe deficiency only few heads emerge. Affected heads develop full grain in the base, shrivelled grains in the middle and a withered necrotic tip called 'rat-tail' (Fig. 71h). The unfilled tip of the head is generally chlorotic and this condition has been named as 'white heads' by Wallace (1961).

Molybdenum: The specific symptom of molybdenum deficiency is the appearance of longitudinal yellow stripe on the middle leaves of wheat plant. Plants become very pale green and limp. Over times plants develop necrosis of the tips and margins of the old and middle aged leaves. Under severe deficiency maturity of plants is delayed and empty white coloured heads are produced.

7.3 CORRECTION OF NUTRIENT DEFICIENCY

Nutrient deficiencies occur in plants as a result of low levels of essential elements in the soil, or when they are present in a form that makes them unavailable to the plant. These physiologic disorders can be neutralized by application of fertilizers, farm yard manures or specific chemical compounds. However, blanket application of nutrient will be wasteful. Therefore, nutrients application is advisable only if soil or plant tissues analysis reveals deficiency of elements. The choice of analytical procedure is based on the nature of an element. For example, soil analysis method for estimation of nitrogen is of limited value because the transformation of N in the soil is greatly influenced by temperature and soil moisture. Soil analysis is not generally useful for micronutrients too. For micronutrients it has proved difficult to find an extractant that accurately estimates the small fraction of the total amount of the nutrient in the soil which is available for uptake by plants, especially across a range of soils. For major macronutrients like nitrogen, phosphorus, test of whole shoot by plants tissues analysis gives a reasonable indication of nutrient status. The status of macronutrients like NPK is kept in balance by addition of appropriate dose of the fertilizers. They are applied directly or through irrigation while micronutrients require either foliar spray or soil application.

The level of nitrogen in soil of India is low. It is made available to the soil in the form of nitrate or ammonia by adding nitrogenous fertilizers *viz.*, ammonium sulphate (20% N), ammonium nitrate (25% N), urea (46% N) and diammonium sulphate. The nitrogen not utilized by the plant, is lost. It forms a gas that escape into the atmosphere, or it dissolves and is carried away by soil moisture. Nitrogen losses can be reduced by incorporating the fertilizer into the soil instead of spreading it on the surface. Applying fertilizer 2–3 times during the crop season, rather than all at once, is another way to raise N-uptake. Joshi (1974) suggested the use of 2% urea as foliar spray for recovery of nitrogen-chlorosis. A 50–60 days crop of *Sesbania* spp. add 85–125 kg N/ha to the soil.

Phosphorus is an immobile element, and it quickly reacts with the soil, which makes it unavailable to plants. Moreover, the level of phosphorus in soil is low to medium. Hence phosphatic fertilizers like single super phosphate (16% P_2O_5) triple super phosphate (46% P_2O_5) etc. are added to correct the dificiency of phosphorus in soil. Full dose of P along with 1/3 N is applied at seedling. Placing phosphorus fertilizer in a band near the seed, can increase efficacy of K-uptake.

Coarse textured soil is generally deficient of potassium. However potash deficiency is not so widespread in the country. Potassium as potassium chloride, potassium nitrate or potassium sulphate is made available to wheat crop by addition of fertilizer like murate of potash. Some fertilizers like DAP has more than one element so its proper dose for application should be based on soil test.

In recent years, the deficiencies of one of the secondary nutrients i.e. sulphur and other micronutrients like zinc, iron, manganese and boron has started emerging in pockets where very intensive cultivation is being practiced. To overcome problem of multiple nutrient deficiencies an integrated nutrient management involving organic sources like crop residue, green manuaing, farm yard manure (FYM) and composts in combination with inorganic elements is recommended (Mishra *et al.*, 2005).

Sulphur deficiency can be corrected by using cosavet @ 15 kg S/ha along with gypsum. Zinc deficiency can be rectified by the application of zinc sulphate (0.5% $ZnS0_4$. $7H_2O$) @ 25 kg/ha along with the fertilizer. One such application will be sufficient for 2–3 years. Zinc deficiency can also be corrected with the foliar application of $ZnSo_4$. For foliar spray 200–300 litres of zinc sulphate, lime mixture in water (5:2.5:1000) should be prepared for an acre of wheat-crop to apply 2–3 times at 15 days interval. Boron deficiency is neutralized by use of boric acid (Na_2 B_{10}, O_{16}, $10H_2O$) or sodium borate (Na B_4O_7, $5H_2O$). Since the excess of these compounds is extremely toxic to plants, the rate should be decided after soil analysis or crop response studies. Ferrus sulphate (0.5–1.0% Fe SO_4) is recommended to correct iron deficiency (Kanwar and Randhawa, 1978). Deficiency of manganese is corrected by applying manganese sulphate (1% $MnSO_4$). It is always better to apply Mn SO_4 on the basis of plant tissues analysis rather than soil analysis as a mean of identification.

REFERENCES

Agarwal, S.C. 1971. Annual Progress Report of ICAR Coordinated Project on Nutrients of soil. 1968-69, 1960-70 and 1970-71.

Arnon, D.I. 1950. Criteria for essentiality of organic micronutrients for plants. **In**: *Trace Elements in Plant Physiology* (Eds. W.D. McElory and B. Glass), Chronica Botanica, Waltham, Mass.

Beevers, L. and R.H. Hageman. 1969. Nitrate reduction in higher plants. *Ann. Rev. Plant Physiol.* **20**: 495-522.

Blank, F. 1947. The anthocyanin pigments of plants. *Bot. Rev.* **13**: 241-317.

Bove, J.M., C. Bove, F.R. Whatley and D.I. Arnon. 1963. Chloride requirement for oxygen evolution in phytogenesis. *Z. Natur far Sch.* **18**B: 683-688.

Carell, E.F. and C.A. Price. 1965. Prophyrins and the iron requirement for chlorophyll formation in *Euglena. Plant Physiol.* **40**: 1-7.

Chapman, H.D. 1966. Calcium. **In**: *Diagnostic Criteria for Plants and Soils* (Eds. H.D. Chapman), Div. Agric. Sci. Univ. California, Berkeley (USA).

Chapman, F.M. and J.L. Mason. 1969. Effect of phosphorus and potassium fertilizers on the agronomic characters of spring wheat and interaction on grain yield. *Can. J. Soil. Sci.* **49**: 343-347.

Chenial, G.M. 1970. Photosystem II and O_2 evolution. *Ann. Rev. Plant Physiol.,* **21**: 247-298.

Epstein, E. 1982. *Mineral Nutrition of Plant: Principles and Perspectives*. John Wiley and Sons, New York.

Evans, H.J. 1956. Role of molybdenum in plant nutrition. *Soil. Sci.* **81**: 199-208.

Fujino, M. 1967. Role of adenosine triphosphatase in stomatal movement. *Sci. Bull. Tech. Educ.* Nagsaki University, Japan **18**: 1-47.

Gauch, H.C. 1972. *Inorganic Plant Nutrition*. Dowden Hutchinson & Ross. Inc. Stroudsburg, 329 p.

Gauch, H.C. and W.M. Dugger Jr. 1954. The physiological action of boron in higher plants: a review and interpretation. *Marylane Agri. Exp. Sta. Bull.* **A-80** (Tech.), 43 p.

Graham, R.D. 1975. Male sterility in wheat plants deficient in copper. *Nature* (London) **254**: 514-515.

It, A. and A. Fujiwara. 1967. Functions of calcium in the cell wall of rice leaves. *Plant Cell Physiol.* **8**: 409-422.

Jones, W.W. 1966. Nitrogen. **In**: *Diagnostic Criteria for Plants and Soils* (Ed. H.D. Chapman), pp. 310-323. University of California, Berkeley.

Joshi, L.M. 1974. Yellowing in 'Kalyansona' variety of wheat. *Indian Phytopath.* **27**: 230-231.

Keisling, T.C., L.F. Thompson and W.R. Slabaugh. 1984. Visual symptoms and tissue manganese toxicity in wheat. *Comm. Soil Sci. Plant Analysis* **15**: 537-540.

Lauchli, A. and Bieleski, R.L. 1983. Inorganic plant nutrition. *Encyclopaedia of Plant Physiology*. New Series, Vol. 15A, Springer-Verlag.

Marschner, H. 1986. *Mineral Nutrition of Higher Plants*. Academic Press, London.

Menegel, K. and E.A. Kirkby. 1982. *Principles of Plant Nutrition*. International Potash Institute, Bern, Switzerland.

Mishra, B., Jag Shoran, R. Chatrath, A.K. Sharma, R.K. Gupta, R.K. Sharma, R. Singh, J. Rane and A. Kumar. 2005. *Cost Effective and Sustainable Wheat Production Technologies*. Directorate of Wheat Research, Karnal, Tech. Bull. No. 8, 36 p.

People, T.R. and D.W. Koch. 1979. Role of potassium in carbon dioxide assimilation in *Medicago sativa* (L.). *Plant Physiol.* **63**: 878-881.

Price, C.A. 1968. Iron compound and plant nutrition. *Ann. Rev. Plant. Physiol.* **19**: 239-248.

Ramamoorthy, B. and S.V. Desai, 1946. *Micronutrient Research in Soils and Plants in India* (Eds. J.S. Kanwar and N.S. Randhawa) ICAR, New Delhi.

Rasmussan, H.P. 1967. Calcium and strength of leaves. I. Anatomy and histochemistry. *Bot. Gaz.* **128**: 219-223.

Scrutton, M.C., M.F. Utter and A.S. Mildvan. 1966. Pyruvate carboxylase VI. The presence of tightly bond manganese. *J. Biol. Chem.* **241**: 3480-3487.

Singh, M. and O.P.S. Bana. 1986. Mineral nutrition of wheat: Physiological functions and deficiency diseases. **In**: *Problems and Progress of Wheat Pathology in South Asia*, (Eds. L.M. Joshi, D.V. Singh and K.D. Srivastava) pp. 277-295. Malhotra Publishing House, New Delhi, 401 p.

Skoog, F. 1940. Relations between zinc and auxin in the growth of higher plants. *Am. J. Bot.* **27**: 939-951.

Snowall, K. and A.D. Robson. 1983. *Symptoms of Nutrient Deficiencies: Subterranean clover and Wheat*. University of Western Australia, Perth.

Snowball, K. and A.D. Robson. 1991. *Nutrient Deficiencies and Toxicities in wheat: A Guide for Field identification*, Mexico, D.F., Mexico. 76 p.

Steward, F.C. 1963. *Plant Physiology*. Vol. III. *Inorganic Nutrition of Plants*, Academic Press, New York, 118 p.

Stocking, C.R. and A. Ongun. 1962. The intracellular distribution of some metallic elements in leaves. *Am. J. Bot.* **49**: 284-289.

Vallee, B.L. and W.E.C. Wacker, 1970. Metalloproteins. **In**: *The Proteins* (2nd ed.) (Ed. H. Neurath) Vol. 5. Academic Press, New York, 192 p.

Wallace, T. 1961. *The Diagnosis of Mineral Deficiencies in Plant by Visual Symptoms-A Colour Atlas and Guide*. Curwen Press Plaistow, London, 125 p.

Chapter

Miscellaneous Leaf and Head Diseases

In cereals some forms of abnormal developments occurs which are not caused by pathogenic organism. Like nutrient deficiencies, these abnormalities are placed under abiotic stresses. For example, disorders like necrosis and chlorosis of leaf and head result from genetic combination. Herbicides used to protect crops against weeds, are injurious to plants and develop abnormalities of plant parts. During periods of high humidity saprophytic fungi often occur on wheat heads and may alter the appearance of invaded tissues, but seldom produce distinctive symptoms.

8.1 HYBRID NECROSIS

Hybrid necrosis in wheat and triticale results from chromosomal instability or genetic combinations. The genetic abnormality was recorded by Caldwell and Compton (1943). In India, hybrid necrosis was reported by Narula *et al.* (1966) and Anand *et al.* (1966).

The initial symptoms of necrosis appear as dull light green spots and mottling on leaves. The spots enlarge and coalese to result in larger zones. Necrotic spots are also spread to leaf sheaths. Based on the degree of necrosis, hybrid necrosis can be classified into following categories:

Lethal Hybrid Necrosis

Quite often the symptoms appear at 2nd and 3rd leaf stage and the affected seedlings die prematurely. In some cases the hybrid can survive till tillering stage but dies when it has put out just two or three tillers.

Sub-lethal Necrosis

In sub-lethal necrosis, symptoms appear in the early tillering phase on wheat leaves. Affected leaves show yellowing due to lack of chlorophyll. Plants are much poorer in growth and even in height and earhead length. Earing is much delayed as compared to the parents. Earheads mature much earlier and consequently produce shrivelled grains.

Genetical studies have shown that necrosis is determined by the interaction of two complementary genes *i.e. Ne1* and *Ne2* (Hermsen, 1963). Sunewaki and Nishikawa (1964) revealed that three genes, *viz., Ne1, Ne2* and *Ne3* located in B, A and D. genomes respectively, are involved with hybrid necrosis. Indigenous *T. aestivum* varieties from India carry the gene *Ne1* (Kochumadhvan *et al.*, 1980). The *Ne2* carriers appear to be absent in *T. sphaerococcum* which was once epidemic in North India and Pakistan. Tomar *et al.* (1986) indicated that *Ne1* gene present in *T. sphaerococcum* interacts weakly. Hybrid chlorosis in the hybrids involving *T. macha, T. dicoccum and T. dicoccoides* is conditioned by Chlorosis gene *Ch1* and in *T. aestivum* by *Ch2* (Tomar *et al.*, 1988; Tomar and Menon, 2001).

8.2 BROWN NECROSIS

Brown necrosis has been observed in some wheat varieties that have derived their resistance to stem rust (*P. graminis tritici*) from Hope and H44 progenies (Zillinsky, 1983). The sign of necrosis appears as dark brown discolouration that progresses on the glumes and peduncles. The brown pigmentaton on glumes may be confused with black chaff (*Xanthomonas translucens*) or glume blotch (*Septoria nodorum*) diseases of wheat.

8.3 HERBICIDAL INJURY

Herbicides pose a hazard to the wheat crop if they are not applied with proper dose, method and timing. Phytotoxic effect of weedicide 2,4-D (2,4-dichlorophenoxyacetic acid) is common in wheat crop. The chemical solution of this weedicide is prepared @ 0.4 kg ai in 750 litres of water for one hectare. It is normally sprayed as a post-emergence treatment to control broad-leaf weeds 5–6 weeks after sowing. Any deviation in the procedure, invites injurious effect of the weedicide. If 2, 4-D is applied too early *i.e.* seedling stage or late at joining stage then wheat growth is suppressed and malformation of spikes is induced. Earheads are deformed due to excessive application of the weedicide. Grain formation in the spike is adversely affected resulting in shrivelled grains with low weight. Features of deformed earheads look like downy mildew (*Sclerophthora macrospora*) of wheat. 2, 4-D should not be used in sensitive varieties like WH 283, HD 2009, Raj 3077 etc. (Yadav, 2008).

8.4 HEAD SCAB

Scab, also known as head blight is a newly emerging problem in the cooler and humid areas in foothills of Punjab and Himachal Pradesh. Although the out break of scab was reported from Arunachal Pradesh, Assam in 1973 (Roy, 1973), but it remained simply a record. Bramha and Singh (1985) recorded head scab in Nilgiris in 1989-90 crop seasons. Subsequently, its sporadic appearance was again recorded in foot hills of Punjab during 1989–90, 1995–96 and 2004–05 crop seasons (Singh *et al.* 1993, Singh and Aujla 1994, Kaur *et al.* 1999, Mishra *et al.*, 2005). Head scab is still a disease of minor importance but it can be a potential disease in future (Sharma, 2008).

Symptoms: Foliar infections are often characterized by zonate lesions which have a water soaked appearance. The Scab phase on the spike is frequently accompanied by a pinkish growth of the mycelium. The entire ear becomes bleached and in severely blighted spikes, a brown discolouration of the glumes is visible (Plate 7B). Small dark perithecia of the fungus may also be seen on the glumes of infected spikelets. Scabby seeds show various degrees of shrivelling and pink discolouration.

Causal Organism: Fusarium head blight of wheat has been linked to at least 17 causal organisms, most records of the disease being associated with five species: *F. graminearum*, *F. culmorum*, *F. avenaceum*, F. *poae* and *F. nivale* (Parry *et al.* 1995). Despite the range of *Fusarium* species implicates the disease, three species predominates internationally *i.e. F. graminearum*, *F. culmorum* and *F. avenaceum.* In India, Fusarium Scab is associated with *Fusarium graminearum* Schwabe [Telemorph *Gibberella zae* (Schwein) Petch]. The macroconidia are produced in orange to peach coloured masses at the base of spikelets or at the edges of glumes or lemmas. They are hyaline, straight or slightly curved and the apical cell curves and narrows sharply to a point. Macroconidial usually measure between 25–50 µm × 2.5–5.0 µm and have three to seven septa. *F. graminearum* regularly produces perithecia in nature on diseases heads. The ascospores are sub-hyaline to light yellow-brown, fusoid, slightly curved with rounded ends three septate and measure 20–25 µm × 3–4 µm. the major mycotoxins produced by the fungus are DON, 15-ADON, NIV, FX and ZEN (Mangan *et al.* 2002).

Disease Cycle: Head scab is a monocyclic disease. Infected crop debris appears to be the most important source of inoculum. Macroconidia that are produced on debris, are believed to be disseminated to wheat ears at or shortly after anthesis by splashing rain or by wind – driven rain. The perfect stage *G. zeae* of the fungus which had overwintered for 2 years on or above the soil surface, serves as an important source of inoculum for infection of wheat ear heads in Canada (Khonga and Sutton, 1988). Thus, it is possible that the perithecia and ascospores produced on crop stubble may survive from one season to another and can be incitant of scab infection. According to Reis (1990), the main source of inoculum in southern Brazil is perithecia that are formed saprophytically on senescened tissues of non-host plants. Free moisture and warm temperature (18°C) favour foliar and spike infection (Saari, 1986). Ascospores inside the perithecia require water as a stimulus to forcibly eject them into air (Tschang *et al.*, 1975).

Management

The strategy for the integrated control of scab relies on host resistance, chemical control and cultural practices. Normally, bread wheats are more resistant than durums. But no resistant varieties are available in India (Mishra *et al.,* 2005). However, in China, where scab is a major problem, the cultivars belonging to 'Ning' system- Ning 7840 and others are resistant (Zong, 1988).

Chemical control of the disease can be achieved with some of the new systemic fungicides such as triadimefon, benomyl, propiconazole and MCB (Bergstrom, 1993). Scab is most severe where wheat is grown into infested debris of previous suspect crop such as wheat maize or rice. Hence, complete burial of infested crop debris reduces inoculum of the pathogen. Early planting is an effective scab avoidance strategy in some parts of the world (Khonga and Sutton, 1988).

8.5 MOULDY EAR HEAD

Mouldy earhead disease is a common occurrence in rainy season crops like maize, sorghum and pearl millet (Wiehe, 1953; Mathur *et al.*, 1967). In case of winter cereals such as wheat, barley and triticale which are invariably harvested in the beginning of the summer season, mouldy head normally does not occur. However, moulds may appear under moist humid environment. The disease is seen particularly in the eastern region of the country or low lying areas where humidity is much high. Srivastava and Chatterji (1972) recorded the occurrence of mouldy head of triticale in Gurdaspur (Punjab).

In India Butler (1918) described *Cladosporium* as a common saprophyte or mild parasite of sooty moulds. Later Arya and Panwar (1955) reported a sooty mould of wheat from Jodhpur (Rajasthan) caused by *C. herbarum.*

Ear heads which are shaded in the field or are weakened otherwise are prone to attack of shooty moulds. Plants that are multi-deficient suffer from lodging, and support the mouldy growth of fungi (Wiese, 1977). Awns, glumes and anther lobes of the ear head are found to be affected by the moulds. If the fungi spread to grains, black point symptoms on grains may appear. Old dead leaves show withering and brownish black mouldy growth.

8.6 LEAF TWIST

Leaf twist is a disease of historical interest because of its occurrence in England during 19th century. It is considered a disease of minor importance in Germany, USA, Canada, France and Switzerland (Weise, 1977). In India the disease was first reported in 1959 from Kashmir valley (Reddy, 1960) and later Munjal and Kaul (1961) confirmed its existence in Handwar region of Kashmir.

As the name of the disease suggests, the fungus *Dilophospora alopecuri* causes twisting of leaf and pedicel below the ear head. The disease is reported to be associated with wheat nematode (*Anguina tritici*). Leaf spotting can occur without the association of nematode but leaf twist stage is extremely rare in its absence. Symptoms on leaves arise as yellow, elliptical to elongated or spindle shaped flecks which later turn brown with black crust in the centre. The lesion mostly arise singly, seldom coalescent. In severe case of infection affected leaves show characteristic twisting.

REFERENCES

Anand, S.C., B.S. Gill and R.P. Jain. 1966. Necrosis in intervarietal crosses of wheat. *Indian J. Genet.* **29**: 131-134.

Arya, H.C. and K.S. Panwar. 1955. Some studies on a virulent strain of *Cladosporium herborum* (Link.) Fr. on wheat. *Indian Phytopath.* **8**: 176-181.

Bergstrom, G.C. 1993. Scab (Head blight). **In**: *Seed borne Diseases and Seed Health Testing of Wheat* (Eds. S.B. Mathur and B.M. Cunfer), pp. 83-94. Danish Government Institute of Seed Pathology for Developing Countries, Copenhegan, Denmark, 168 p.

Brahma, R.N. and S.D. Singh. 1985. Occurrance of scab of wheat in the Nilgiri hills. *Curr. Sci.*, **54**:1184-1185.

Butler, E.J. 1918. *Fungi and Diseases in Plants*. Thacker & Spinck and Co. Calcutta, 547 p.

Caldwell, R.M. and L.E. Crompton. 1943. Complementry lethal genes in wheat causing a progressive lethal necrosis of seedlings. *Heredity* **34**: 66-70.

Hermsen, J.G. 1963. Hybrid necrosis as a problem for wheat breeder. *Euphytica* **12**: 1-16.

Kaur, S., S.S. Chahal and N. Singh. 1999. Effect of temperature on the development of the pathogen and head blight disease in wheat. *Plant Dis. Res.* **14**: 191-194.

Khonga, E.B. and J.C. Sutton. 1988. Inoculum production and survival of *Gibberella zeae* in maize and wheat residues. *Canad. J. Plant Pathol.* **10**: 232-239.

Kochumadhavan, M., S.M.S. Tomar and P.N.N. Nambisan. 1980. Investigations on hybrids necrosis in bread wheat. *Indian J. Genet.* **40**: 496-502.

Magan, N.R., Hope, A. Colleate and E.S. Baxter. 2002. Relationship between growth and mycotoxin production by *Fusarium* species, biocides and environment. *European Jour. Plant Path.* **108**: 685-690.

Mathur, S.B., R. Sharma and L.M. Joshi. 1967. Mouldy-head disease of sorghum, isolation of associated fungi on blotter agar. *Proc. Intern Seed Test. Assoc.* **32**: 639-645.

Mishra, B., Jag Shoran, R. Chatrath, A.K. Sharma, R.K. Gupta, R.K. Sharma, R. Singh, J. Rane and A. Kumar. 2005. *Cost Effective and Sustainable Wheat Production Technologies,* DWR, Karnal, *Tech. Bull.* No. 8, 36 p.

Munjal, R.L. and T.N. Kaul. 1961. Dilophospora leaf spot of wheat in India. *Indian Phytopath.* **14**: 13-15.

Narula, P.N., S. Singh and P.S.L. Srivastava. 1966. Hybrid necrosis in bread wheat. *Curr. Sci.* **35**: 547-548.

Parry, D.W., P. Jenkins and L. McLeod. 1995. Fusarium ear blight (scab) in small grain cereals-a review. *Plant Pathology* **44**:207-238.

Reddy, D.B. 1960. Outbreak and new records. *FAO Plant Prot. Bull.* **8**: 113-114.

Reis, E.M. 1990. Perithecial formation of *Gibberella zeae* on senescent stems of grasses under natural conditions. *Fitopathologia Brasilera* **15**: 52-54.

Roy, A.K. 1973. Ear blights and scab of wheat in Arunachal Pradesh. *Curr. Sci.* **43**: 162.

Saari, E. 1986. Wheat diseases problems in South-East Asia. **In**: *Problems and Progress of Wheat Pathology in South Asia* (Eds. L.M. Joshi, D.V. Singh and K.D. Srivastava), pp. 31-40. Malhotra Publ. House, New Delhi, 401 p.

Sharma, A.K. 2008. IPM in wheat. **In**: *A Compendium of Lectures on Integrated Pest Management* (Eds. A.K. Sharma and D.P. Singh). DWR, Compendium No. 2. Directorate of Wheat Research, Karnal, India, 258 p.

Singh, D.V., K.D. Srivastava, R. Aggarwal and P. Bahadur. 1993. Wheat Disease problems: The changing scenario. **In**: *Pest and Pest Management in India* (Eds. H.C. Sharma and V. Rao), pp. 116-120. Plant Protection Association of India, Hyderabad, 312 p.

Singh, P.J. and S.S. Aujla. 1994. Effect of lodging on the development of head scab of wheat. *Indian Phytopath.* **47**: 56-57.

Srivastava, K.D. and S.C. Chatterji. 1972. Mouldy head of triticale. *Indian Phytopath.* **25**: 591-593.

Tomar, S.M.S. and M.K. Menon. 2001. Genes for disease resistance in wheat, IARI, New Delhi, 152 p.

Tomar, S.M.S., M. Kochumadhavan and P.N.N. Nambisan. 1986. Hybrid lethality and resistance to wheat pathogens in free threshing hexaploid wheats. *Indian J. Genet.* **46**: 375-381.

Tomar, S.M.S., M. Kochumadhavan, P.N.N. Nambisan and B.C. Joshi. 1988. Hybrid necrosis and chlorosis in wild emmer, *Triticum dicoccoides* Korn. pp. 165-168. **In**: *Proc.* 7th *Int. Wheat Genet. Symp.* (Eds. T.E. Miller and R.M.D. Koebner), Combridge, England.

Tschanz, A.T., R.K. Horst and P.E. Nelson. 1975. Ecological aspects of ascospore discharge in *Gibberella zeae. Phytopathology* **65**: 597-599.

Wiehe, P.O. 1953. Mycological Paper No. 53, CMI, Key, U.K.

Weise, M.V. 1977. *Compendium of wheat Diseases*. American Phytopathological Society, APS Press, St. Paul, Minnesota, 106 p.

Yadav, D.B. 2008. Weed problems and management approaches in wheat based system **In**: *A Compendium on Lectures on Integrated Pest Management in Wheat based cropping systems*, DWR Compendium No. 2, Directorate of wheat Research, Karnal, 258 p.

Zillinsky, F.J. 1983. *Common Diseases of Small Grain Cereals*. A Guide to identification, CIMMYT, Mexico, 140 p.

Zong, W.Y. 1988. Screening techniques and sources of resistance to Fusarium head blight. **In**: *Wheat Production Constraints in Tropical Environment* (Ed. A.R. Klatt), pp. 239-250. CIMMYT, Mexico, 410 p.

Chapter

Genomics and Genetic Engineering for the Management of Wheat Diseases

Plant biotechnology has made significant advancements during the past decade, and several crops are now grown commercially. Moreover, plant breeding continues to improve wheat production but is limited by the gene pool required for developing varieties for biotic stress tolerance. Recombinant methods have the potential to meet the future demands but require robust transformation technology that is directly applicable to established and adapted commercial cultivars. Biotechnology can be defined in many different ways, but for the purpose of this chapter, all areas that use recent molecular approaches including genomics, next-generation strategies and genetic engineering strategies will be discussed.

9.1 GENOMICS OF DISEASE RESISTANCE

9.1.1 Mapping Wheat Genome for Resistance

The resistance against most non-adapted pathogens (non-host resistance) and the ability to reduce the disease severity of an adapted pathogen (basal resistance) lets a plant to complete its life cycle. Preformed structural and biochemical barriers and induced defence both contribute to tolerance or resistance against pathogens. Like animals plants also have the capacity to recognize a potential pathogen and mount defence responses against it. Most of the pathogens are recognized by their conserved structural or chemical components and plants mount defence responses but pathogens have also evolved to suppress this basal defence strategy by their virulence effectors. These effectors are recognized in resistant plant through resistance (R) genes. Thus, throughout the co-evolution of plant and microbes this 'tug of war' is going on. These resistance mechanisms follow a "gene-for-gene" model, whereby specific plant recognition of a pathogen gene product yields near-absolute resistance to pathogens containing the recognized gene. However, this form of resistance is vulnerable to rapid counter-evolution by the pathogen. This generates a continuous and resource-intensive cycle of plant resistance gene discovery, crop breeding, and eventual resistance gene "defeat" by the pathogen, which has driven interest in identifying durable forms of resistance to pathogens. In most plant-pathogen systems, durable resistance has been identified through quantitative trait locus (QTL) mapping. Although identified QTL provide quantitative resistance to a range of pathogen species and genotypes therein, the molecular bases of these broad quantitative resistances have remained largely unknown.

Genetic mapping of wheat has provided the basis for identifying QTL that control complex traits. The International Triticeae Mapping Initiative (ITMI) has overseen the coordination of resources from different organizations world-wide to provide genetic mapping information (http://www.scri.sari.ac.uk/ITMI/). The developments in molecular genetics in wheat have been relatively slow, especially when compared to other crops, such as maize, rice or tomato, due to wheat's ploidy level, the size and complexity of its genome, the very high percentage of repetitive sequences and the low level of polymorphism. However, due to the large number of disease resistances controlled by major genes, the mapping of such genes has dominated the research activities in wheat genomics.

A large number of disease genes were identified and tagged in wheat and listed in Table 18. QTL studies for various diseases have been conducted in wheat, leading to mapping of QTL on different chromosomes. In most of these studies, either single marker regression approach or QTL interval mapping has been utilized. Gupta *et al.* (2008) have recently reviewed the prospects of wheat genomics including the disease resistance. The availability of markers associated with disease resistance also allowed development of major programs on marker assisted selection (MAS), and facilitated map-based cloning of a number of genes/QTL. Despite the large number of markers, few of those markers are close enough to the genes of interest to be useful in breeding applications.

Table 18. A list of disease gene/QTL tagged/mapped in wheat.

Disease/ Nematode	Gene/QTL (chromosome)	Mapping population	Reference(s)
Disease			
(i) Leaf rust resistance	*Lr9* (6BL)	NILs	Schachermayr *et al.*, 1994
	Lr1 (5DL)	F2	Feuillet *et al.*, 1995
	Lr24 (3DL)	F2	Schachermayr *et al.*, 1995
	Lr10 (1AS)	F2	Schachermayr *et al.*, 1997
	Lr28 (4AL)	F2:3	Naik *et al.*, 1998
	Lr3 (6BL)	F2	Sacco *et al.*, 1998
	Lr35 (2B)	F2	Seyfarth *et al.*, 1999
	Lr47 (7A)	BC1F2	Helguera *et al.*, 2000
	LrTr (4BS)	F2	Aghaee-Sarbarzeh *et al.*, 2001
	Lr19 (7DL)	Deletion lines	Prins *et al.*, 2001
	Lr39 (=*Lr41*)(2DS)	F2	Raupp *et al.*, 2001
	Lr37 (2AS)	NILs	Seah *et al.*, 2001
	Lr20 (7AL)	F2	Neu *et al.*, 2002
	Lr19 (7D)	F2	Cherukuri *et al.*, 2003
	Lr21/*Lr40* (1DS)	F2	Huang *et al.*, 2003
	Lr1 (5DL)	F2:3 families	Ling *et al.*, 2004
	Lr28 (-)	F2:3	Cherukuri *et al.*, 2005
	Lr34 (7D)	RILs	Spielmeyer *et al.*, 2005
	Lr52 (*LrW*) (5B)	F2	Hiebert *et al.*, 2005
	Lr16 (2BS)	DH	McCartney *et al.*, 2005
	Lr19 (7DL)	F2	Gupta *et al.*, 2006
	Lr24 (3DL)	F2	Gupta *et al.*, 2006
	Lr34 (7DS)	RILs	Bossolini *et al.*, 2006
	Lr22a (2DS)	F2	Hiebert *et al.*, 2007
	Lr1 (5DL)	RILs	Qiu *et al.*, 2007
	Unknown (5B)	F2:3 lines	Obert *et al.*, 2005
	QTL (7D, 1BS)	RILs	Schnurbusch *et al.*, 2004
	QTL (2D, 2B)	F2	Leonova *et al.*, 2007
	QTL (7DS, linked to *Lr34*)	RILs	Singh *et al.*, 2000
(ii) Stripe rust resistance	*Yr15* (1B)	F2	Sun *et al.*, 1997
	YrH52 (1B)	F2	Peng *et al.*, 1999

Table 18. *Contd.*

Disease/ Nematode	Gene/QTL (chromosome)	Mapping population	Reference(s)
Disease			
	Yrns-B1 (3BS)	F3 lines	Börner *et al.*, 2000
	Yr15 (1B)	F2 lines	Peng *et al.*, 2001
	Yr28 (4DS)	RILs	Singh *et al.*, 2000
	Yr9 (1B/1R)	BC7F2:3	Shi *et al.*, 2001
	Yr17 (2A)	NILs	Seah *et al.*, 2001
	Yr26 (1BS)	F2 lines	Ma *et al.*, 2001
	Yr10 (1B)	F2 lines	Wang *et al.*, 2002
	Yr5 (2B)	BC7F3	Sui *et al.*, 2009
	Yr18 (7D)	RILs	Spielmeyer *et al.*, 2005
	Yr36 (6B)	RILs	Uauy *et al.*, 2005
	YrCH42 (1B)	F2	Li *et al.*, 2006
	YrZH84 (7BL)	F2, F3	Li *et al.*, 2006
	Yr34 (5AL)	DH	Bariana *et al.*, 2006
	Yr26 (1B)	F2:3 lines	Wang *et al.*, 2008
	QTL (2D, 5B, 2B, 2A)	RILs	Mallard *et al.*, 2005
	QTL (2AL, 2AS, 2BL, 6BL)	DH	Christiansen *et al.*, 2006
(iii) Stem rust resistance	*Sr22* (7A)	F2	Paull *et al.*, 1994
	Sr38 (2AS)	NILs	Prins *et al.*, 2001
	Sr2 (3BS)	F3 lines	Spielmeyer *et al.*, 2003
(iv) Fusarium head blight resistance	*Fhb2* (6BS)	RILs	Cuthbert *et al.*, 2007
	QTL (5A, 3B, 1B)	DH	Buerstmayr *et al.*, 2002
	QTL (3BS, 3A, 5B)	RILs	Bourdoncle *et al.*, 2003
	QTL (3B)	Advanced lines	Blanco *et al.*, 2003
	QTL (3B, 6B, 2B)	RILs	Lin *et al.*, 2004; Lin *et al.*, 2006
	QTL (6D, 4A, 5B)	RILs	Paillard *et al.*, 2004
	QTL (3A, 5A)	DH	Steiner *et al.*, 2004
	QTL (1B, 3B)	RILs	Zhang *et al.*, 2004
	QTL (2B)	RILs	Gilsinger *et al.*, 2005
	QTL (3B)	DH	Jia *et al.*, 2005
	QTL (6AL, 1B, 2BL, 7BS)	RILs	Schmolke *et al.*, 2005
	QTL (1BL)	RILs	Haberle *et al.* 2009
	QTL (3A)	RICLs	Chen *et al.*, 2007
	QTL (4D)	DH	Draeger *et al.*, 2007
	QTL (3BS, 5AS, 2DL)	RILs	Jiang *et al.*, 2007

Table 18. *Contd.*

Disease/ Nematode	Gene/QTL (chromosome)	Mapping population	Reference(s)
Disease			
	QTL (1BS, 1DS, 3B, 3DL, 5BL,7BS, 7AL)	RILs	Klahr *et al.*, 2007
	QTL (7E)	RILs	Shen and Ohm *et al.*, 2007
(v) Scab resistance	QTL (2AS, 2BL, 3BS)	RILs	Zhou *et al.*, 2002
	QTL (3BS)	F3:4 lines	Zhou *et al.*, 2003
(vi) Powdery mildew resistance	*Pm2* (5DS)	F2	Hartl *et al.*, 1995
	Pm18 (5DS)	F2	Hartl *et al.*, 1995
	Pm12 (6B)	F2	Jia *et al.*, 1996
	Pm21 (6AL)	BC lines	Liu *et al.*, 1999
	Pm3g (1A)	DH	Sourdille *et al.*, 1999
	Pm24 (1DS)	F2:3 lines	Huang *et al.*, 2000
	Pm26 (2BS)	RSI	Rong *et al.*, 2000
	Pm6 (2BL)	NILs	Tao *et al.*, 2000
	Pm27 (6B)	F2	Järve *et al.*, 2000
	Pm8/Pm17 (1BL)	F3 families	Mohler *et al.*, 2001
	Pm3 (1AS)	RILs	Bougot *et al.*, 2002
	Pm1 (7AL)	F2	Neu *et al.*, 2002
	Pm29 (7D)	F2 & F4 lines	Zeller *et al.*, 2002
	Pm30 (5BS)	BC2F2 lines	Liu *et al.*, 2002
	Pm13 (3S)	AL	Alberto *et al.*, 2003
	Pm5e (7BL)	F2	Huang *et al.*, 2003
	Pm4a (2A)	F2	Ma *et al.*, 2004
	PmU (7AL)	F2	Qiu *et al.*, 2005
	Pm34 (5D)	F2:3 lines	Miranda *et al.*, 2006
	PmY39 (2B)	BC3F4:5	Zhu *et al.*, 2006
	Pm35 (5DL)	F2:3 lines	Miranda *et al.*, 2007
	Pm5d (7BL)	F3 lines	Nematollahi *et al.*, 2008
	Pm36	F3	Blanco et al., 2008
	Pm42 (2BS)	F2	Hua *et al.*, 2009
	Pm12 (6B)	BC3F2	Song *et al.*, 2007
	MlRE, QTL (6A, 5D)	F3 lines	Chantret *et al.*, 2000
	MlG (6AL)	BC2F3	Xie *et al.*, 2003
	mlRD30 (7AL)	F2	Singrün *et al.*, 2004
	Mlm2033, Mlm80 (7A)	F2	Yao *et al.*, 2007

Table 18. *Contd.*

Disease/ Nematode	Gene/QTL (chromosome)	Mapping population	Reference(s)
Disease			
	QTL (5A, 7B, 3D)	RILs	Keller *et al.*, 1999
	QTL (1B, 2A, 2B)	F2:3 lines	Liu *et al.*, 2001
	QTL (2B, 5D, 6A)	DH	Mingeot, 2002
	QTL (2B)	F2	Bougot *et al.*, 2006
	QTL (1BL, 2AL, 2BL)	RILs	Tucker *et al.*, 2007
(vii) Common bunt resistance	*Bt-10* (-)	F2	Laroche *et al.*, 2000
	QTL (1B, 7A)	DH	Fofana *et al.*, 2008
(viii) Tan spot and *Stagonospora nodorum* blotch resistance	QTL (1A, 4A, 1B, 3B)	RILs	Faris *et al.*, 1997
	QTL (5B, 3B)	Inbred, CS lines	Singh *et al.*, 2007
	tsn3a, tsn3b, tsn3c (3D)	F2:3 lines	Tadesse *et al.*, 2007
(ix) *Septoria tritici* blotch resistance	*Stb5* (7D)	SCRI	Arraiano *et al.*, 2001
	QTL (3A)	DH	Eriksen *et al.*, 2003
	QTL (1D, 2D, 6B)	RILs	Simón *et al.*, 2004
(x) Barley yellow dwarf tolerance	QTL (12 chromosomes)	RILs	Ayala *et al.*, 2002
(xi) Leaf and glume blotch resistance	QTL (4B, 7B, 5A)	RILs	Aguilar *et al.*, 2005
(xii) Wheat streak mosaic virus resistance	*Wms1* (4D)	F2	Talbert *et al.*, 1996
	WSSMV (2DL)	RILs	Khan *et al.*, 2000
(xiii) Yellow mosaic virus resistance	*YmYF* (2D)	F2	Liu *et al.*, 2005
(xiv) Eyespot (straw breaker foot rot) resistance	*Pch2* (7AL)	F2	de la Peòa *et al.*, 1997
	Pch1 (7A)	F3 lines	Huguet-Robert *et al.*, 2001
	Pch1, Ep-D1 (7D)	TC	Groenewald *et al.*, 2003
Nematodes			
(i) Cereal cyst nematode resistance	*Cre1* (2B)	NILs, F2	Williams *et al.*, 1994
	Cre5 (2AS)	NILs	Jahier *et al.*, 2001
	Cre6 (5A)	F2	Ogbonnaya *et al.*, 2001

Table 18. *Contd.*

Disease/ Nematode	Gene/QTL (chromosome)	Mapping population	Reference(s)
Nematodes			
	QTL (1B)	DH	Williams *et al.*, 2006
(ii) Root-knot-nematode resistance	*Rkn-mn1* (3BL)	BC3F2, F3 lines	Barloy *et al.*, 2000
(iii) Root-lesion	*Rlnn1* (7AL)	DH	Williams *et al.*, 2002.

RILs: recombinant inbred lines; DH: double haploid; RICLs: recombinant inbred chromosome lines; SCRI: single-chromosome recombinant lines; AL: addition lines; BILs: backcross inbred lines; NILs: near isogenic lines.

9.1.2 Isolation of Disease Genes by Map-based Cloning

A number of genes for disease resistance have been recently cloned by using map-based cloning strategy (Table 19). Some of the genes are currently being cloned and characterized. The first genes to be isolated by map-based cloning were three resistance genes against fungal diseases. These genes are leaf rust genes, *Lr21* and *Lr10* and powdery mildew *Pm3b* (Feuillet *et al.,* 2003; Huang *et al.,* 2003; Yahiaoui *et al.,* 2004). However, recently two different CC-NBS-LRR genes were discovered for the *Lr10*-mediated leaf rust resistance (Loutre *et al.*, 2009). A candidate for *Lr19*, one of the few widely effective genes conferring resistance to the leaf rust was isolated recently; however, its significance is yet to determined (Gennaro *et al.*, 2009). Krattinger *et al.* (2009) describe the cloning of the wheat *Lr34* QTL that has been used to confer resistance to multiple rusts and a mildew in the field for nearly 50 years. The locus harbours the gene *Lr34*, which encodes an ATP-binding cassette (ABC) transporter. ABC transporters, or pleiotropic drug resistance transporters, move diverse chemical compounds, including plant natural products, across membranes. Fu *et al.* (2009) find that another QTL, *Yr36*, which provides quantitative resistance to stripe rust, harbours the gene *WHEAT KINASE START 1*(*WKS1*). The WKS1 protein contains a steroidogenic acute regulatory protein-related lipid transfer domain (START) and a functional enzymatic (kinase) domain, suggesting a role for plant lipids in cell signalling mechanisms that confer disease resistance. Therefore, that mechanisms underlying quantitative resistance are not simply weak alleles of genes involved in specific gene-for-gene resistance, as has been proposed. Rather, each plant species may contain multiple, possibly independent, mechanisms of quantitative resistance.

Table 19. Disease genes cloned or likely to be cloned.

Gene/QTL	Disease	Reference(s)
Lr34/Yr18/Pm38	Multiple pathogens (leaf rust, stripe rust, and powdery mildew)	Krattinger *et al.*, 2009
Lr1	Leaf rust resistance	Ling *et al.*, 2003; Cloutier *et al.*, 2007
Lr10	Leaf rust resistance	Feuillet *et al.*, 2003; Loutre et al., 2009
Lr19	Leaf rust resistance	Gennaro *et al.*, 2009
Lr21	Leaf rust resistance	Huang *et al.*, 2003
Yr36 (WKS1)	Wheat stripe rust	Fu *et al.*, 2009
Yr5	Wheat stripe rust	Ling *et al.*, 2005
Pm3b	Powdery mildew resistance	Yahiaoui *et al.*, 2004
Qfhs.Ndsu-3bs	Fusarium head blight resistance	Liu *et al.*, 2005
Sr2	Stem rust resistance	Kota *et al.*, 2006

9.1.3 Functional Genomics Approaches

Expressed sequence tags and transcript profiling: To know the functions of all the genes in a plant genome is the most challenging task. Moreover, the combination of existing knowledge and resources with modern biotechnology and functional genomics is providing the opportunity to study the genetic, biochemical and physiological basis of complete traits. New initiatives to analyze the expressed portion of the wheat genome are going to increase our knowledge of the genes that are linked to various diseases. The expressed sequence tag (EST) is tag-based studies of gene expression and they are one of the important strategies of functional genomics. More than 1200000 wheat ESTs have been deposited in GenBank. These ESTs are presently being used for a variety of activities including development of functional molecular markers, preparation of transcript maps, and construction of cDNA arrays. Several other transcript profiling strategies like serial analysis of gene expression (SAGE), massively parallel signature sequencing (MPSS), and microarrays, are now available in many crops for the estimation of mRNA abundance for large number of genes simultaneously. The functional genomics approach provides a set of candidate genes that can be used for understanding the biology of wheat-pathogen interactions and for the development of perfect or diagnostic marker(s) to be used in map-based cloning of genes and MAS.

Gene silencing (RNAi) Approaches: Most of the techniques currently available to study gene function in plants involve knockouts of target genes followed by the phenotypic characterization of the mutant plants. In

Arabidopsis and rice, large gene knockout collections have been assembled from chemical mutagenesis and T-DNA or transposon insertional mutagenesis. In wheat, a powerful reverse genetics approach was recently implemented through the combination of EMS-mediated mutagenesis and TILLING technology. Because of the tolerance of polyploid wheat to high mutation rates, this method is very efficient in identifying mutations in the target genes. The discovery of RNA mediated gene silencing (RNAi) creates a viable alternative strategy for gene functional analysis through the simultaneous knockdown if expression of multiple related gene copies. Even though there are a limited number of RNAi studies in wheat, some general trends are emerging from its first applications that have successfully modified important agronomic and quality traits (Silivia *et al.*, 2006; Fu *et al.*, 2007). By using an RNAi gene-silencing assay, the critical role of monolignol biosynthesis was observed in cell wall apposition (CWA)-mediated defense against powdery mildew fungus penetration into diploid wheat (Bhuiyan *et al.* 2009). Silencing monolignol genes led to super-susceptibility of wheat leaf tissues to an appropriate pathogen, *Blumeria graminis* f. sp. *tritici* (*Bgt*). and compromised penetration resistance to a non-appropriate pathogen, *B. graminis* f. sp. *hordei* (Bhuiyan *et al.* 2009).

Virus-induced Gene silencing: To functionally characterize disease genes, the development of virus-induced gene silencing (VIGS) system for wheat that allows us to switch-off or "knockout" chosen genes so that we may infer gene function based on the knockout phenotypes. VIGS is very useful for functional genomics in wheat for two reasons: (1) most cultivated wheat is hexaploid, which greatly complicates gene identification through mutant analysis and (2) wheat is highly recalcitrant to transformation, which makes gene identification strategies based on creating transgenic plants very difficult. A virus-induced gene-silencing system was developed for hexaploid wheat and its use in functional analysis of the Lr21-mediated leaf rust resistance pathway (Scofield *et al.*, 2005). Recently, virus-induced gene silencing system was used to simultaneously silence two wheat genes using barley stripe mosaic virus (BSMV) related to leaf rust and Fusarium head blight (Cakir and Scofield, 2008). The BSMV-VIGS system has the capability for the rapid assessment of disease resistance candidate gene function, however it is not a simple process and care must be taken to ensure that phenotypic changes observed during VIGS are a consequence of gene silencing rather than some other aspect of virus infection.

Next Generation Technologies and Future of Wheat-pathogen interaction research: The technological advances in DNA sequencing over the past five years have changed our approaches to gene expression analysis, fundamentally altering the basic methods used and in most cases driving a shift from hybridization-based approaches to sequencing-based approaches. Quantitative, tag-based studies of gene expression were one of the earliest applications of these next-generation technologies, but the tremendous depth of sequencing facilitates *de novo* transcript discovery, which replaces traditional expressed sequence tag (EST) sequencing (Simon *et al.*, 2009). Using next-generation sequencing technologies (NGS) it is possible to isolate entire transcriptome more efficiently and economically and in greater depth (Varshney *et al.* 2009). The recent advancement regarding wheat genome is available on web site of international wheat genome sequencing consortium (http://www.wheatgenome.org). The large numbers of genetic markers are possible through NGS technologies in wheat which are going to facilitate disease resistance mapping and making marker-assisted breeding more feasible in future.

9.2 GENETIC ENGINEERING OF WHEAT

Cereals, including the hexaploid wheat remain a challenge for genetic transformation for many years. However, the progress towards efficient cereal transformation has been slow, mainly due to difficulties encountered in the establishment of embryogenic cell culture methods and a lack of efficient DNA delivery systems. Currently, transgenes are typically introduced using particle bombardment and *Agrobacterium*-mediated transformation of dissected explants. Particle bombardment has been used to transform embryogenic wheat calli and dissected scutella, and T-DNA transfer by *Agrobacterium* has been achieved with embryos, pre-cultured immature embryos, and embryogenic calli (Patnaik and Khurana, 2002; Zale *et al.*, 2009). Efforts to generate selectable marker-free wheat have also been demonstrated in past (Permingeat *et al.*, 2003). Very recently, the floral dip method was demonstrated for stable low-copy number transformants in wheat without the use of tissue culture (Zale *et al.*, 2009).

With the development of plant transformation techniques, it is possible to generate disease resistant plants other than the conventional breeding. Genetic engineering also allows the expression of foreign genes from distant unrelated species as well as the modification of the usual pattern of expression of an already present gene. The availability of potential anti-microbial peptides, defence-related proteins, key regulatory protein genes and defence signalling components have greatly enhanced the possibility of engineering wheat for resistance against pathogens. Most of the works on genetic engineering of wheat for resistance against biotic stress have focussed on developing protection against fungal pathogens. Introduction of genes encoding for chitinases from barley resulted in increased resistance against *Erysiphe graminis* (Bliffeld *et al.,* 1999). They have reported the adverse effect of a ribosome inactivating protein on plant regeneration and development. The genes encoding for thaumatin like protein (TLP) and stilbene synthase in transgenic wheat have been shown to improve resistance of T1, T2 progeny plants against the fungal pathogens (Leckband and Lorz, 1998; Chen *et al.,* 1999). Over expression of defence genes α-1-purothionin, thaumatin-like protein and β-1,3-Glucanse in transgenic wheat gave enhanced resistance against Fusarium head blight (Mackintosh *et al.*, 2007). Most of the introduced genes confer increased resistance to the corresponding pathogens in the transgenic plants. An ethylene response factor (ERF) gene from a wheat relative *Thinopyrum intermedium*, TiERF1, was overexpressed in transgenic wheat lines showed resistance against *Rhizoctonia cerealis* which causes wheat sharp eyespot disease (Chen *et al.*, 2008). Recently, the transgenic wheat expressing a barley class II chitinase gene has shown to enhance resistance against *Fusarium graminearum* (Shin *et al.*, 2008). The constitutive expression of a defence-related peroxidase in transgenic wheat was shown to potentiate cell death in pathogen-attacked leaf epidermis and exhibit enhanced resistance to wheat powdery mildew fungus *Blumeria graminis* f.sp. *tritici* (Schweizer, 2008).

REFERENCES

Aghaee, M., Sarbarzeh, H. Singh and H.S. Dhaliwal. 2001. A microsatellite marker linked to leaf rust resistance transferred from *Aegilops triuncialis* into hexaploid wheat. *Plant Breeding*. **120**: 259-261.

Aguilar, V., P. Stamp, M. Winzeler, *et al*. 2005. Inheritance of field resistance to *Stagonospora nodorum* leaf and glume blotch and correlations with other morphological traits in hexaploid wheat (*Triticum aestivum* L.). *Theor. Appl. Genet.* **111**: 325-336.

Alberto, D. Renato, T. O. Antonio, C. Carla, P. Marina and P. Enrico. 2003. Genetic analysis of the *Aegilops longissima* 3S chromosome carrying the *Pm13* resistance gene. *Euphytica.* **130**: 177-183.

Arraiano, L.S., A.J. Worl, C. Ellerbrook and J. K. M. Brown. 2001. Chromosomal location of a gene for resistance to *Septoria tritici* blotch (*Mycosphaerella graminicola*) in the hexaploid wheat 'Synthetic 6x'. *Theor. Appl. Genet.* **103**: 758-764.

Ayala, L., M. Henry, M. van Ginkel, R. Singh, B. Keller and M. Khairallah. 2002. Identification of QTLs for BYDV tolerance in bread wheat. *Euphytica.* **128**: 249-259.

Bariana, H. S., N. Parry, I.R. Barclay, *et al.* 2006. Identification and characterization of stripe rust resistance gene *Yr34* in common wheat. *Theor. Appl. Genet.* **112**: 1143-1148.

Barloy, J. Lemoine, F. Dredryver and J. Jahier. 2000. Molecular markers linked to the *Aegilops variabilis*-derived root-knot nematode resistance gene *Rkn-mn1* in wheat. *Plant Breeding.* **119**: 169-172.

Bhuiyan N.H., G. Selvaraj, Y. Wei and J. King. 2009. Role of lignification in plant defense. *Plant Signal Behavior.* **4**: 158-159.

Blanco, A. del, R. C. Frohberg, R.W. Stack,W. A. Berzonsky and S.F. Kianian. 2003. Detection of QTL linked to Fusarium head blight resistance in Sumai 3-derived North Dakota bread wheat lines. *Theor. Appl. Genet.* **106**: 1027-1031.

Blanco, A., A. Gadaleta, A. Cenci, A. V. Carluccio, A. M. M. Abdelbacki and R. Simeone. 2008. Molecular mapping of the novel powdery mildew resistance gene *Pm36* introgressed from *Triticum turgidum var. dicoccoides* in durum wheat. *Theor. Appl. Genet.* **117**: 135-142.

Bliffeld, M., Mundy, J, Potrykus, I. and Futterer, J. 1999. Genetic engineering of wheat for increased resistance to powdery mildew disease. *Theor. Appl. Genet.*. ***98**: 1079-1086.*

Börner, A., M.S. Röder, O. Unger and A. Meinel. 2000. The detection and molecular mapping of a major gene for nonspecific adult-plant disease resistance against stripe rust (*Puccinia striiformis*) in wheat. *Theor. Appl. Genet.* **100**: 1095-1099.

Bossolini, E., S. G. Krattinger and B. Keller. 2006. Development of simple sequence repeat markers specific for the *Lr34* resistance region of wheat using sequence information from rice and *Aegilops tauschii*. *Theor. Appl. Genet.* **113**: 1049-1062.

Bougot, Y.J., Lemoine, M.T. Pavoine, D. Barloy and G. Doussinault. 2002. Identification of a microsatellite marker associated with *Pm3* resistance alleles to powdery mildew in wheat. *Plant Breeding*. **121**: 325-329.

Bougot, Y.J., Lemoine, M.T. Pavoine, *et al.*, 2006. A major QTL effect controlling resistance to powdery mildew in winter wheat at the adult plant stage. *Plant Breeding*. **125**: 550-556.

Bourdoncle, W. and H.W. Ohm. 2003. Quantitative trait loci for resistance to Fusarium head blight in recombinant inbred wheat lines from the cross Huapei 57-2/Patterson. *Euphytica*. **131**: 131-136.

Brunner, S., P. Srichumpa, N. Yahiaoui and B. Keller. 2005. Positional cloning and evolution of powdery mildew resistance gene at *Pm3* locus of hexaploid wheat. **In**: *Proceedings of the Plant & Animal Genome XIII Conference*, p. 73, Town & Country Convention Center, San Diego, Calif., USA.

Buerstmayr, H., M. Lemmens, L. Hartl, *et al.* 2002. Molecular mapping of QTLs for Fusarium head blight resistance in spring wheat. I. Resistance to fungal spread (type II resistance). *Theor. Appl. Genet.* **104**: 84-91.

Cakir C. and S.R. Scofield. 2008. Evaluating the ability of the barley stripe mosaic virus-induced gene silencing system to simultaneously silence two wheat genes. *Cereal Research Communications*. **36**: 217-222.

Chantret, N., P. Sourdille, M. R¨oder, M. Tavaud, M. Bernard and G. Doussinault, 2000. Location and mapping of the powdery mildew resistance gene *MlRE* and detection of a resistance QTL by bulked segregant analysis (BSA) with microsatellites in wheat. *Theor. Appl. Genet.* **100**: 1217-1224.

Chen Liang, ZengYan Zhang, HongXia Liang, HongXia Liu, LiPu Du, Huijun Xu and Zhiyong Xn. 2008. Over expression of TiERF1 enhances resistance to sharp eyespot in transgenic wheat. *Jour. of Exp. Bot.* **59**: 4195-4204.

Chen, W.P., P.D. Chen, D.J. Liu, R. Kynast, B. Friebe, R. Velazhahan, S. Muthukrishnan and B.S. Gill. 1999. Development of wheat scab symptoms is delayed in transgenic wheat plants that constitutively express a rice thaumatin-like protein gene. *Theor. Appl. Genet.,* **99**: 755-760.

Chen, X., J.D. Faris, J. Hu, *et al.* 2007. Saturation and comparative mapping of a major Fusarium head blight resistance QTL in tetraploid wheat. *Molecular Breeding.* **19**: 113-124.

Cherukuri, D.P., S.K. Gupta, A. Charpe *et al.* 2003. Identification of a molecular marker linked to an *Agropyron elongatum* derived gene *Lr19* for leaf rust resistance in wheat. *Plant Breeding*, **122**: 204-208.

Cherukuri, D.P., S.K. Gupta, A. Charpe *et al.* 2005. Molecular mapping of *Aegilops speltoides* derived leaf rust resistance gene *Lr28* in wheat. *Euphytica.* **143**: 19-26.

Christiansen, M.J., B. Feenstra, I.M. Skovgaard and S.B. Andersen. 2006. Genetic analysis of resistance to yellow rust in hexaploid wheat using a mixture model for multiple crosses. *Theor. Appl. Genet.* **112**: 581-591.

Cloutier, S., B.D. McCallum, C. Loutre *et al.* 2007. Leaf rust resistance gene *Lr1*, isolated from bread wheat (*Triticum aestivum* L.) is a member of the large *Psr567* gene family. *Plant Molecular Biology.* **65**: 93-106.

Cuthbert, P.A., D.J. Somers and A. Brulé-Babel. 2007. Mapping of *Fhb2* on chromosome 6BS: a gene controlling Fusarium head blight field resistance in bread wheat (*Triticum aestivum* L.). *Theor. Appl. Genet.* **114**: 429-437.

de la Peòa, R.C., T.D.Murray and S.S. Jones. 1997. Identification of an RFLP interval containing *Pch2* on chromosome 7AL in wheat. *Genome.* **40**: 249-252.

Draeger, R., N. Gosman, A. Steed, *et al.* 2007. Identification of QTLs for resistance to Fusarium head blight, DON accumulation and associated traits in the winter wheat variety Arina. *Theor. Appl. Genet.* **115**: 617-625.

Eriksen, L., F. Borum and A. Jahoor. 2003. Inheritance and localization of resistance to *Mycosphaerella graminicola* causing *Septoria tritici* blotch and plant height in the wheat (*Triticum aestivum* L.) genome with DNA markers. *Theor. Appl. Genet.* **107**: 515-527.

Faris, J.D., J.A. anderson, L.J. Francl and J.G. Jordahl. 1997. RFLP mapping of resistance to chlorosis induction by *Pyrenophora tritici-repentis* in wheat. *Theor. Appl. Genet.* **94**: 98-103.

Feuillet, C., M. Messmer, G. Schachermayr and B. Keller. 1995. Genetic and physical characterization of the *Lr1* leaf rust resistance locus in wheat (*Triticum aestivum* L.). *Molecular and General Genetics.* **248**: 553-562.

Feuillet, S. Travella, N. Stein, L. Albar, A. Nublat and B. Keller. 2003. Map-based isolation of the leaf rust disease resistance gene *Lr10* from the hexaploid wheat (*Triticum aestivum* L.) genome. *Pro. Nat. Acad. Sci.,* **100**: 15253-15258.

Fofana, D. G. Humphreys, S. Cloutier, C. A. McCartney and D. J. Somers. 2008. Mapping quantitative trait loci controlling common bunt resistance in a doubled haploid population derived from the spring wheat cross RL4452 × AC Domain. *Molecular Breeding.* **21**: 317-325.

Fu Daolin, C. Uauy, A. Distelfeld, A. Blechl, L. Epstein, X. Chen, H. Sela, T. Fahima and J. Dubcovsky. 2009. A Kinase-START gene confers temperature-dependent resistance to wheat stripe rust. *Science.* **323**: 1357-1359.

Fu Daolin, C. Uauy, A. Blechl and J. Dubcovsky. 2007. RNA interference for wheat functional gene analysis. *Transgenic Res.* **16**: 689-701.

Gennaro andrea, Robert M.D. Koebner and Carla Ceoloni. 2009. A candidate for *Lr19,* an exotic gene conditioning leaf rust resistance in wheat. *Funct. Integr Genomics.* **9**: 325-334.

Gilsinger, J., L. Kong, X. Shen and H. Ohm. 2005. DNA markers associated with low Fusarium head blight incidence and narrow flower opening in wheat. *Theor. Appl. Genet.* **110**: 1218-1225.

Groenewald, J.Z., A.S. Marais and G.F. Marais. 2003. Amplified fragment length polymorphism-derived microsatellite sequence linked to the *Pch*1 and *Ep-D*1 loci in common wheat. *Plant Breeding.* **122**: 83-85.

Gupta, P.K., R.R. Mir, A. Mohan and J. Kumar. 2008. Wheat genomics: Present status and future prospects. *International Journal of Plant Genomics.* **2008**: 896451.

Gupta, S.K., A. Charpe, K.V. Prabhu and Q.M.R. Haque. 2006. Identification and validation of molecular markers linked to the leaf rust resistance gene *Lr*19 in wheat. *Theor. Appl. Genet.* **113**: 1027-1036.

Häberle, Jennifer, Josef Holzapfel, Günther Schweizer and Lorenz Hartl. 2009. a major QTL for resistance against Fusarium head blight in European winter wheat. *Theor Appl Genet.* **119**: 325-332.

Hartl, L., H. Weiss, U. Stephan, F.J. Zeller and A. Jahoor, 1995. Molecular identification of powdery mildew resistance genes in common wheat (*Triticum aestivum* L.). *Theor. Appl. Genet.* **90**: 601-606.

Helguera, M., I.A. Khan and J. Dubcovsky. 2000. Development of PCR markers for the wheat leaf rust resistance gene *Lr*47. *Theor. Appl. Genet.* **100**: 1137-1143.

Hiebert, C.W., J.B. Thomas and B.D. McCallum. 2005. Locating the broad-spectrum wheat leaf rust resistance gene *Lr*52(*LrW*) to chromosome 5B by a new cytogenetic method. *Theor. Appl. Genet.* **110**: 1453-1457.

Hiebert, C.W., J.B. Thomas, D.J. Somers, B.D. McCallum and S.L. Fox. 2007. Microsatellite mapping of adult-plant leaf rust resistance gene *Lr*22*a* in wheat. *Theor. Appl. Genet..* **115**: 877-884.

Hua, Wei., Ziji Liu, Jie Zhu, Chaojia Xie, Tsomin Yang, Yilin Zhou, Xiayu Duan, Qixin Sun and Zhiyong Liu. 2009. Identification and genetic mapping of *pm42,* a new recessive wheat powdery mildew resistance gene derived from wild emmer (*Triticum turgidum var. dicoccoides*). *Theor. Appl. Genet.* **119**: 223-230.

Huang, L., S.A. Brooks, W. Li, J.P. Fellers, H.N. Trick and B.S. Gill. 2003. Map-based cloning of leaf rust resistance gene *Lr*21 from the large and polyploid genome of bread wheat. *Genetics* **164**: 655-664.

Huang, X. Q., L.X. Wang, M.X. Xu and M.S. Röder. 2003. Microsatellite mapping of the powdery mildew resistance gene *Pm*5*e* in common wheat (*Triticum aestivum* L.). *Theor. Appl. Genet.* **106**: 858-865.

Huang, X.Q., S.L.K. Hsam, F.J. Zeller, G. Wenzel and V. Mohler. 2000. Molecular mapping of the wheat powdery mildew resistance gene *Pm*24 and marker validation for molecular breeding. *Theor. Appl. Genet.* **101**: 407-414.

Huguet-Robert, V., F. Dedryver, M.S. Röder, *et al.* 2001. Isolation of a chromosomally engineered durum wheat line carrying the *Aegilops ventricosa Pch*1 gene for resistance to eyespot. *Genome* **44**: 345-349.

Jahier, J., P. Abelard, A.M. Tanguy, *et al.* 2001. The *Aegilops ventricosa* segment on chromosome 2AS of the wheat cultivar 'VPM1' carries the cereal cyst nematode resistance gene *Cre*5. *Plant Breeding* **120**: 125-128.

Järve, K., H.O. Peusha, J. Tsymbalova, S. Tamm, K.M. Devos and T.M. Enno. 2000. Chromosomal location of a *Triticum timopheevii*-derived powdery mildew resistance gene transferred to common wheat. *Genome* **43**: 377-381.

Jia, G., P. Chen, G. Qin, *et al*. 2005. QTLs for Fusarium head blight response in a wheat DH population of Wangshuibai/Alondra's'. *Euphytica* **146**: 183-191.

Jia, J., K.M. Devos, S. Chao, T.E. Miller, S.M. Reader and M.D. Gale. 1996. RFLP-based maps of the homologous group-6 chromosomes of wheat and their application in the tagging of *Pm*12, a powdery mildew resistance gene transferred from *Aegilops speltoides* to wheat. *Theor. Appl. Genet.* **92**: 559-565.

Jiang, G.-L., Y. Dong, J. Shi and R.W. Ward. 2007. QTL analysis of resistance to Fusarium head blight in the novel wheat germplasm CJ 9306—II: resistance to deoxynivalenol accumulation and grain yield loss. *Theor. Appl. Genet.* **115**: 1043-1052.

Keller, M., B. Keller, G. Schachermayr, *et al*. 1999. Quantitative trait loci for resistance against powdery mildew in a segregating wheat × spelta population. *Theor. Appl. Genet.* **98**: 903-912.

Khan, A., G.C. Bergstrom, J.C. Nelson and M.E. Sorrells. 2000. Identification of RFLP markers for resistance to wheat spindle streak mosaic bymovirus (WSSMV) disease. *Genome* **43**: 477-482.

Klahr, G. Zimmermann, G. Wenzel and V. Mohler. 2007. Effects of environment, disease progress, plant height and heading date on the detection of QTLs for resistance to Fusarium head blight in an European winter wheat cross. *Euphytica* **154**: 17-28.

Kota, R.S., W. Spielmeyer, R.A. McIntosh and E.S. Lagudah. 2006. Fine genetic mapping fails to dissociate durable stem rust resistance gene *Sr*2 from pseudo-black chaff in common wheat (*Triticum aestivum* L.). *Theor. Appl. Genet.* **112**: 492-499.

Laroche, T. Demeke, D.A. Gaudet, B. Puchalski, M. Frick and R. McKenzie. 2000. Development of a PCR marker for rapid identification of the *Bt*-10 gene for common bunt resistance in wheat. *Genome* **43**: 217-223.

Leckband G. and H. Lorz. 1998. Transformation and expression of a stilbene synthase gene of *Vitis vinifera* L. in barley and wheat for increased fungal resistance. *Theor. Appl. Genet.* **96**: 1004-1012.

Leonova, N., L.I. Laikova, O. M. Popova, O. Unger, A. Börner and M. S. Röder. 2007. Detection of quantitative trait loci for leaf rust resistance in wheat-*T. timopheevii* / *T. tauschii* introgression lines. *Euphytica* **155**: 79-86.

Li He-Ping, Jing-Bo Zhang, Run-Ping Shi, Tao Huang, Rainer Fischer and Yu-Cai Liao. 2008. Engineering of Fusarium head blight resistance in wheat by expression of a fusion protein containing a *Fusarium*-specific antibody and an Antifungal Peptide. *Molecular Plant-Microbe Interactions.* **21**: 1242-1248.

Li, G.Q., Z.F. Li, W.Y. Yang, *et al*. 2006. Molecular mapping of stripe rust resistance gene *YrCH*42 in Chinese wheat cultivar Chuanmai 42 and its allelism with *Yr*24 and *Yr*26. *Theor. Appl. Genet.* **112**: 1434-1440.

Lin, F., S.L. Xue, Z.Z. Zhang, *et al*. 2006. Mapping QTL associated with resistance to Fusarium head blight in the Nanda 2419 × Wangshuibai population—II: Type I resistance. *Theor. Appl. Genet.* **112**: 528-535.

Lin, F., Z.X. Kong, H.L. Zhu, *et al*. 2004. Mapping QTL associated with resistance to Fusarium head blight in the Nanda 2419 × Wangshuibai population—I: Type II resistance. *Theor. Appl. Genet.* **109**: 1504-1511.

Ling, H.-Q., J. Qiu, R.P. Singh and B. Keller. 2004. Identification and genetic characterization of an *Aegilops tauschii* ortholog of the wheat leaf rust disease resistance gene *Lr*1. *Theor. Appl. Genet.* **109**: 1133-1138.

Ling, H.-Q., Y. Zhu and B. Keller. 2003. High-resolution mapping of the leaf rust disease resistance gene *Lr*1 in wheat and characterization of BAC clones from the *Lr*1 locus. *Theor. Appl. Genet.,* **106**: 875-882.

Ling, P., X. Chen, D.Q. Le and K. G. Campbell. 2005. Towards cloning of the *Yr*5 gene for resistance to wheat stripe rust resistance, in *Proceedings of the Plant & Animal Genomes XIII Conference*, Town & Country Convention Center, San Diego, Calif., USA.

Liu, S., C.A. Griffey and M.A.S. Maroof. 2001. Identification of molecular markers associated with adult plant resistance to powdery mildew in common wheat cultivarMassey. *Crop Science* **41**: 1268-1275.

Liu, S., M.O. Pumphery, X. Zhang, *et al.* 2005. Towards positional cloning of Qfhs.ndsu-3BS, a major QTL for Fusarium head blight resistance in wheat, in *Proceedings of the Plant & Animal Genome XIII Conference,* p. 71, Town & Country Convention Center, San Diego, Calif., USA.

Liu, W., H. Nie, S. Wang, *et al.* 2005. Mapping a resistance gene in wheat cultivar Yangfu 9311 to yellow mosaic virus, using microsatellite markers. *Theor. Appl. Genet.,* **111**: 651-657.

Liu, Z., Q. Sun, Z. Ni and T. Yang. 1999. Development of SCAR markers linked to the *Pm*21 gene conferring resistance to powdery mildew in common wheat. *Plant Breeding* **118**: 215-219.

Liu, Z., Q. Sun, Z. Ni, E. Nevo and T. Yang. 2002. Molecular characterization of a novel powdery mildew resistance gene *Pm*30 in wheat originating from wild emmer. *Euphytica* **123**: 21-29.

Loutre Caroline, T. Wicker, S. Travella, P. Galli, S. Scofield, T. Fahima, C. Feuillet and B. Keller. 2009. Two different CC-NBS-LRR genes are required for *Lr10*-mediated leaf rust resistance in tetraploid and hexaploid wheat. *The Plant Journal. (In Press)*

M.R. Simón, F.M. Ayala, C.A. Cordo, M.S. R¨oder and A. Börner. 2004. Molecular mapping of quantitative trait loci determining resistance to *Septoria tritici* blotch caused by *Mycosphaerella graminicola* in wheat. *Euphytica* **138**: 41-48.

Ma, J., R. Zhou, Y. Dong, L. Wang, X. Wang and J. Jia. 2001. Molecular mapping and detection of the yellow rust resistance gene *Yr26* in wheat transferred from *Triticum turgidum* L. using microsatellite markers. *Euphytica* **120**: 219-226.

Ma, Z.-Q., J.-B. Wei and S.-H. Cheng. 2004. PCR-based markers for the powdery mildew resistance gene *Pm4a* in wheat. *Theor. Appl. Genet.* **109**: 140-145.

Mackintosh C.A., J. Lewis, L.E. Radmer, S. Shin, S.J. Heinen, L.A. Smith, M.N. Wyckoff, R. Dill-Macky, C.K. Evans, S. Kravchenko, G.D. Baldridge, R.J. Zeyen and G.J. Muehlbauer. 2007. Overexpression of defense response genes in transgenic wheat enhances resistance to Fusarium head blight. *Plant Cell Rep.* **26**: 479-488.

Mallard, S., D. Gaudet, A. Aldeia, *et al.* 2005. Genetic analysis of durable resistance to yellow rust in bread wheat. *Theor. Appl. Genet.* **110**: 1401-1409.

McCartney, C.A., D.J. Somers, B.D. McCallum, *et al.* 2005. Microsatellite tagging of the leaf rust resistance gene *Lr16* on wheat chromosome 2BSc. *Molecular Breeding* **15**: 329-337.

Mingeot, D., N. Chantret, P.V. Baret, *et al.* 2002. Mapping QTL involved in adult plant resistance to powdery mildew in the winter wheat line RE714 in two susceptible genetic backgrounds. *Plant Breeding* **121**: 133-140.

Miranda, L.M., J.P. Murphy, D. Marshall and S. Leath. 2006. *Pm34*: a new powdery mildew resistance gene transferred from *Aegilops tauschii* Coss. to common wheat (*Triticum aestivum* L.). *Theor. Appl. Genet.* **113**: 1497-1504.

Miranda, L.M., J.P. Murphy, D. Marshall, C. Cowger and S. Leath. 2007. Chromosomal location of *Pm35*, a novel *Aegilops tauschii* derived powdery mildew resistance gene introgressed into common wheat (*Triticum aestivum* L.). *Theor. Appl. Genet.* **114**: 1451-1456.

Mohler, V., S.L.K. Hsam, F.J. Zeller and G. Wenzel. 2001. An STS marker distinguishing the rye-derived powdery mildew resistance alleles at the *Pm8* / *Pm17* locus of common wheat. *Plant Breeding* **120**: 448-450.

Naik, S., K.S. Gill, V.S. Prakasa Rao, *et al.* 1998. Identification of a STS marker linked to the *Aegilops speltoides*-derived leaf rust resistance gene *Lr28* in wheat. *Theor. Appl. Genet.*. **97**: 535-540.

Nematollahi, G., V. Mohler, G. Wenzel, F.J. Zeller and S.L.K. Hsam. 2008. Microsatellite mapping of powdery mildew resistance allele *Pm5d* from common wheat line IGV1-455. *Euphytica* **159**: 307-313.

Neu, C., N. Stein and B. Keller. 2002. Genetic mapping of the *Lr20-Pm1* resistance locus reveals suppressed recombination on chromosome arm 7AL in hexaploid wheat. *Genome* **45**: 737-744.

Obert, D.E., A.K. Fritz, J.L.Moran, S. Singh, J.C. Rudd and M.A. Menz. 2005. Identification and molecular tagging of a gene from PI 289824 conferring resistance to leaf rust (*Puccinia triticina*) in wheat. *Theor. Appl. Genet.* **110**: 1439-1444.

Ogbonnaya, F.C., S. Seah, A. Delibes, *et al.* 2001. Molecular genetic characterisation of a new nematode resistance gene in wheat. *Theor. Appl. Genet.* **102**: 623-629.

Paillard, S., T. Schnurbusch, R. Tiwari, *et al.* 2004. QTL analysis of resistance to Fusarium head blight in Swiss winter wheat (*Triticum aestivum* L.). *Theor. Appl. Genet.* **109**: 323-332.

Patnaik, D. and P. Khurana. 2001. Wheat Biotechnology: a mini review. *EJB Electronic Journal of Biotechnology* **4**: 74-102.

Paull, J.G., M.A. Pallotta, P. Langridge and T.T. The. 1994. RFLP markers associated with *Sr22* and recombination between chromosome 7A of bread wheat and the diploid species *Triticum boeoticum*. *Theor. Appl. Genet.* **89**: 1039-1045.

Peng, J.H., T. Fahima, M.S. Röder, *et al.* 1999. Microsatellite tagging of the stripe-rust resistance gene *YrH52* derived from wild emmer wheat, *Triticum dicoccoides* and suggestive negative crossover interference on chromosome 1B. *Theor. Appl. Genet.* **98**: 862-872.

Peng, J.H., T. Fahima, M.S. Röder, *et al.* 2001. High-density molecular map of chromosome region harboring stripe-rust resistance genes *YrH52* and *Yr15* derived from wild emmer wheat, *Triticum dicoccoides*. *Genetica.* **109**: 199-210.

Permingeat Hugo R., M.L. Alvarez, D.L. Gerardo, C.R. Ravizzini and R.H. Vallejos. 2003. Stable wheat transformation obtained without selectable markers. *Plant Molecular Biology* **52**: 415-419.

Prins, R., J.Z. Groenewald, G.F. Marais, J.W. Snape and R.M.D. Koebner. 2001. AFLP and STS tagging of *Lr19*, a gene conferring resistance to leaf rust in wheat. *Theor. Appl. Genet.* **103**: 618-624.

Qiu, J.-W., A.C. Schürch, N. Yahiaoui, *et al.* 2007. Physical mapping and identification of a candidate for the leaf rust resistance gene *Lr1* of wheat. *Theor. Appl. Genet.* **115**: 159-168.

Raupp, W.J., S. Singh, G.L. Brown-Guedira and B.S. Gill. 2001. Cytogenetic and molecular mapping of the leaf rust resistance gene *Lr39* in wheat. *Theor. Appl. Genet.* **102**: 347-352.

Rong, J.K., E. Millet, J. Manisterski and M. Feldman. 2000. A new powdery mildew resistance gene: introgression from wild emmer into common wheat and RFLP-based mapping. *Euphytica* **115**: 121-126.

Sacco, F., E.Y. Su´arez and T. Naranjo. 1998. Mapping of the leaf rust resistance gene *Lr3* on chromosome 6B of Sinvalocho MA wheat. *Genome* **41**: 686-690.

Schachermayr, G.M., M.M. Messmer, C. Feuillet, H. Winzeler, M. Winzeler and B. Keller. 1995. Identification of molecular markers linked to the *Agropyron elongatum* derived leaf rust resistance gene *Lr24* in wheat. *Theor. Appl. Genet.* **90**: 982-990.

Schachermayr, G., C. Feuillet and B. Keller. 1997. Molecular markers for the detection of the wheat leaf rust resistance gene *Lr10* in diverse genetic backgrounds. *Molecular Breeding* **3**: 65-74.

Schachermayr, G., H. Siedler, M.D. Gale, H. Winzeler, M. Winzeler and B. Keller. 1994. Identification and localization of molecular markers linked to the *Lr9* leaf rust resistance gene of wheat. *Theor. Appl. Genet.* **88**: 110-115.

Schmolke, M., G. Zimmermann, H. Buerstmayr, *et al*. 2005. Molecular mapping of Fusarium head blight resistance in the winter wheat population Dream/Lynx. *Theor. Appl. Genet.* **111**: 747-756.

Schnurbusch, T., S. Paillard, A. Schori, *et al*. 2004. Dissection of quantitative and durable leaf rust resistance in Swiss winter wheat reveals a major resistance QTL in the *Lr34* chromosomal region. *Theor. Appl. Genet.* **108**: 477-484.

Schweizer, P. 2008. Tissue-specific expression of a defence-related peroxidase in transgenic wheat potentiates cell death in pathogen-attacked leaf epidermis. *Mol. Plant Pathol.* **9**: 45-57.

Scofield, S.R., H. Li, A.S. Brandt and B.S. Gill. 2005. Development of virus-induced gene-silencing system for hexaploid wheat and its use in functional analysis of the *Lr21*-mediated leaf rust resistance pathway. *Plant Physiology.* **138**: 2165-2173.

Seah, S., H. Bariana, J. Jahier, K. Sivasithamparam and E.S. Lagudah. 2001. The introgressed segment carrying rust resistance genes *Yr17*, *Lr37* and *Sr38* in wheat can be assayed by a cloned disease resistance gene-like sequence. *Theor. Appl. Genet.* **102**: 600-605.

Seyfarth, R., C. Feuillet, G. Schachermayr, M. Winzeler and B. Keller. 1999. Development of a molecular marker for the adult plant leaf rust resistance gene *Lr35* in wheat. *Theor. Appl. Genet.* **99**: 554-560.

Shen, X. and Ohm, H. 2007. Molecular mapping of *Thinopyrum*derived *Fusarium* head blight resistance in common wheat. *Molecular Breeding* **20**: 131-140.

Shi, Z.X., X.M. Chen, R.F. Line, H. Leung and C.R. Wellings. 2001. Development of resistance gene analog polymorphism markers for the *Yr9* gene resistance to wheat stripe rust. *Genome* **44**: 509-516.

Shin S., C.A. Mackintosh, J. Lewis, S.J. Heinen, L. Radmer, R. Dill-Macky, G.D. Baldridge, R.J. Zeyen and G.J. Muehlbauer. 2008. Transgenic wheat expressing a barley class II chitinase gene has enhanced resistance against *Fusarium greminearum*. *Journal of Experimental Botany.* **9**: 2371-2378.

Simón G.K., E.S. Lagudah, W. Spielmeyer, R.P. Singh, J.H.-Espino, H. McFadden, E.L. Selter and B. Keller. 2009. A putative ABC transporter confers durable resistance to multiple fungal pathogens in wheat. *Science* **323**: 1360-1363.

Simon S.A., J. Zhai, R. Sekhar, Nandety, K.P. McCormick, J. Zenga, D. Mejia and B.C. Meyers. 2009. Short-read sequencing technologies for transcriptional analyses. *Ann. Rev. Plant Biol.* **60**: 305-333.

Singh, P.K., M. Mergoum, T.B. Adhikari, S.F. Kianian and E.M. Elias. 2007. Chromosomal location of genes for seedling resistance to tan spot and *Stagonospora nodorum* blotch in tetraploid wheat. *Euphytica* **155**: 27-34.

Singh, R.P., J.C. Nelson and M.E. Sorrells. 2000. Mapping *Yr28* and other genes for resistance to stripe rust in wheat. *Crop Science* **40**: 1148-1155.

Singrün, Ch., S.L.K. Hsam, F.J. Zeller, G. Wenzel and V. Mohler. 2004. Localization of a novel recessive powdery mildew resistance gene from common wheat line RD30 in the terminal region of chromosome 7AL. *Theor. Appl. Genet.* **109**: 210-214.

Song, W., H. Xie, Q. Liu, *et al.* 2007. Molecular identification of *Pm12*-carrying introgression lines in wheat using genomic and EST-SSR markers. *Euphytica* **158**: 95-102.

Sourdille, P., P. Robe, M.-H. Tixier, G. Doussinault, M.-T. Pavoinc and M. Bernard. 1999. Location of *Pm3g*, a powdery mildew resistance allele in wheat, by using a monosomic analysis and by identifying associated molecular markers. *Euphytica* **110**: 193-198.

Spielmeyer, W., P.J. Sharp and E.S. Lagudah. 2003. Identification and validation of markers linked to broad-spectrum stem rust resistance gene *Sr2* in wheat (*Triticum aestivum* L.). *Crop Science* **43**: 333-336.

Spielmeyer, W., R.A. McIntosh, J. Kolmer and E.S. Lagudah. 2005. Powdery mildew resistance and *Lr34/Yr18* genes for durable resistance to leaf and stripe rust cosegregate at a locus on the short arm of chromosome 7D of wheat. *Theor. Appl. Genet.* **111**: 731-735.

Steiner, B., M. Lemmens, M. Griesser, U. Scholz, J. Schondelmaier and H. Buerstmayr. 2004. Molecular mapping of resistance to Fusarium head blight in the spring wheat cultivar Frontana. *Theor. Appl. Genet.* **109**: 215-224.

Sui X.X., M.N. Wang and X.M. Chen. 2009. Molecular mapping of a stripe rust resistance gene in spring wheat cultivar zak. *Phytopathology.* **99**: 1209-1215.

Sun, G.L., T. Fahima, A.B. Korol, *et al.* 1997. Identification of molecular markers linked to the *Yr15* stripe rust resistance gene of wheat originated in wild emmer wheat, *Triticum dicoccoides. Theor. Appl. Genet.* **95**: 622-628.

Tadesse, W., M. Schmolke, S.L.K. Hsam, V. Mohler, G. Wenzel and F.J. Zeller. 2007. Molecular mapping of resistance genes to tan spot [*Pyrenophora tritici-repentis* race 1] in synthetic wheat lines. *Theor. Appl. Genet.* **114**: 855-862.

Talbert, L.E., P.L. Bruckner, L.Y. Smith, R. Sears and T.J. Martin. 1996. Development of PCR markers linked to resistance to wheat streak mosaic virus in wheat. *Theor. Appl. Genet.* **93**: 463-467.

Tao, W.J., D. Liu, J.Y. Liu, Y. Feng and P. Chen. 2000. Genetic mapping of the powdery mildew resistance gene *Pm6* in wheat by RFLP analysis. *Theor. Appl. Genet.* **100**: 564-568.

Travella S., T.E. Klimm and B. Keller. 2006. RNA Interference-based gene silencing as an efficient tool for functional genomics in hexaploid bread wheat. *Plant Physiology* **142**: 6-20.

Tucker, D.M., C.A. Griffey, S. Liu, G. Brown-Guedira, D.S. Marshall and M.A.S. Maroof. 2007. Confirmation of three quantitative trait loci conferring adult plant resistance to powdery mildew in two winter wheat populations. *Euphytica* **155**: 1-13.

Uauy, C., J.C. Brevis, X. Chen, *et al.* 2005. High-temperature adult-plant (HTAP) stripe rust resistance gene *Yr36* from *Triticum turgidum* ssp. *dicoccoides* is closely linked to the grain protein content locus Gpc-B1. *Theor. Appl. Genet.* **112**: 97-105.

Varshney R.K., S.N. Nayak, G.D. May and S.A. Jackson. 2009. Next-generation sequencing technologies and their implications for crop genetics and breeding. *Cell.* **27**: 522-530.

Wang, L., J. Ma, R. Zhou, X. Wang and J. Jia. 2002. Molecular tagging of the yellow rust resistance gene *Yr10* in common wheat, P.I. 178383 (*Triticum aestivum* L.). *Euphytica* **124**: 71-73.

Wang, Y. Zhang, D. Han, *et al.* 2008. SSR and STS markers for wheat stripe rust resistance gene *Yr26. Euphytica.* **159**: 359-366.

Williams, K.J., J.M. Fisher and P. Langridge. 1994. Identification of RFLP markers linked to the cereal cyst nematode resistance gene (*Cre*) in wheat. *Theor. Appl. Genet.* **89**: 927-930.

Williams, K.J., K.L. Willsmore, S. Olson, M. Matic and H. Kuchel. 2006. Mapping of a novel QTL for resistance to cereal cyst nematode in wheat. *Theor. Appl. Genet.* **112**: 1480-1486.

Williams, K.J., S.P. Taylor, P. Bogacki, M. Pallotta, H. S. Bariana and H. Wallwork. 2002. Mapping of the root lesion nematode (*Pratylenchus neglectus*) resistance gene *Rlnn1* in wheat. *Theor. Appl. Genet.* **104**: 874-879.

Xie, Q. Sun, Z. Ni, T. Yang, E. Nevo and T. Fahima. 2003. Chromosomal location of a *Triticum dicoccoides*-derived powdery mildew resistance gene in common wheat by using microsatellite markers. *Theor. Appl. Genet.* **106**: 341-345.

Y.C. Qiu, R.H. Zhou, X.Y. Kong, S.S. Zhang and J.Z. Jia. 2005. Microsatellite mapping of a *Triticum urartu* Tum. derived powdery mildew resistance gene transferred to common wheat (*Triticum aestivum* L.). *Theor. Appl. Genet.* **111**: 1524-1531.

Yahiaoui, N., P. Srichumpa, R. Dudler and B. Keller. 2004. Genome analysis at different ploidy levels allows cloning of the powdery mildew resistance gene *Pm3b* from hexaploid wheat. *Plant Journal* **37**: 528-538.

Yao, G., J. Zhang, L. Yang, *et al.* 2007. Genetic mapping of two powdery mildew resistance genes in einkorn (*Triticum monococcum* L.) accessions. *Theor. Appl. Genet.* **114**: 351-358.

Zale Janice M., S. Agarwal, S. Loar and C.M. Steber. 2009. Evidence for stable transformation of wheat by floral dip in *Agrobacterium tumefaciens. Plant Cell Rep.* **28**: 903-913.

Zeller, F.J., L. Kong, L. Hartl, V. Mohler and S.L.K. Hsam. 2002. Chromosomal location of genes for resistance to powdery mildew in common wheat (*Triticum aestivum* L. em Thell.) 7. Gene *Pm29* in line Pova. *Euphytica* **123**: 187-194.

Zhang, X., M. Zhou, L. Ren, *et al.* 2004. Molecular characterization of *Fusarium* head blight resistance from wheat variety Wangshuibai. *Euphytica* **139**: 59-64.

Zhou, W., F.L. Kolb, G. Bai, G. Shaner and L.L. Domier. 2002. Genetic analysis of scab resistance QTL in wheat with microsatellite and AFLP markers. *Genome.* **45**: 719-727.

Zhou, W.-C., F.L. Kolb, G.-H. Bai, L.L. Domier, L.K. Boze and N.J. Smith. 2003. Validation of a major QTL for scab resistance with SSR markers and use of marker-assisted selection in wheat. *Plant Breeding* **122**: 40-46.

Zhu, Z., R. Zhou, X. Kong, Y. Dong and J. Jia. 2006. Microsatellite marker identification of a *Triticum aestivum—Aegilops umbellulata* substitution line with powdery mildew resistance. *Euphytica* **150**: 149-153.

Chapter

Integrated Disease Management

The development and use of pesticides in crop protection is a success story. However, public concern about pesticides has increased over the last two decades, probably as part of a general response to the rising levels of pollution and a growing popular awareness of the fragility of the environment, but also as a direct- consequence surrounding the use and withdrwal of some pesticides; notably DDT and the publication of "Silent Spring" (Carson, 1962). These events forced the scientists to reconsider the crop protection strategies and suggest alternative. The integrated control of pests was first formalized at California University in 1959 (Stern *et al.*, 1959). In 1965 the FAO panel experts recognized the concept of integrated pest management and defined IPM as "pest management system that in the context of the associated environment and population dynamics of the pest species, utilizes all suitable techniques and methods in as compatible manner as possible and maintains the pest populations below those causing economic injury" (F.A.O., 1966). In this background the integrated pest management came under focus as ecology oriented protection strategy.

The IPM stresses the rational use of combination of pest control techniques. There is an inter-relationship of crop, biotic influence, crop management and agro-ecosystems (Fig. 50). Since the IPM is based on ecological principles operating in an eco-system, any imbalance in the systems is likely to influence the productivity of the crop variety, while proper management system will increase the productivity. Hence the concept of integrated disease management implies that the agro-ecosystem

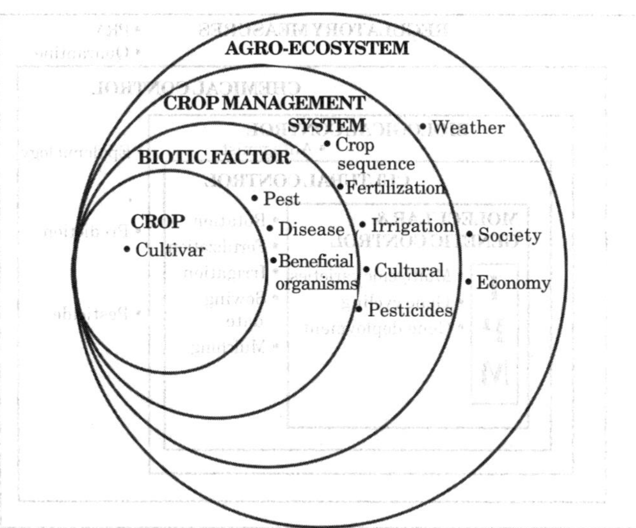

Figure 50: Inter-relationship of crop, biotic factors, crop management system and agro-ecosystem

should be adjusted or maintained at a sub-economic level of population so that extensive injury is avoided and productivity is sustained. Therefore, instead of being called pest control, the term pest management has been used. Infect, usage of pesticides is a top priority for control of a disease, while management focuses on other methods of pest control. The IPM strategy tolerates certain level of pest infestation which may be even slightly higher than the prescribed economic threshold level (Singh *et al.*, 2006).

The present day IPM is knowledge based. It requires information on risk and losses caused by pests and diseases; epidemiology and population dynamics; forecasts and control threshold; potential of beneficial organisms and influence of the IPM components on the agro-ecosystem. Thus, for an effective and efficient integrated management system high level managerial skill on the part of researchers, extension personnels and farmers is needed (Singh *et al.*, 2006).

10.1 COMPONENTS OF IPM

In the practical IPM system, six basic principles of plant disease control, *viz.*, avoidance of pathogen, exclusion of inoculum, eradication, protection, host resistance and therapy are usually opted for integration. IPM as a pest management system utilizes all suitable approaches involving genetical, cultural, biotechnological. biological, regulatory and chemical practices (Fig. 51) which are blended in a compatible manner to reduce and maintain pest population at levels below those causing economic injury. Among these options selective breeding and genetic mainipulation of host plants for

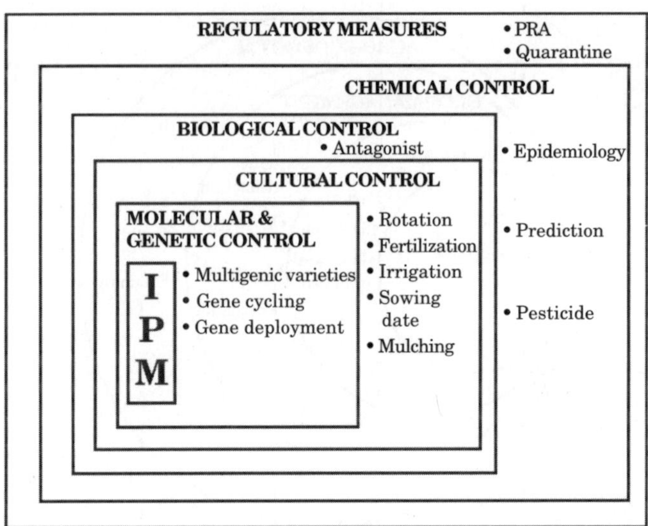

Figure 51: Main components of integrated pest management of wheat.

resistance to pests and diseases is of paramount importance. For example, gene deployment strategy has minimised the risk of rust epidemics in wheat growing regions of the country. Cultural methods which are considered useful in IPM system include crop rotation variation in planting dates, fertilization, mulching, irrigation and tillage. Several antagonisistic bioagents capable of suppressing plant diseases are part of integrated management strategy. Molecular aided information of host has drawn attention as an IPM approach in recent times. Regulatory practices through quarantine and pest risk analysis (PRA) are equally important in IPM system. The chemical control in IPM includes applications of insecticides/fungicides at the action level (Singh and Srivastava, 1992).

10.2 IPM MODULE IN WHEAT

IPM is a recent concept in wheat but some cultural practices used by the farmers in India, are centuries old. However, the scientifically designed programmes are only a few decades old. The first decade of 20th century witnessed improvement in productivity by mere utilization of yield potential of breeding material with no consideration for pest and disease resistance (Howard and Howard, 1910). In 1935, Dr. B.P. Pal and Prof. K.C. Mehta initiated a collaborative programme to breed rust resistant varieties. The joint approach between breeder and pathologist further strengthened the productivity of NP series of wheats with resistance to individual rust and combined resistance to leaf and stem rusts (Randhawa, 1979). In 1964, the Indian Council of Agricultural Research started the All Indian Coordinated Wheat Improvement Programme. Under this programme scientists specializing in different disciplines came together to work for multidisciplinary research in wheat. Resultantly valuable data on control measures of pests and diseases could be generated through coordinated approach (Table 20).

Based on some of the above mentioned techniques, the Directorate of Wheat Research has synthesized, evaluated and validated an IPM module on large scale at farmers' fields (Sharma and Babu, 2006). The IPM package which is spread over the entire growth period of the crop, is being promoted to target pests like rusts, loose smut, powdery mildew, termites and aphids in north-western plains zone in India (Sharma, 2008). This package consists of the following schedule:

Table 20. Some techniques evolved for management of wheat pathogens in India

Technique	Disease	Pathogen	References
Gene deployment	Rusts	*Puccinia graminis tritici* *P. recondita* *P. striiformis*	Bahadur and Nagarajan, 1985, Bahadur *et al.,* 1994, Joshi and Nagarajan, 1978
Cultural practice	Root rot	*Sclerotium rolfsii*	Padwick, 1939
Crop rotation	Seedling blight	*Rhizoctonia solani*	Kulkarni and Hegde, 1978
Mulching	Flag smut	*Urocystis agropyrii*	Bedi, 1957
	Karnal bunt	*Neovossia indica*	Singh *et al.,* 1992
Chemical control	Rusts	*P. graminis tritici* *P. recondita* *P. striiformis*	Srivastava *et al.,* 1983, Aujla *et al.,* 1993
	Loose smut	*Ustilago nuda tritici*	Nene *et al.,* 1971, Bahadur and Sinha, 1978, Aggarwal *et al.,* 1993
	Powdery mildew	*Erysiphe graminis*	Aggarwal *et al.,* 1997
	Karnal bunt	*N. indica*	Singh, 2008
Biological control	Loose smut	*Ustilago nuda tritici*	Aggarwal *et al.,* 1991
Epidemiological strategies	Rusts	*P. graminis tritici* *P. recondita* *P. striiformis*	Joshi *et at.,* 1985 Nagarajan and Joshi, 1985, Nagarajan *et al.,* 2006
	Karnal bunt	*N. indica*	Sidhartha *et al.,* 1995
Survey and surveillance	All diseases	—	Joshi, 1975

At the time of sowing: In North Indian conditions, treatment of wheat seed with recommended fungicides like carboxin (Vitavax 75 WP @ 2.5 g/kg seed), carbendazim (Bavistin 50 WP @ 2.5 g/kg seed) or tebuconazole (Raxil 2DS @ 1.25 g/kg seed) is advocated. If the level of loose smut pathogen (*Ustilago nuda tritici*) in the seed is low (*i.e.* less than 5%), then seed treatment with a combination of bio control agent – *Trichoderma viride* (4 g/kg seed) with half the recommended dose of carboxin (1.25 g/kg seed) is applied. It is realized that the treatment with *T. viride* helps is better germination and improves vigour of the plants. The seed treatment with fungicide also needs to be done in area having problem of flag smut and foot rot. In areas having termite problem, the seed can be treated with insecticide endosulfan @ 7 ml/kg seed, along with carboxin/carbendazim/tebuconazole and *T. viride* which are compatible with each other.

Seedling stage: To avoid damage due to termites, the broadcast of insecticide treated soil along with first irrigation is recommended. For this, endosulfan @ 0.8 kg a.i./ha is mixed in 40–50 kg of sand or fine soil along with 4–5 litres of water. This treated soil is broadcasted in the field with sufficient moisture (1–2 days) after first irrigation.

Boot leaf stage: In Karnal bunt (*Neovossia indica*) endemic area, the spray of propiconazole (Tilt 25EC @ 0.1%) at ear-head emergence stage can be given to protect the wheat crop from air-borne infection of the pathogen. However, the spray can be avoided in case there are no rains during second week of February, *i.e.* coinciding with the ear head emergence stage. Propiconazole a broad spectrum fungicide controls not only Karnal bunt but is also effective against rusts, powdery mildew and blights. One spray of *T. viride* given at crop growth stage of 31–40, followed by another spray at 41–49, can provide non-chemical management of Karnal bunt.

One spray of imidacloprid (Confidor 200 SL) @ 100 ml/ha be given on border rows upto five meter width to control foliar aphids, which colonize the wheat plants on borders before entering the main field.

Ear-head stage: Loose smut of wheat appears at ear-head stage. In some varieties boot leaf shows yellowing due to loose smut infection. At this stage field inspection and roughing of infected plants is suggested to check the spread of the fungus. The diseased ear-heads should be burnt.

Storage: The studies conducted at Wheat Directorate have indicated that pre-harvest spray of either malathion or cypermethrin can protect the grain from insect injury in storage up to six months of harvest. Proper sanitation and cleaniness should be ensured in the store room. Also prophylactic spray of malathion @ 10 ml/litre water at recommended dose in 100 m^2 area proves helpful in storage area prior to the storage.

REFERENCES

Aggarwal, R., K.D. Srivastava, D.V. Singh and P. Bahadur. 1991. Possible control of loose smut of wheat. *Indian J. Biol. Control.* **6**: 111-112.

Aggarwal, R., K.D. Srivastava and D.V. Singh (1993). Raxil - a potent fungicide to control loose smut of wheat. *Indian Phytopath.* **46**: 172-173.

Aggarwal, R., P. Bahadur and S.K. Jain. 1997. Efficacy of some systemic fungicides against powdery mildew (*Erysiphe graminis* f. sp. *tritici*) of wheat. *Indian J. Plant Prot.* **25**: 84-87.

Aujla, S.S., A.K. Basandarai and G.S. Rattan. 1993. Chemical control of leaf rust (*Puccinia recondita* Rob. ex. Desm. f. sp. *tritici*) of wheat. *Indian J. Mycol. Plant Pathol.* **23**: 209-213.

Bahadur, P. and S. Nagarajan. 1985. Gene deployment for management of stem rust of wheat (*Puccinia graminis*) Pers. f. sp. *tritici* Erikss. & Henn.) in India. *Proc. Indian Acad. Sci.* **95**: 29-33.

Bahadur, P., D.V. Singh and K.D. Srivastava. 1994. Mangement of wheat rusts – a revised strategy for gene deployment. *Indian Phytopath.* **47**: 41-47.

Bahadur, P. and V.C. Sinha. 1978. Efficacy of bavistin for controlling loose smut of wheat. *Pesticides* **12**: 31-32.

Bedi, K.S. 1957. Fighting the flag smut. *Indian Fmg.* **7**: 25-27.

Carson, R. 1962. *Silent Spring*. Houghton- Miffin W., Boston, USA, 368 p.

FAO 1966. *Proceedings of the FAO Symposium on Integrated Pest Control*, 1965, Vol. 1-3, Rome

Joshi, L.M. 1975. Surveys on wheat rusts in India. The rust situation since 1972. *Cereal Rust Bull.* **3**: 7-9.

Joshi, L.M. and S. Nagarajan. 1978. Regional deployment of *Lr* genes for brown rust management. **In**: *Genetics and Wheat Improvement* (Ed. A.K. Gupta). Oxford and IBH. Publ. Co., New Delhi, 268 p.

Joshi, L.M., K.D. Srivastava and D.V. Singh. 1985. Monitoring of wheat-rusts in the Indian subcontinent. *Proc. Indian Acad. Sci.* (Plant Sci.). **94**: 387-407.

Howard, A. and G.L.C. Howard. 1910. *Wheat in India: Its Production, Varieties and Improvement*. Thacker- Spink and Co. Culcutta, 288 p.

Kulkarni, S. and R.K. Hegde. 1978. Diseases of wheat and their control in Karnataka. *Kisan World* **5**: 32.

Nagarajan, S. and L. M. Joshi. 1985. Epidemiology in the Indian subcontinent. **In**: *The Cereal Rusts*, Vol. II (Eds. A.P. Roelfs and W.R. Bushnel), pp. 371-402. Academic Press, New York.

Nagarajan, S., S.K. Nayar and J. Kumar. 2006. Checking the *Puccinia* species that reduce the productivity of wheat (*Triticum* spp.) in India–A review. *Proc. Indian Natn. Sci. Acad.* **72**: 239-247.

Nene, Y.L., S.C. Saxena and S.S.L. Srivastava. 1971. Influence of fungicidal seed treatment on the incidence of loose smut and rusts of wheat *Pesticides* **6**: 11-14.

Padwick, G.W. 1939. Report of the Imperial Mycologist. *Sci. Rep. Agric. Sci.* 231-236.

Randhawa, M.S. 1979. *A History of Indian Council of Agricultural Research*, ICAR, New Delhi, 510 p.

Sidhartha, V.S., D.V. Singh, K.D. Srivastava and R. Aggarwal (1995). Some epidemiological aspects of Karnal bunt of wheat. *Indian Phytopath.* **48**: 419-426.

Sharma, A.K. 2008. IPM in wheat. **In**: *A Compendium of Lectures on Integrated Pest Management in Wheat based Cropping Systems,* (Eds. A.K. Sharma and D.P. Singh), pp. 104-111. DWR Compendium No. 2, Directorate of Wheat Research, Karnal, India, 258 p.

Sharma, A.K. and K.S. Babu. 2006. Status and prospectus of integrated pest management strategies in selected crops: Wheat. **In**: *Integrated Pest Management – Principles and Applications* (Eds. A. Singh, O.P. Sharma, and D.K. Garg). Vol II, CBS Publishers and Distributors New Delhi, 85-101 pp.

Singh, A., O.P. Sharma and D.K. Garg. 2006. *Integrated Pest Management: Principles and Application* Vol. 1: *Principles*. CBS Publishers and Distributors, New Delhi.

Singh, D.P. 2008. Disease problems of wheat and management approaches. **In**: *A Compendium of Lectures on Integrated Pest Management in Wheat based Cropping Systems*, (Eds. A.K. Sharma and D.P. Singh), pp. 77-84. DWR Compendium No. 2, Directorate of Wheat Research, Karnal, India, 258 p.

Singh, D.V. and K.D. Srivastava. 1992. Wheat diseases control through IPM strategies. **In**: *Farming Systems and Integrated Pest Management* (Eds. J.P. Verma and A. Varma), pp. 159-176. Malhotra Publishing House, New Delhi, 332 p.

Srivastava, K.D., D.V. Singh and L.M. Joshi. 1983. Efficacy of triadimefon in the control of leaf rust of wheat. *Indian Phytopath*. **36**: 712-715.

Stern, V.M., R.F. Van den Bosch, Robert and K.S. Hagen. 1959. The integration of chemical and biological control of the spotted alfalfa aphid. I. The integrated control concept. *Hilgardia,* **26**: 81-101.

Index

C

D

E

F

G

H

I

J

K

L

M

N

O

P

R

S

X

Y

Z